STATISTICAL MECHANICS

STATISTICAL MECHANICS

Shang-Keng Ma

Translated by
M. K. Fung

World Scientific

Philadelphia • Singapore

Published by

World Scientific Publishing Co Pte Ltd.
P. O. Box 128, Farrer Road, Singapore 9128
242, Cherry Street, Philadelphia PA 19106-1906, USA

Library of Congress Cataloging in Publication Data
Main entry under title:

Ma, Shang-Keng
   Statistical Mechanics

   Includes Index
   1. Statistical Mechanics. I. Title
QC174.8.M24   1985   530.1'3   85-2299
ISBN 9971-966-06-9
ISBN 9971-966-07-7 (pbk)

Copyright © 1985 by World Scientific Publishing Co Pte Ltd.

All rights reserved. This book, or parts thereof, may not be reproduced in any form or by any means, electronic or mechanical, including photocopying, recording or any information storage and retrieval system now known or to be invented, without written permission from the Publisher.

Printed in Singapore by Singapore National Printers (Pte) Ltd.

To

Chiansan (千山) *and* Tienmu (天目)

PREFACE

This book serves both as a graduate text and as a reference. Statistical mechanics is a fundamental branch of science which has made substantial progress in the past decade, opening up many new areas of research. The aim of this book is not to give a complete introduction to this field, but rather to give a clear exposition of the most basic principles, and to use relatively up-to-date material as examples of application. Concepts as well as applications are thus emphasised.

The statistical mechanics of today is so fragmented into subfields and so extensively applied to various phenomena that the reader cannot hope to become an expert by just reading one or two textbooks. This text merely aims at helping the reader establish a solid foundation on the principles, clarify some concepts, as well as learn some simple 'tricks' so that he can cope with new problems himself. At the time of writing, there are already many textbooks around including many classics. But they now look somewhat old-fashioned both conceptually as well as in content. On the other hand, many new books are too specialised. Scientists are now now heavily devoted to research, and they expend less effort in writing books. In addition, new knowledge accumulates rapidly, so much so that it is extremely difficult to write a textbook which can stand the test of time. This book is a short one, serving the need of students for the time being. I hope that it can of use to contemporary readers.

Statistical mechanics is a branch of theoretical physics, distinguished by its extensive applications in physics, astronomy, chemistry, materials science as well as biology. It is a convenient tool to link the molecular structure and properties with the macroscopic thermodynamic and electromagnetic behavior. The application of statistical mechanics can be classified according to its degree of difficulty as "elementary" and "advanced". The elementary part is essentially the ideal gas, including the quantum ideal gas. It includes all cases when the interaction between particles is unimportant; the free electron model is one of these cases. Its application is rather extensive. The advanced part discusses the interaction, which plays an important role in many problems, for example in the phenomena of phase transitions. Naturally, the advanced part is much more difficult. Except for some special cases, approximation and numerical calculation are the only methods of solution. The simple and effective approximations usually simplify the problem to certain models of an ideal gas. The mean field approximation is an important example.

In my view, advanced statistical mechanics is a product of the emergence of solid state physics. Although the relation between statistical mechanics and phase transitions is forty years old, most of the applications and developments were made within the last twenty years. The field especially has been fertile in the last ten years. Most of the applications are on solid state, because the phenomena in solid state are numerous, the experimental techniques have become increasingly sophisticated and the physical systems to be investigated are not as complicated as biological systems. Hence the theoretical analysis can cope with such systems. Many new concepts, in fact, originate from solid state physics.

The greater part of this book is on the discussion of examples. The elementary examples are used to clarify the concepts, while the advanced examples are used to discuss the phenomena and methods. In fact, the basic principles and the applications are inseparable. All the examples in this book are used, directly or indirectly, to illustrate some basic concepts. The examples are selected for their extreme simplicity. The material is somewhat biased towards phenomena in solid state physics and I have tried to include some modern, interesting and inspiring topics, in the hope of providing the reader with a glimpse of the modern developments in statistical mechanics.

One motivation for writing this book is to air some of my personal views. My views on some of the basic concepts differ from the conventional ones. I think that statistical mechanics, as it is at present, is an ill-proportioned subject, with many successful applications but relatively little understanding of the basic principles. Yet most textbooks treat statistical mechanics as a complete science.

Roughly speaking, statistical mechanics is based on some rules of calculation. We can proceed forward and discuss the application of the rules of calculation, or we can turn back to discuss their origin. (These rules or calculation can be said to be the Boltzmann's formula for the entropy, which is equivalent to the statement that a state occurs with a probability proportional to $e^{-H/T}$, H being the energy of the state and T the temperature). There are many examples of applications, but works on the quest for the origin are relatively rare, especially at the textbook level. One reason is that, over the years, the applications have been so successful that a substantial degree of confidence is now attached to these rules of calculation. Although the origin may not be clear, the problem seems almost irrelevant because the rules are so useful and successful.

There is another reason. At present we still do not know how to start from mechanics and proceed to understand these rules of calculation; these rules are still an assumption. Many books start from the rules of calculation, ignoring their origin, and treat statistical mechanics purely as a matter of mathematical technique. This viewpoint has its validity. However, some textbooks attempt to make up for the lack of a firm conceptual foundation by giving fancy arguments to set up these rules as established laws. If students do not understand these fancy arguments, it does not matter, so long as they can still use the rules to calculate. This was my own experience as a student.

Of course, many successful applications do not require a profound and exact understanding. This is characteristic of scientific development. Nevertheless, if the range of applications is extended, problems will sooner or later appear. A deeper understanding will become urgent. Today the development of statistical mechanics has reached the point of delving into its origin.

The basic assumption in statistical mechanics is in fact an assumption of independence or randomness. If we can understand this assumption, then it would be helpful in understanding other stochastic phenomena, such as random number generation in the computer. The converse is also true.

We think that the proper attitude to science and to knowledge in general is expressed by the Chinese saying, "Recognise what you know, and recognise what you do not know." Hence this book pays much attention to all the weak points in statistical mechanics, the limitations on its rules of calculation, as well as its questionable and ambiguous points. Although this book does not provide the answer to the important questions, I have at least tried to state the questions clearly.

From the experience of learning mechanics and electromagnetism, the students are accustomed to precisely stated laws. Yet I constantly remind the reader of all those questionable points in statistical mechanics. It may engender a feeling of unease. I hope that this feeling of unease will make the reader become more cautious, cultivating the habit of not accepting too much on faith.

This book is divided into seven parts totalling thiry chapters. These seven parts are

Part I	Equilibrium	Chapters 1 to 4
Part II	Hypothesis	Chapters 5 to 9
Part III	Probability	Chapters 10 to 13
Part IV	Applications	Chapters 14 to 19
Part V	Dynamics	Chapters 20 to 22
Part VI	Theoretical basis	Chapters 23 to 26
Part VII	Condensation	Chapters 27 to 30

Part I is a warm-up. It reviews some common concepts and terms, as well the theory of the ideal gas (including the quantum ideal gas).

Part II, III and VI emphasise the basic concepts. Statistical mechanics is established on a daring assumption. The meaning of this assumption is by no means clear. This book discusses this assumption from the point of view of the underlying molecular motion, avoiding abstract concepts such as "ensembles". My view is that the concept of ensembles is unnecessary and indeed not compatible with reality. Probability is regarded here as a way of handling statistical data. It is a tool, not a theory.

Part V discusses some non-equilibrium phenomena, with a view to elucidate the maintenance of equilibrium. Metastable states and numerical simulation are included in this part.

Part IV and VII analyse various phenomena, as applications of the theory. Included are some discussions on frozen impurities, superfluidity and the two dimensional Coulomb gas. The contents are not a report of the results in the literature, but are my own understanding of these problems expressed for the reader in the most direct fashion I know. My view on these problems may not be the most appropriate and the treatment given here reflect my personal prejudices. The reader must read these chapters with this in mind.

The discussion in this book are not too rigorous; that is to say, many conclusions are not rigorously proved, but only illustrated by simple examples. I do not mean to say that rigorous proof is not important, but it is beyond the scope of my expertise. The mathematical prerequisite for this book is not high, and the necessary calculational techniques are few. There are in fact no complicated calculations in the book at all. Complicated calculations are no doubt very important, but I think that the necessary techniques can be acquired only through actual practice, and it is of no great use to include a lot of complicated calculations here in the book.

Some very important but specialised topics have not been covered in this book.

These include critical phenomena and the renormalisation group (although the phenomena of phase transitions occupy many pages in this book). Non-equilibrium phenomena are little talked about. Examples are biased to problems in solid state. There is also little on thermodynamics and in general, abstract formal things are avoided. Some materials in the book appeared in journals but not in current books; for example the relation between the virial coefficient and the collision time in Chapter 14, the echo phenomena in Chapter 24 and the calculation of entropy from the trajectory in Chapter 25. Many points of emphasis and approach differ from that of standard textbooks. For example, we emphasise the relation of the observation time with equilibrium, the importance of the metastable state and the intimate relation of superfluidity with the conservation of the winding number.

No profound mathematical prerequisite is necessary for reading this book. The mathematics used in undergraduate quantum mechanics is sufficient. The reader should be acquainted with mechanics, quantum mechanics, thermodynamics and electromagnetism. Rather, this book requires the reader to have a cautious and independent mind. Statistical mechanics is a difficult subject in basic physics and this book is not easy to study. Problems are an integral part of the contents. If the reader does not do the problems, he cannot learn the techniques of application, and he may even miss the basic concepts.

The reader should remember that the contents of this book as well as those of other books are not what you understand. Your own understanding of statistical mechanics is acquired through your own thinking and work. This book as well as others are for your reference. You should not be led along blindly.

I have another motive in writing this book. Scientific texts in Chinese are extremely scarce today, and I hope this book may be of some help in alleviating this situation. I deeply feel that basic sciences should be taught in one's own tongue, and textbooks should be written by scientists who are native speakers of their languages. In this century, scientists have become accustomed to writing research reports in the languages of the West. If asked to write a book in Chinese, the answer would often be that one is unable to do so, whereas the truth is that one does not try. The standard of my Chinese is humble, and writing this book was a bold attempt. In the beginning, I found the writing difficult, but I gained fluency as I went along. Indeed science is not literature, and there should be no problem in writing scientific books in one's own language, provided the contents are there. If this book is not easy to read, it is mainly due to the complexity of the content. That the presentation in Chinese is short of fluency is but a secondary factor.

Most parts of this book were written during my stay at the National Tsing Hua University in 1977 and 1981. Many professors and the students have offered me advice, help and encouragement, especially Professors Y Y Lee (李怡嚴), Y Shen (單越), E Yen (閻愛德), W T Ni (倪維斗), Y M Shih (石青民), H M Huang (黃幸美), T L Lin (郎棣), H C Yen (顏晃徹), T T Chen (陳逼) and graduate students C L Lin (林其隆), F K Yang (楊芳鏗), J L Chen (陳俊良), and D G Lin (林達觀). I express my gratitude to them here. The administrative and secretarial staff in the Graduate Institute of the National Tsing Hua University have given me much help, especially all the directors and Miss Y M Chang (張月梅). I record my thanks to them here.

Moreover, I have to thank my parents. They have always expressed their full support for my work. That this book could be finished at all owes much to their encouragement and support; but I have seldom doen my filial duty. I therefore dedicate this book (Chinese edition) to my parents.

This book has much room for improvement. I hope the reader will not hesitate to give his views.

Spring 1982　　　　　　　　　　　　　　　　　　　　　　　　　　S K Ma

Publisher's Acknowledgement

The publisher is deeply grateful to Professor M K Fung for translating the Chinese text and to K Young, J Chen, K C Chang, C H Lai, J Martinez, Y Koh and J Tay for checking the translation and galley proofs. Since Professor Ma passed away before the English edition could be completed their cooperation with us in this effort was indispensable. We hope that the spirit and depth of Professor Ma's views on Statistical Mechanics are well communicated in this book.

CONTENTS

Preface   vii

Introduction   xxi

PART I   EQUILIBRIUM   1

1   Equilibrium   2

1.1   Equilibrium and Observation Time   2
1.2   Equilibrium and Molecular Motion   4
1.3   Large Scale Change and Local Equilibrium   4
1.4   Structure of Matter   6
1.5   Supplement and Conclusions   8
     Problems   9

2   The Basic Concepts of Thermodynamics   10

2.1   Equilibrium and Reversible Processes   10
2.2   Work and Heat   11
2.3   Adiabatic Process and Entropy   13
2.4   Absolute Temperature and Entropy   16
2.5   The Ideal Gas   19
2.6   The Second Law of Thermodynamics   24
2.7   The Efficiency and Speed of a Heat Engine   24
     Problems   28

3 The Law of Detailed Balance — 30

- 3.1 The Energy Distribution of the State — 31
- 3.2 The Energy Distribution of Fermions — 35
- 3.3 Conservation Laws, Chemical Potential and the Constant Term in the Entropy — 39
- 3.4 Distribution of Bosons and Black Body Radiation — 41
- 3.5 Chemical Reactions and the Law of Concentrations — 43
- Problems — 45

4 Electrons in Metals — 48

- 4.1 Dense Ideal Gas — 48
- 4.2 Heat Capacity — 51
- 4.3 Influence of Periodic Structure — 55
- 4.4 The Structural Change Caused by Electrons — 60
- Problems — 63

PART II HYPOTHESIS — 65

5 The Basic Assumption of Entropy and Molecular Motion — 66

- 5.1 Region of Motion — 66
- 5.2 Assumption for the Calculation of the Entropy — 67
- 5.3 Fermions — 70
- 5.4 The Ideal Gas — 72
- 5.5 Some Special Features of High Dimensional Space — 76
- 5.6 Peak Integration and the Law of Large Numbers — 78
- 5.7 The Law of Equipartition of Energy — 80
- 5.8 Perspectives of the Region of Motion — 81
- 5.9 Trajectory and the Region of Motion — 85
- Problems — 86

6 Some Elementary Applications of the Basic Assumption — 88

- 6.1 Systems and Subsystems — 88
- 6.2 Vibrations — 94
- 6.3 The Debye Model — 98
- 6.4 Phonons and Second Sound — 102

6.5	General Conclusions, Average Values and the Normal Distribution	111
	Problems	116

7 Rules of Calculation 118

7.1	Various Thermodynamic Potentials	118
7.2	The Various Rules of Calculation	122
	Problems	130

8 Illustrative Examples 133

8.1	Time Scale and the Region of Motion	133
8.2	Gas of Hydrogen Molecules	139
8.3	One-Dimensional Model	145
	Problems	150

9 Yang-Lee Theorem 153

9.1	Macroscopic Limit (i.e. Thermodynamic Limit)	154
9.2	Generalisation of the Theorem	159
	Problems	161

PART III PROBABILITY 163

10 Probability and Statistics 164

10.1	The Use and Misuse of Probability	164
10.2	Distributions and the Statistics of the Average Value	166
10.3	The General Definition of Probability	174
	Problems	178

11 Independence and Chaos 180

11.1	The Definition and Consequence of Independent Phenomena	180
11.2	Test of Independence	186
11.3	Random Sequences	188
11.4	Scattering Experiments and the Measurement of the Correlation Functions	190
	Problems	195

12	**Sum of Many Independent Variables**	**198**
12.1	Normal Distributions	198
12.2	The Central Limit Theorem (Rudimentary Discussion)	200
12.3	Higher Order Averages and the Cumulant	203
12.4	The Cumulant Expansion Theorem	207
12.5	Central Limit Theorem (General Exposition)	209
12.6	Repeated Trials and the Determination of Probability	211
12.7	Random Motion and Diffusion	214
12.8	Fluctuation of Macroscopic Variables	217
12.9	Fluctuations and an Extension of the Basic Assumption	219
	Problems	222
13	**Correlation Functions**	**225**
13.1	Response and Fluctuation	225
13.2	Density Correlation Function	230
13.3	The Fermion Gas	235
13.4	The Response and Correlation Functions in Quantum Mechanics	238
	Problems	242

PART IV APPLICATIONS 245

14	**Corrections to the Ideal Gas Law**	**246**
14.1	Intermolecular Interactions	246
14.2	Corrections to the Ideal Gas Law	248
14.3	Time Delay in Collisions	250
14.4	Quantum Mechanical Calculation	254
14.5	Bound State and the Levinson Theorem	257
	Problems	259
15	**Phase Equilibrium**	**261**
15.1	Gaseous and Liquid Phases	261
15.2	The Growth of Water Droplets	265
15.3	Latent Heat and the Interaction Energy of Molecules	268
15.4	Melting and the Lindemann Formula	270
15.5	Definitions of μ_1 and μ_2	271
	Problems	274

16	**Magnetism**	**277**
16.1	Paramagnetism	277
16.2	Magnetic Susceptibility	281
16.3	Diamagnetism of Charged Particles	283
16.4	Spin-spin Interaction	291
	Problems	295
17	**Ising Model**	**297**
17.1	Ising Ferromagnetism in One Dimension	297
17.2	Proof of the Existence of Ising Ferromagnetism in Two Dimensions	301
17.3	Other Ising Models	307
	Problems	310
18	**Impurities and Motion**	**312**
18.1	Solutions and Osmotic Pressure	312
18.2	Effective Interaction Energy	314
18.3	Low Density Case	316
18.4	Mobile and Stationary Impurities	318
18.5	The Amorphous State	323
	Problems	325
19	**Electrostatic Interaction**	**329**
19.1	Short Distance and Long Distance are Both Important	329
19.2	The Plasma Gas and Ionic Solutions	330
19.3	Electrons in Metals	333
19.4	Electron Crystals	335
19.5	Two-Dimensional Coulomb Gas Model	337
	Problems	341

PART V DYNAMICS		**345**
20	**The Equation of Motion For Gases**	**346**
20.1	Flow and Collision	346
20.2	Case of No Collisions — Plasma Oscillations	349
20.3	Zero Sound	351
20.4	Collisions and Diffusion	354

xviii STATISTICAL MECHANICS

20.5	Collisions and Sound Waves, Viscosity and Heat Conduction	357
20.6	H-theorem	363
	Problems	364

21 The Diffusion Equation — 367

21.1	Simple Examples	367
21.2	The Metastable State	372
21.3	Transformation to the Wave Equation	374
21.4	Derivation of the Diffusion Equation	376
21.5	Two-State Cluster Model	380
	Problems	386

22 Numerical Simulation — 388

22.1	Numerical Solution of Molecular Motion	389
22.2	Random Sequences	391
22.3	Monte Carlo Simulation	391
22.4	Conceptual Problems to be Noted	394
	Problems	397

PART VI THEORETICAL BASES — 399

23 Laws of Thermodynamics — 400

23.1	Adiabatic Processes	400
23.2	Adiabatic Process and Entropy	403
23.3	The Second Law of Thermodynamics	405
23.4	The Third Law of Thermodynamics	408
23.5	The Amorphous State and the Third Law	411
	Problems	413

24 Echo Phenomena — 415

24.1	Demonstration of the Viscous Liquid Bottle	416
24.2	Analysis of the Demonstration	417
24.3	Spin Echo Experiment	419
24.4	The Plasma Echo	422
	Problems	424

25	**Entropy Calculation from the Trajectory of Motion**	425
25.1	Number of Coincidences and the Size of the Region	426
25.2	Independence of Different Parts of a System	429
25.3	Correlation Time	431
25.4	Process of Computation	432
25.5	Entropy of the Metastable State	434
25.6	The Third Law of Thermodynamics	435
	Problems	437
26	**The Origin of the Basic Assumption**	440
26.1	The Basic Assumption	441
26.2	Ergodicity and Ensembles	442
26.3	Mixing and Independence	445
26.4	Probability and Experiments	448
26.5	Instability of the Trajectory	450
26.6	The General Problem of Independence	454
26.7	Difficulties Encountered in Applications	455
	Problems	455

PART VII CONDENSATION 457

27	**Mean Field Solutions**	458
27.1	Mean Field Theory	458
27.2	Total Magnetic Moment	461
27.3	Thermodynamical Potential and the Coexistence of Different Phases	464
27.4	The Ising Lattice Gas	466
27.5	The Van der Waals Equation	470
27.6	Common Properties of Condensation Phenomena	473
	Problems	476
28	**Fluctuation of the Boundary Surface**	480
28.1	The Liquid Surface Model	481
28.2	The Boundary Surface of the Ising Model	485
28.3	Crystal Surface — the Coulomb Gas Model	488
	Problems	493

29	**Models with Continuous Symmetry**	**495**
29.1	The Planar Vector Model	495
29.2	Density Fluctuations of a Crystal	498
29.3	Quantum Vector Model	502
29.4	Continuous Symmetry and Soft Modes	505
29.5	Defects in the Condensation	508
	Problems	514

30	**Theory of Superfluidity**	**518**
30.1	Quantum Lattice Gas	519
30.2	The Ground State and Low Temperature Model	522
30.3	State of Flow and Winding Number	525
30.4	Stability of Superfluidity	528
	Problems	535

References   **541**

Index   **545**

INTRODUCTION

1. On Teaching and Reading

This book is meant for two-semester course for graduate students of physics, materials science and chemistry.

On the whole, the first three parts are elementary, while the last four are more difficult. In each chapter, we start from the elementary, leading gradually to more advanced topics. The choice of material to be covered in the course would have to depend on the need and level of the students, and the opinion of the teacher. I suggest below some points to be noted in teaching and reading.

(a) It is important to discuss and anlyse different points of view. A dogmatic pedagogical approach would be most unwise. It would not be appropriate to skip any material in Parts I-III. Part VI is also general, and should not be omitted.

(b) Mathematical derivations should not be neglected. We have already avoided complicated calculation so that students should be fully capable of repeating every derivation in the book. The objective of the derivation is to instill an appreciation of the approximations, hypotheses and consequences.

(c) Each chapter ends with a set of problems. These include many exercises of a conventional variety, as well as many intended for discussions. The latter may not have 'standard' answers and often I do not know the answers myself. These discussion problems are important, and should not be neglected.

(d) The references listed are not strictly necessary as a part of the course. They provide direction of further study, or merely identify the sources of the tables and figures. My own view is that graduate courses should stress independent thinking about a problem before consulting references, with references serving only as a source of specific information. It is indeed necessary to have clear concepts in order to benefit from the literature. Of course, references are useful, especially in broadening one's views. Unfortunately, for students who are not native speakers of English, there is an unavoidable language barrier. (Nevertheless, I must emphasise that with a clear understanding of the concepts and a proper foundation, books and articles in English should present no great difficulty. Without clear concepts, the best ability in English would be of little use.) Therefore it is also the task of the teacher to introduce some extra material where possible.

(e) Difficulties encountered in the course can often be traced to a poor foundation in elementary concepts. The new material is seldom very difficult in itself, nor can the problems be blamed to the lack of reference books. It is extremely important to review elementary concepts periodically.

2. Units and Constants

For simplicity, we set Boltzmann's constant to be 1, so kT becomes T. The temperature T is regarded as an energy, and energy can be measured in temperature units. Planck's constant $\hbar$ is also set equal to 1. These constants are nevertheless restored wherever necessary, in order to exhibit their roles. The following units and constants may be used for conversion.

Boltzmann's constant $k = 1.38 \times 10^{-16}$ erg/K
$= 8.62 \times 10^{-5}$ eV/K

Avogadro's number $N = 6.02 \times 10^{23}$ mole^{-1}

Gas constant $R = Nk = 1.98$ cal deg^{-1} mole^{-1}

[Note: 1 eV $\approx 10^4$ K; 1 cal/mole ≈ 0.5 K/molecule]

Planck's constant $\hbar = 1.05 \times 10^{-27}$ erg sec $= 6.577 \times 10^{-16}$ ev.sec.

$h = 2\pi\hbar$

[Note: 1 eV/$h = 2.42 \times 10^{14}$ sec^{-1}]

Proton mass $= 1.67 \times 10^{-24}$ gm

Electron mass $= 9.11 \times 10^{-28}$ gm

Electron charge $= -e$

$e^2 = (4.80 \times 10^{-10})^2$ erg cm

Bohr radius $a = 0.529$ Å

Binding energy of hydrogen $= e^2/2a = 13.6$ eV

Bohr magneton $\mu_B = e\hbar/2mc = 9.27 \times 10^{-21}$ erg/gauss.

3. Mathematical Formulas

The formulas listed below are not meant to take the place of a proper table. They may nevertheless be useful when a table is not at hand.

(a) δ-function

$$\theta(x) = 1, \quad x > 0,$$
$$= \tfrac{1}{2}, \quad x = 0,$$
$$= 0, \quad x < 0.$$

$$\frac{d}{dx}\theta(x) = \delta(x),$$

$$\int_a^b \delta(x)\,dx = \theta(b) - \theta(a).$$

$$\delta(ax) = \frac{\delta(x)}{|a|}$$

$$\operatorname{sgn} x = \frac{x}{|x|},$$

$$\frac{d}{dx}\operatorname{sgn} x = 2\delta(x).$$

$$\delta(x-a) = \frac{1}{2\pi}\int_{-\infty}^{\infty} dk\, e^{ik(x-a)}$$

$$\sum_{n=-\infty}^{\infty} \delta(x-a-nb) = \frac{1}{b}\sum_{m=-\infty}^{\infty} e^{i2\pi m(x-a)/b} \quad \text{(Poisson sum formula)}.$$

(b) Useful formulas

$$\Gamma(n) = (n-1)!$$

$$\int_0^{\infty} dx\, x^n e^{-x} = \Gamma(n+1) = n!$$

$$\Gamma(1+n) = n\Gamma(n) .$$

$$\Gamma(\tfrac{1}{2}) = \sqrt{\pi} = 1.772 .$$

$$\Gamma(\tfrac{3}{2}) = \tfrac{1}{2}\sqrt{\pi} .$$

$$\ln \Gamma(1+x) = -Cx + O(x^2) ,$$

$$C = 0.5772 = -\int_0^\infty dx\, e^{-x} \ln x .$$

$$N! = N^N e^{-N} (2\pi N)^{1/2} \left[1 + \frac{1}{12N} + \frac{1}{288N^2} + \ldots \right].$$

$$\int_0^\infty \frac{dx}{e^x + 1} = \ln 2$$

$$\int_0^\infty \frac{x\,dx}{e^x + 1} = \frac{\pi^2}{12} .$$

$$\int_0^\infty \frac{x^2\,dx}{e^x + 1} = \frac{3}{2} \zeta(3), \ \zeta(3) = 1.202 .$$

$$\int_0^\infty \frac{x\,dx}{e^x - 1} = \frac{\pi^2}{6} .$$

$$\int_0^\infty \frac{x^2\,dx}{e^x - 1} = 2\zeta(3) .$$

$$\int_0^\infty \frac{x^3\,dx}{e^x - 1} = \frac{\pi^2}{15} .$$

$$\int_{-\infty}^\infty dx\, e^{-x^2/2a^2} e^{-ikx} = \sqrt{2\pi}\, a\, e^{-1/2 k^2 a^2} .$$

$$\tanh^{-1} x = \frac{1}{2} \ln \frac{1+x}{1-x} = x + \frac{x^3}{3} + \ldots, \quad |x| < 1 .$$

$$\coth^{-1} x = \frac{1}{x} + \frac{1}{3x^3} + \ldots, \quad |x| > 1 .$$

(c) Useful numbers

$\ln 2 = 0.693$,

$\ln 10 = 2.30$,

$\sqrt{\pi} = 1.77$.

$6! = 720$, $9! = 362\,880$, $12! \approx 4.8 \times 10^8$.

$2^{10} = 1024$, $2^{15} = 32\,768$, $2^{20} \approx 10^6$.

PART I
EQUILIBRIUM

This part is divided into four chapters which introduces the basic concepts of equilibrium and provides some simple examples. The first two chapters review the concept of equilibrium and the laws of thermodynamics, with special emphasis on the facts that equilibrium is not an instantaneous state and the observation time is important. The outstanding feature of thermodynamics is the appearance of entropy. Chapter 3 discusses the law of detailed balance, pointing out the interaction between molecules as the origin of equilibrium. Chapter 4 discusses electrons in metals, reviewing some basic knowledge in solid state physics and some elementary calculations in statistical mechanics. This part lays the preparatory ground work. Various basic concepts will be analysed again in the following parts.

Chapter 1

EQUILIBRIUM

This chapter discusses the fundamental concept of equilibrium with emphasis on the relation between equilibrium and the observation time. On the time scale of molecular motion, equilibrium is a state characterizing the system over a long observation time and not an instantaneous state. To discuss equilibrium we must consider the approximate length of the observation time.

1.1. Equilibrium and Observation Time

Basic concepts always originate from simple phenomena. Equilibrium is the most basic concept of thermodynamics and statistical mechanics. We use a few simple examples to illustrate the meaning of equilibrium and to discuss some of its related problems.

Equilibrium refers to a state which is unchanging. For example, an ancient painting hanging on the wall is an unchanging state. However, to say this state is unchanging is only an approximation and is not absolute. From the date of its creation to the present, this ancient painting has undergone many changes. For example, because of the chemical reaction with air, some pigments have changed and the paper has become pale. Consequently, when one speaks of equilibrium, one must take into consideration the problem of the observation time. If the observation time is short, e.g., one second, half a second, or two or three days, or even one year, this painting can be said to be in an equilibrium state, because during the observation time its change is rather minute. Suppose now we have an extremely long observation time, e.g., one hundred years or two hundred years, then this painting cannot be regarded as being in an equilibrium state, because its change is rather obvious and cannot be neglected, not to mention the change in the hanger and the attached wall. What change is of importance and what may be neglected depends on the aims of our analysis. Generally speaking, to discuss equilibrium we must specify the observation time. The specifications of this observation time depend on the requirement and interest of the observer.

Consider another example. If we pour some boiling water from a thermos flask into a tea cup, after a few seconds the water in the cup becomes an unchanging state at rest, and this is an equilibrium state. The volume of the water in the cup can be measured from the height of the water and the cross-section area of the cup. The temperature of the water can be measured by a thermometer. Within one or two seconds, these measured quantities will not change significantly, and thus during this observation time the water is in equilibrium. If the time is too long, the temperature will obviously change and the water cannot be regarded as being in an equilibrium state. After an hour, the temperature of the water will be equal to the room temperature. If the room temperature is constant for several hours, then the temperature of the water will accordingly remain constant. Therefore, if the observation time is within this range of several hours, the cup of water can be regarded as being in equilibrium. However, water molecules will continually evaporate from the water surface. In several hours the volume of the water will not change significantly, but if the time is long enough, such as two or three days, all the water will evaporate. So if the observation period is over two or three days, this cup of water cannot be regarded as being in an equilibrium state. After a few days, when all the water has evaporated, then again this empty cup can be regarded as being in an equilibrium state. However, strictly speaking, even this cup is not in an absolute unchanging state, because the molecules of the cup can evaporate, although in a few years' time there will be no change observed.

These examples point out that the observation time cannot be too large, otherwise equilibrium will be meaningless. There are no absolutely unchanging things in this world, but we can use different methods to prolong the period of equilibrium. For example, we can put the ancient painting in a glass case, or pour the water into a thermos flask rather than a cup. But absolute isolation is impossible and there is always some transfer of energy. Notice that there is no permanent container, and any substance under tension or shear is a metastable state, and is not absolutely stable. If the observation time is not too long, and the state of the system does not change, we call this an equilibrium state. This definition is somewhat ambiguous, and a more precise definition and better understanding will emerge from an analysis of more examples.

1.2. Equilibrium and Molecular Motion

Equilibrium is a macroscopic phenomenon. "Macroscopic" refers to a length and time scale much larger than those characterizing the molecules. The time scales, discussed in the previous section, are macroscopic times, i.e., times much longer than the time of molecular motion. From the microscopic point of view, i.e., on the scale of the size of the molecules, it must be recognised that the painting of the previous section, is composed of many molecules, which vibrate continually. Similarly, the water molecules in the cup as discussed above are executing very complicated motion and none of the molecules is stationary. The macroscopic phenomena and properties that are observed are the time average of these very complicated molecular motion during observation. From the microscopic point of view, equilibrium is a very special state of motion in which, within the observation time, the macroscopic properties are unchanged. The aim of statistical mechanics is to understand the relation between equilibrium and molecular motion.

As equilibrium is an unchanging state, any phenomena changing with time can be called nonequilibrium phenomena. The conduction of heat and electricity, the propagation of sound, blowing winds and falling rain drops are examples of nonequilibrium phenomena. To understand nonequilibrium phenomena, it is essential to understand the molecular motion. To understand equilibrium, we must analyse the factors maintaining equilibrium and also the factors causing changes. Hence to understand equilibrium, we must also consider nonequilibrium problems.

In the previous section, we have emphasised the intimate relationship between equilibrium and the observation time in order to develop a correct viewpoint about the range of applicability of statistical mechanics. Most textbooks start with the concept of absolute equilibrium. The so called "absolute equilibrium" is an idealisation, referring to the state of an absolutely isolated system and an infinitely long observation time. There are also more abstract definitions of equilibrium. Such approaches have their merits in providing a simpler and clearer introductory discussion. However, the reader is somewhat hampered, since many problems in statistical mechanics become nonequilibrium problems under this idealised definition. In contrast, we introduce the notion of an observation time right from the beginning. This may appear to be cumbersome at first, but it will provide a greater flexibility in the range of applications.

1.3. Large Scale Change and Local Equilibrium

In the previous section, we have emphasised that equilibrium is a large scale phenomenon. On the scale of a molecule, it is not an instantaneous property,

EQUILIBRIUM 5

but a property associated with a long observation period. On the other hand, from the large scale considerations, the observation time cannot be too large, for otherwise equilibrium cannot be maintained. Now we use the equations of fluid mechanics as an example to further emphasise the importance of the scale.

The following equations describe gas motion on a large scale.

$$\frac{\partial \rho}{\partial t} + \nabla \cdot (\rho \mathbf{v}) = 0 \, ,$$

$$\rho \frac{\partial \mathbf{v}}{\partial t} + \rho (\mathbf{v} \cdot \nabla) \mathbf{v} = -\nabla p \, , \qquad (1\text{-}1)$$

where $\rho(\mathbf{r}, t)$ is the mass density at position $\mathbf{r}$ and time t, $\mathbf{v}(\mathbf{r}, t)$ is the flow velocity, $p(\mathbf{r}, t)$ is the pressure and $\rho\mathbf{v}$ is the gas flow, i.e., the momentum density. The first equation expresses conservation of mass and the second is Newton's law relating the change of momentum to the applied force. We neglect viscosity for simplicity. The meaning of (1-1) can be seen more clearly by integrating the equations

$$\frac{\partial}{\partial t} \int \rho \, d^3 r + \int d\boldsymbol{\sigma} \cdot \mathbf{v} \rho = 0 \, ,$$

$$\frac{\partial}{\partial t} \int \rho \mathbf{v} \, d^3 r + \int d\boldsymbol{\sigma} \cdot \mathbf{v} \rho \mathbf{v} = -\int d\boldsymbol{\sigma} p \, .$$

The integral $d^3 \mathbf{r}$ is over all points in a certain region and $d\boldsymbol{\sigma}$ is an integration over the surface of the region.

Notice that the scales of time and length in (1-1) are understood to be macroscopic and the differential dt is regarded as a rather long macroscopic observation time, i.e.,

$$dt \gg \tau \, , \qquad (1\text{-}2)$$

where τ is the time scale of the molecular motion which may be taken to be the mean free time, i.e., the free flight time between two consecutive collisions. The differential volume $d^3 \mathbf{r} = dx\,dy\,dz$ is likewise a rather large volume, larger than the length scale of molecular motion

$$dx \gg \lambda \, , \qquad (1\text{-}3)$$

where λ is the mean free path, $\lambda \sim \tau v$ where v is the average speed of the moluecules. Within an observation time dt, the gas in a volume $d^3 \mathbf{r}$ is approximately in equilibrium, i.e., local equilibrium. But if the time is much longer than dt or the volume is much larger than $d^3\mathbf{r}$, then changes can occur and nonequilibrium appears, as described by (1-1).

Equations (1-2) and (1-3) point out the range of application of (1-1) — the changes of p, $\mathbf{v}$ and ρ must be rather smooth on a scale dictated by τ and λ, whose magnitudes depend on specific situations.

A point to be noted is that (1-2) and (1-3) are not very precise. It is no doubt better to have larger dt and $d^3\mathbf{r}$, but what is the lower limit: 10τ, 10λ or 1000τ, 1000λ? In principle this can be answered from microscopic analysis, i.e., to deduce the large scale equations of motion from the molecular motion, but in fact these problems are very difficult, as illustrated by some examples which we shall discuss later.

1.4. Structure of Matter

Under different external conditions, matter has different equilibrium properties. The external conditions are described by physical variables such as temperature, pressure and magnetic field, and to a certain extent can be created and controlled in the laboratory. The so-called equilibrium properties of matter denote the structural as well as the thermal, electrical, or elastic properties etc. The main task of statistical mechanics is to understand how these properties emerge from the molecular motion.

Now we briefly review some special features of the structure of matter. Matter in the usual states can be regarded as a collection of nuclei and electrons. Different compositions and motions result in different structures and properties. If one or several nuclei attract several electrons and form an entity (i.e., atom or molecule), but each individual entity does not combine with others, but moves independently, we have a fluid. If the nuclei cannot move freely, we have a solid. If the arrangement of the atoms is random, we have amorphous substance or a glassy state. If the arrangement is orderly, we have a crystal. If there are free electrons in the solid, then it is a conductor. Otherwise it is an insulator. Low temperature liquid helium (near absolute zero) is a superfluid without viscosity. Many metals and alloys at low temperatures are superconductors with zero resistivity.

Each electron has a spin with a magnetic dipole moment. In a magnetic field the spins may align with the field and this is paramagnetism. If this alignment does not require an external magnetic field, then it is ferromagnetism. If the alignment is alternately up and down, it is called antiferromagnetism. If the arrangement is random, it is an amorphous magnet or spin glass.

The above are only some possible structures. There are millions of different molecules, crystals and plastics. Materials science is an active field and synthetic

Table 1-1. Some examples of the change of the properties of matter with temperature (at one atmosphere pressure).

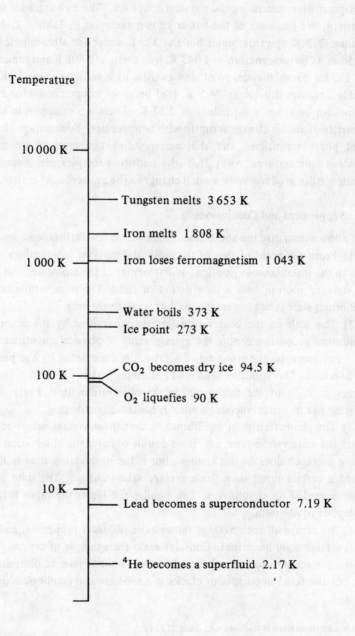

Temperature

10 000 K

— Tungsten melts 3 653 K

— Iron melts 1 808 K

1 000 K — Iron loses ferromagnetism 1 043 K

— Water boils 373 K
— Ice point 273 K

100 K — CO_2 becomes dry ice 94.5 K

O_2 liquefies 90 K

10 K —

— Lead becomes a superconductor 7.19 K

— ^{4}He becomes a superfluid 2.17 K

materials are constantly being developed and utilized.[a]

The structure and properties of matter change with temperature, pressure and other parameters describing the physical condition. These changes are numerous in variety. We list some of the better known examples in Table 1-1. From the ice point 273 K upwards: water boils at 373 K under one atmospheric pressure, iron loses its ferromagnetism at 1 043 K, iron melts at 1 808 K and tungsten melts at 3 653 K. From the ice point downwards: oxygen liquefies at 90 K, carbon dioxide becomes dry ice at 94.5 K, lead becomes a superconductor at 7.19 K and helium becomes a superfluid at 2.17 K. These are examples in which the properties of matter change abruptly with temperature. Such abrupt changes are called phase transitions, and the corresponding temperatures are called the transition temperatures. Away from the transition temperature, a small change in temperature produces only a small change in the properties of matter.

1.5. Supplement and Conclusions

We now summarise the above discussions with a few further remarks:

(1) From the point of view of the molecular motion, "equilibrium" does not refer to the instantaneous positions and velocities of the molecules, but rather to the state of motion over a long period of time. The observation time of an equilibrium state is not instantaneous, but is relatively long.

(2) The scale of the observation time is determined by the observer. The equilibrium properties denote the average value of physical quantities over this time. For example, the structure of a crystal is determined by the positions of all the nuclei. This structure is an equilibrium property. The position means the average position of the nuclei within the observation time. Every nucleus is vibrating and its instantaneous position is continually changing.

(3) The characteristic of equilibrium is that these average values do not vary within the observation time, e.g., if we double or halve the observation time, the various average values do not change. But if the observation time is allowed to exceed a certain limit, then these average values change. The state is then no longer regarded as equilibrium. The smaller the limit, the more fuzzy is the concept of equilibrium.

(4) We regard all such average values as equilibrium properties, and not just the few large scale quantities. Consider again the example of crystal. Not only are the average positions of all the molecules in the crystal equilibrium properties, but the fixed impurities or cracks in a solid are also equilibrium properties.

---

[a] See books on materials science, e.g., Tang (1971).

(5) Some equilibrium properties are not related to the molecular motion in an obvious way. Temperature and entropy (which we shall discuss carefully later) are the prominent examples.

(6) The above discussion of equilibrium is not complete. A more precise definition and a better understanding of equilibrium require additional concepts such as detailed balance and the independence of the motions of parts. Moreover, the distinction between large and small scale, i.e., between the macroscopic and the microscopic, cannot always be well defined.

In short, the topics and concepts we shall analyse in equilibrium are not always well defined. They are often complicated and fuzzy.

Problems

1. Carbon has two common crystal structures: graphite and diamond. On everyday time scales the two have very well-defined equilibrium properties.
Question: Under an infinite time scale, which is more stable?
(Answer: Graphite is more stable than diamond. The reader can refer to solid state physics data and find out their energies.)

2. There are many crystalline forms of sulphur and the structures of its liquid and gas phases are very interesting. The reader should consult a suitable book on this subject.

3. Figure 1-1 shows a cup filled with liquid helium in a box at low temperature. If the temperature is lower than 2.17 K the liquid helium in the cup will flow out until the liquid levels are the same inside and outside. Consult reference to understand this phenomenon.

4. If the temperature is slightly higher than 2.17 K, can the liquid helium flow out of the cup?

5. If instead of helium, there is water in the cup, and the box is at room temperature, what happens?

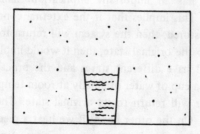

Fig. 1-1

Hint: The main difference in Problems 3, 4 and 5 lies in the different time scales.

Chapter 2

THE BASIC CONCEPTS OF THERMODYNAMICS

Thermodynamics is mainly concerned with the exchange of energy (heat and work) and the equilibrium properties. It does not consider the structure of matter, but analyses only the macroscopic behaviour. This chapter reviews the basic assumptions and concepts as a preparation for the discussion of statistical mechanics.

2.1. Equilibrium and Reversible Processes

The basic assumption in thermodynamics is that under a specific set of environmental conditions, a given system has definite equilibrium properties. The environmental conditions are determined by the external factors such as temperature, pressure and magnetic field, etc. Once these factors are fixed, the equilibrium state is defined. This assumption is valid under many situations, but there are exceptions.

This assumption has an important implication, namely the existence of reversible processes. This implies that if the external conditions of a system are changed and then restored, then the system will return to the original state. If it does not return to the original state, then it would imply that under the same conditions we could have different states and the process is irreversible. For example, if we heat a cup of water originally at room temperature and then cool it, this cup of water will return to its original state. Therefore this particular process is reversible. On the other hand, if we heat a piece of iron to very high temperatures and then suddenly cool it, many properties of this piece of iron will change. Different rates of heating and cooling can result in different structures, so this process is irreversible. To understand this irreversible process and the property of iron, we must study the detailed change of the structure.

We now discuss only reversible processes. Generally speaking, smooth changes are usually reversible, while abrupt changes are usually irreversible. Of course, smoothness or abruptness depends on the time scale. Roughly speaking, if the time scale of change is much longer than the time scale of the motion of the molecules, then the process can be called smooth.

2.2. Work and Heat

The energy of an object may be changed by external factors, e.g., we can stretch an object, compress it, or add an external magnetic field, etc. These external factors can do work on the object and change its energy. The work done is given by the product of force and displacement (i.e., change of coordinates). For example, the work done, dW, by a tension f pulling an iron rod by a distance dX, is

$$dW = f \, dX \, . \tag{2-1}$$

The bar on d means that dW is only an infinitesimal quantity and not the exact differential of the function W. As a second example, let consider the work done on an object by an external magnetic field h. The increase in energy of the object is given in terms of the increase in the magnetic moment dM

$$dW = h \, dM \, . \tag{2-2}$$

If we fix one end of a rod and apply a torque η to the other end, turning the rod through an angle $d\theta$, the energy increases by

$$dW = \eta \, d\theta \, . \tag{2-3}$$

We give two more examples of work in thermodynamic systems: the pressure p is the "force" that causes a change in the volume V, and the corresponding work is

$$dW = -p \, dV \, . \tag{2-4}$$

The negative sign in dV denotes the decrease in volume due to compression. The chemical potential μ is the "force" that causes a change in the number of particles N. For example, a beaker of water contains N dissolved oxygen molecules. The oxygen molecules above the water provide a "force" on the number N below, and this is described by a chemical potential μ. If dN oxygen molecules are pushed into the water, then the work done is

$$dW = \mu \, dN \, . \tag{2-5}$$

Similarly nitrogen and water vapour in the air have their own chemical potentials μ' and μ''. In general, each species of particles is characterised by its own chemical potential.

The chemical potential is not a familiar concept, so we add a few remarks here. The chemical potential is a "force" increasing the number of the particles, and is very important in the discussion of chemical reactions. The chemical potentials of various molecules determine the direction of the reaction. Suppose A, B combine to form C, and C decomposes to A and B

$$A + B \leftrightarrow C \ .$$

The presence of C exerts a "force" to increase the population of A and B; similarly A and B tend to increase the population of C. In equilibrium the chemical potentials on the two sides must balance,

$$\mu_A + \mu_B = \mu_C \ .$$

We shall discuss this in detail when talking about chemical reactions.

The notion of chemical potential is analogous to that of population pressure in social science. Early marriages, high birth rates and longevity increase the population pressure. Increase of employment rates and improvement of living conditions will decrease the population pressure. The development of a city creates more jobs and its population pressure is lower than the surrounding country and people will rush into the city. Once the urban population increases, the living condition deteriorates, jobs are fewer and the population pressure in the city rises, thus causing a decrease in the flow of population into the city.

The length X, the magnetic moment M, the turning angle θ, the particle number N are all "coordinates" characterizing the state and/or the configuration of the system, and they are called generalised coordinates. The tension f, the magnetic field h, the torque η, and the chemical potential μ are external "forces" capable of changing the coordinates and hence doing work.

These generalised coordinates are large scale coordinates or thermodynamic coordinates. The small scale coordinates expressing the detailed motion of the molecules do not appear in thermodynamics. Of course, the small scale coordinates exist and are very important. Energy can flow out of or into a system without the above generalised coordinates showing any change. Such energy flow is heat and is the result of the change of the small scale motion. This shows that the small scale motion cannot be ignored completely. But we need not analyse the coordinates of every molecule. The most outstanding feature of thermodynamics is to introduce a new generalised coordinate to express the small scale motion in a collective way. This new generalised coordinate is denoted by the symbol S, and the "force" responsible for increasing this coordinate is the absolute temperature T. The increase in energy is

$$dQ = T\, dS \, . \tag{2-6}$$

Why does T turn out to be the absolute temperature? This can be said to be a coincidence, or can be considered as a choice. We now carefully discuss this problem, reviewing some terms and definitions.

The sum of the energy changes in (2-1) to (2-5) is called the work done dW, because this is the work done by external "forces" to change the large scale coordinates. We call (2-6) the heat dQ because this is the heat flowing in, due to the change of the small scale molecular motions. The change of total energy dE is the sum of the two:

$$dE = dW + dQ \, . \tag{2-7}$$

This is the first law of thermodynamics. This law is a statement of the conservation of energy and tells us that there are two ways of changing the energy of work done on the body or heat flowing into the body.

2.3. Adiabatic Process and Entropy

If $dQ = 0$, then during the process of energy change, no heat flows into or out of the system and this is called an adiabatic process. Absolute heat insulation is, of course, impossible. According to experience, some materials, such as stone or asbestos, are rather good insulators. Others such as metals are very bad insulators. In addition to insulating materials, the amount of heat flow is also controlled by the time duration of the process. If the process is very quick, then the heat flow is small. But if time is prolonged, it is not easy to ensure negligible heat exchange. Starting from the concept of an adiabatic process, we shall develop step-by-step the concept of entropy. To be concise, we first consider a single coordinate X of a physical system. The horizontal axis in Fig. 2-1 denotes X and the vertical axis is the total energy E. Each point in the (X, E) plane represents an equilibrium state, and the curves are adiabatic. In other words, from A, we can reach B or any other points on the same curve by an adiabatic process. Similarly we can start from A' and reach B' or any other points on the curve $A'B'$ by an adiabatic process. To change from one curve to another requires heat exchange.

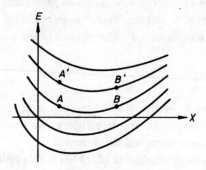

Fig. 2-1 Adiabatic curves in the X-E plane.

These adiabatic curves can be represented by the lines of constant values of a function $g(X, E)$:

$$g(X, E) = \sigma \quad . \quad \text{广延？} \tag{2-8}$$

Each curve is determined by a constant σ. This situation is like the contour lines in maps, where points with the same height are joined together. Here $g(X, E)$ is the "height" at the point (X, E). Of course, we can use another function to represent these curves. The only condition is that the value of this function is constant on the same line, e.g., $g(X, E)^3 = \sigma$ is also possible. These curves are in fact the solution of the differential equation $dE = f dX$ where σ is the integration constant. Each σ determines a curve. The conclusion of the above discussion is that we have introduced a new quantity σ which will be called the entropy. Note that its definition is still to a certain degree arbitrary.

From (2-8) we can solve for E, formally

$$E = E(X, \sigma) \quad . \tag{2-9}$$

That is to say, the energy E of the system is a function of the coordinate X and the entropy σ. If the entropy is a constant, (2-9) determines the adiabatic process characterised by σ. If X is constant, then the change of energy is equal to the heat flowing in:

$$dQ = dE$$

$$= \left(\frac{\partial E}{\partial \sigma}\right)_X d\sigma \quad . \tag{2-10}$$

Combining this with the first law given by (2-7), we get

$$dE = f\,dX + \tau\,d\sigma \;,$$

$$f = \left(\frac{\partial E}{\partial X}\right)_\sigma \;, \qquad \tau = \left(\frac{\partial E}{\partial \sigma}\right)_X \;,$$

$$d\sigma = \frac{dQ}{\tau} \;. \tag{2-11}$$

The above analysis can be slightly generalised to include all the coordinate changes:

$$dE = \sum_k f_k\,dX_k + \tau\,d\sigma \;,$$

$$f_k = \left(\frac{\partial E}{\partial X_k}\right)_\sigma \;,$$

$$\tau = \left(\frac{\partial E}{\partial \sigma}\right)_{X_1, X_2, \ldots} \;, \tag{2-12}$$

where f_k, X_k denote the various "forces" and coordinates in examples (2-1) to (2-5).

Hence entropy can be considered as a new coordinate like X_k, and τ is a new force like f_k.

Two questions remain to be discussed:

(1) The curves in Fig. 2-1 do not intersect. In fact we have already assumed this. If there were an intersection such as point C in Fig. 2-2(a), then f can have two different values at C. But $f = dE/dX$ is the slope of the curve and is physically the external "force" at equilibrium, i.e., the "force" inside the body equal and opposite to the external "force". For example, the pressure p is a "force" acting on the body, and is also the pressure inside the body acting against the external "force", thus maintaining equilibrium. To say that under the same equilibrium condition we have two different values of f is contradictory to experience. If each equilibrium state has only one unique value of f, then the intersection in Fig. 2-2(a) is not possible. A line curving back like that in Fig. 2-2(b) is also impossible, because the value of f would then be $+\infty$ or $-\infty$. These physical considerations imply that entropy must be a monotonic increasing function of E.

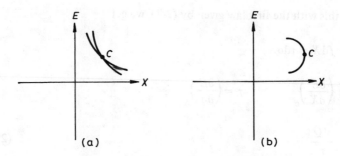

Fig. 2-2 (a) Intersection of two adiabatic curves.
(b) An adiabatic line curving back.

Notice that the basic assumption of thermodynamics (i.e., the equilibrium state is completely specified by a small number of coordinates) is most important. If there are other coordinates whose changes are unknown to us, then Figs. 2-2(a) and (b) would be possible and the above definition of entropy is unreliable. The validity of this basic assumption depends on the problem under consideration. A hidden worry in thermodynamics is: How many coordinates are necessary to determine the equilibrium state? How do we know that we have not missed any?

(2) The above definition leaves some room for arbitrariness. If we find $g(X, E)$ so that σ is defined by (2-8), we can still define another function $\bar{\sigma}$ such that

$$\bar{\sigma} = A(\sigma) ,$$

$$\bar{\tau} = \tau \Big/ \left(\frac{d\bar{\sigma}}{d\sigma}\right) ,$$

$$dQ = \bar{\tau} \, d\bar{\sigma} , \qquad (2\text{-}13)$$

where $A(\sigma)$ must increase with σ but is otherwise arbitrary. This condition is very weak. Hence we have the freedom to choose a suitable σ as the entropy.

2.4. Absolute Temperature and Entropy

In the last section we discussed an adiabatic process (see Fig. 2-1) and we introduced a new coordinate and a new "force", i.e., the entropy σ and the variable $\tau = \left(\dfrac{\partial E}{\partial \sigma}\right)_X$, see (2-9) – (2-12). The reader may ask: why not simply use the temperature as a coordinate? Why introduce the concept of entropy?

Actually all the coordinates plus temperature can determine the equilibrium state. In principle there is nothing wrong in using temperature as another coordinate. We can then use an isothermal process to draw curves in place of those in Fig. 2-1. But then the distinction between heat and work will not be so clear. Moreover, if temperature appears as a coordinate, entropy will appear like a force. Now we consider some special features of equilibrium to further specify the definition of entropy so that τ becomes the absolute temperature T, and σ becomes the entropy S as commonly understood. To treat T as the "force" and entropy as the coordinate yields a convenient conceptual framework.

A system in equilibrium has an important property: each and every part of it is in equilibrium. If we focus on one part of the system, then the remaining parts become the surrounding. Therefore, the so-called physical variables describing the surrounding must also describe the various parts. The most important variable of the surrounding is the temperature, because no matter how complicated or how inhomogeneous the structure of the system is, if it is in equilibrium, then the temperature is the same throughout. (Of course, if some part is isolated so that heat exchange cannot occur, then the temperature may be different.) What is the relevance of this to σ and τ?

Consider a physical system as composed of two parts 1 and 2. We define σ and τ as above and similarly σ_1, τ_1 and σ_2, τ_2 for the two parts. Since an equilibrium state is specified by the temperature and the other coordinates, it follows that $\sigma, \tau, \sigma_1, \tau_1$, and σ_2, τ_2 are also functions of the temperature θ and all the coordinates. We use θ to denote temperature measured on any convenient scale: Celsius, Fahrenheit, etc. Let X_{1k}, $k = 1, 2, 3, \ldots$ be all the coordinates of part 1. Similarly X_{2k} are all the coordinates of part 2. Suppose in a certain process, a certain amount of heat dQ enters this system. Out of this, dQ_1 flows into part 1 and dQ_2 into part 2. Then $dQ = dQ_1 + dQ_2$, i.e.,

$$\tau d\sigma = \tau_1 d\sigma_1 + \tau_2 d\sigma_2 \ . \tag{2-14}$$

This differential formula shows that σ is a function of σ_1 and σ_2, i.e., although σ changes with X_{1k}, X_{2k} and θ, this change is determined through the changes of σ_1 and σ_2. (σ_1 and σ_2 are functions of X_{1k} and X_{2k} respectively, and also functions of θ.) Therefore, we can say

$$\sigma = \sigma(\sigma_1, \sigma_2) \ . \tag{2-15}$$

Note that there is no other variable within the parentheses. Of course the differential of σ with respect to σ_1 or σ_2 can also be a function only of σ_1 and σ_2:

$$\left(\frac{\partial \sigma}{\partial \sigma_1}\right)_{\sigma_2} = \frac{\tau_1}{\tau} = U_1(\sigma_1, \sigma_2) ,$$

$$\left(\frac{\partial \sigma}{\partial \sigma_2}\right)_{\sigma_1} = \frac{\tau_2}{\tau} = U_2(\sigma_1, \sigma_2) , \qquad (2\text{-}16)$$

where τ_1, is a function of θ and X_{1k} only, and is not related to X_{2k}. Therefore, the relation between $\tau = \tau_1/U_1(\sigma_1, \sigma_2)$ and X_{2k} is only through σ_2. For the same reason, from the second equation of (2-16) we know that τ is related to X_{1k} only through σ_1. Hence

$$\tau = \tau(\theta, \sigma_1, \sigma_2) . \qquad (2\text{-}17)$$

Substituting this result into (2-16), we discover that τ_1, τ_2 are functions of σ_1, σ_2 and θ. Therefore,

$$\tau_1 = \tau_1(\sigma_1, \theta) ,$$

$$\tau_2 = \tau_2(\sigma_2, \theta) . \qquad (2\text{-}18)$$

Note that τ_1 cannot be a function of σ_2 because τ_1 is only related to X_{1k}. Equation (2-18) also shows that

$$\tau = \tau(\sigma, \theta) \qquad (2\text{-}19)$$

because the property of the parts is shared by the entire system. (Part 1 can be considered to represent almost the entire system.)

Now we note that the ratios τ_1/τ and τ_2/τ in (2-16) are not functions of θ, but τ_1, τ_2, τ themselves are directly related to θ. The only possibility is

$$\tau_1 = T(\theta) S_1'(\sigma_1) ,$$

$$\tau_2 = T(\theta) S_2'(\sigma_2) ,$$

$$\tau = T(\theta) S'(\sigma) . \qquad (2\text{-}20)$$

That is, their relations with θ is through a certain function $T(\theta)$. The functions S_1', S_2' and S' may be different, but τ_1, τ_2 and τ contain the same $T(\theta)$. Hence τ_1/τ, τ_2/τ will not depend on θ as a variable. Since any two systems can be combined into a single one, once $T(\theta)$ is determined, it can be applied universally to any body.

Now we modify the definition of entropy. We define $S(\sigma)$ as the entropy in the manner of (2-13), $\bar{\sigma} = S$, $dA/d\sigma = S'(\sigma)$,

$$S(\sigma) = \int S'(\sigma) \, d\sigma \quad,$$

$$dQ = \tau \, d\sigma = T \, dS \quad, \tag{2-21}$$

and call $T(\theta)$ the absolute temperature. The unit of $T(\theta)$ has yet to be chosen, because in (2-20) if we multiply $T(\theta)$ by a constant C, and divide S'_1, S'_2 and S' by C, the equations are unchanged. If the ice point is fixed at $T = 273$ K, then one obtains the usual absolute temperature scale. We can also use other units as well. Once the unit of T is established, the unit of S is fixed. But S still has an undetermined integration constant, which we shall discuss later.

If τ is replaced by T and σ by S in (2-14), then

$$dS = dS_1 + dS_2 \quad, \tag{2-22a}$$

where T has been cancelled on both sides.

This result can immediately be generalised: divide the system into m parts, then

$$dS = \sum_{i=1}^{m} dS_i \tag{2-22b}$$

Integrating each dS_i we get S_i with an undetermined integration constant. How do we choose the integration constants? As dS is the sum of the dS_i, it is convenient to require S to be sum of the S_i:

$$S = \sum_{i=1}^{m} S_i \quad. \tag{2-22c}$$

If we impose (2-22c), then the integration constants of each dS_i cannot be chosen arbitrarily. We now treat (2-22c) as an imposed condition. It is easily satisfied in actual calculations, but there are some details which we shall illustrate in the examples in the next section. The additivity of the entropy is not straightforward or obvious. The above analysis of the entropy is rather abstract. Now let us consider an example.

2.5. The Ideal Gas

Now we apply the above analysis to the case of an ideal gas. Our aim is not to derive the properties of the ideal gas, but to illustrate the abstract ideas of the preceding section.

There is a simple relationship for the pressure p, the total energy E and the volume V of an ideal gas, i.e.,

$$p = \frac{2E}{3V} \ . \tag{2-23}$$

Its origin is as follows. Let $\mathbf{v}$ be the velocity of a molecule and v_x be the component perpendicular to one wall of the container. Each collision results in a momentum change (impulse) of $2mv_x$ and the pressure is the impulse per unit time per unit area on the wall:

$$\begin{aligned} p &= \frac{1}{2} \int d^3\mathbf{v}\, f(\mathbf{v})\, v_x \cdot 2mv_x \\ &= \int d^3\mathbf{v}\, f(\mathbf{v})\, mv_x^2 \\ &= \frac{2}{3} \int d^3\mathbf{v}\, f(\mathbf{v}) \cdot \frac{1}{2} mv^2 \\ &= \frac{2}{3} \frac{E}{V} \ , \end{aligned} \tag{2-24}$$

where $d^3\mathbf{v}\, f(\mathbf{v})$ is the density of the molecules with velocity $\mathbf{v}$ and $d^3\mathbf{v}\, f(\mathbf{v})\, v_x$ is the flux in the x direction. The factor $\frac{1}{2}$ is due to the fact that on the average, only half of the molecules move towards the wall, while the other half move away from the wall. The distribution of the directions of the velocities is isotropic and hence v_x^2 in the integral has been replaced by $(\frac{1}{3})(v_x^2 + v_y^2 + v_z^2) \equiv (\frac{1}{3})v^2$. The validity of (2-24) does not depend on the distribution function $f(\mathbf{v})$.

Suppose the number of molecules N is constant. (We have not yet introduced entropy.) The curves drawn from the adiabatic processes are determined by

$$dE = -p\, dV \ . \tag{2-25}$$

Substituting in (2-24), we get

$$\frac{dE}{E} + \frac{2}{3}\left(\frac{dV}{V}\right) = 0 \ , \tag{2-26}$$

with the solution

$$EV^{2/3} = \sigma \ , \tag{2-27}$$

where σ is an integration constant. Each value of σ determines a curve in the

(V, E) plane (see (2-8)). Since σ defines the entropy,

$$\tau = \left(\frac{\partial E}{\partial \sigma}\right)_V = V^{-2/3} \qquad (2\text{-}28)$$

Now we have the coordinates V and σ and their values specify the equilibrium state of this ideal gas. The total energy of the gas is $E = \sigma V^{-2/3}$. The concept of temperature has not yet appeared. The discussion in Section 2-4 showed that given a measured temperature θ, then the absolute temperature $T(\theta)$ is fixed. However, we have to determine θ first. The pressure and the energy are precise concepts in mechanics and indeed this enabled us to find their relationship (2-24). However, the temperature is different; its determination is easy, but its relation with the motion of the molecules is not clear. We can, nonetheless, use experimental results. According to experiments, for a dilute gas of N molecules, the combination pV/N is unchanged at a fixed temperature. This fact yields the relation

$$\frac{pV}{N} = f(\theta) \qquad (2\text{-}29)$$

where θ can be measured on any scale. The function $f(\theta)$ can be determined experimentally on the scale chosen.

As pointed out in the last section, τ must be the absolute temperature $T(\theta)$ multiplied by a function $S'(\sigma)$ of σ (see (2-20)). Therefore, from the τ in (2-28) and the experimental result (2-29) of the ideal gas, we can derive the absolute temperature. The steps are as follows. From (2-27) and (2-23), we can calculate pV and together with (2-29), we obtain

$$f(\theta) = \frac{2}{3}\left(\frac{\sigma}{N}V^{-2/3}\right) \qquad (2\text{-}30)$$

Hence, (2-28) can be written as

$$\tau = f(\theta)\frac{3}{2}\left(\frac{N}{\sigma}\right) \qquad (2\text{-}31)$$

Let us define the absolute temperature as $f(\theta)/k = T(\theta)$, then we obtain from (2-29) and (2-20)

$$kT = \frac{pV}{N} , \qquad (2\text{-}31')$$

$$S'(\sigma) = \frac{3}{2}\left(\frac{Nk}{\sigma}\right) \qquad (2\text{-}32)$$

The absolute temperature can be measured by p, V and N in this way. The relation of the existing temperature scale and the absolute scale is established by the experimental function $f(\theta)$. Here k is a constant determined by the unit of T. The combination pV/N has the same unit as energy; if T has the usual unit, then

$$k = 1.38 \times 10^{-16} \text{ erg K}^{-1} \ . \tag{2-33}$$

This is the Boltzmann constant. If we simply use the unit of the energy to measure T, then

$$k = 1 \ . \tag{2-34}$$

In the following chapters, we shall use $k = 1$ for simplicity. The choice of k is a choice of units and is not unique.

From (2-32), we obtain

$$S = \int S'(\sigma) \, d\sigma = \frac{3}{2} N \ln \sigma + \text{const.} \tag{2-35}$$

The constant in the equation can be a function of N. Since (2-25) has not taken dN into account, N is thus a constant. If we substitute (2-27) into (2-35) and write the constant in (2-35) as

$$N \ln(C^{3/2} N^{-5/2}) \ ,$$

then

$$S = N \ln \left[(V/N)(CE/N)^{3/2} \right] \ , \tag{2-36}$$

is the entropy of an ideal gas, where E/N is the average energy of one molecule and N/V is the density. The constant C is independent of N because S must be additive, so under a fixed temperature and density, S must be proportional to N.

Now we mention the problem of the entropy of mixing. Suppose we have two containers (1 and 2) of gas at the same temperature and pressure, with N_1 and N_2 molecules and occupying volumes V_1 and V_2. If we now connect the two volumes, the gases will mix. The pressure and temperature will remain the same after mixing. In fact we do not observe any change (assume that the two gases have the same colour).

Let $V = V_1 + V_2$, $N = N_1 + N_2$, $E = E_1 + E_2$. The entropy of each gas before mixing can be calculated from (2-36)

$$S_1 = \frac{3}{2} N_1 \ln \epsilon C + N_1 \ln \frac{V_1}{N_1} ,$$

$$S_2 = \frac{3}{2} N_2 \ln \epsilon C + N_2 \ln \frac{V_2}{N_2} ,$$

where

$$\epsilon \equiv \frac{E_1}{N_1} = \frac{E_2}{N_2} = \frac{E}{N} ,$$

$$\frac{V_1}{N_1} = \frac{V_2}{N_2} = \frac{V}{N} . \tag{2-37}$$

The sum of the two entropies is

$$S = S_1 + S_2 = \frac{3}{2} N \ln \epsilon C + N \ln \frac{V}{N} . \tag{2-38}$$

After mixing, the energy is unchanged but each gas occupies the two containers, of total volume V, so their entropy is

$$S_1' = \frac{3}{2} N_1 \ln \epsilon C + N_1 \ln \frac{V}{N_1} ,$$

$$S_2' = \frac{3}{2} N_2 \ln \epsilon C + N_2 \ln \frac{V}{N_2} . \tag{2-39}$$

Their sum $S' = S_1' + S_2'$ is larger than S:

$$S' - S = N_1 \ln \frac{N}{N_1} + N_2 \ln \frac{N}{N_2} . \tag{2-40}$$

This excess comes from mixing, and is called the entropy of mixing.

We have said before that any system can be divided into several parts and we can calculate the entropies separately and then add them up to obtain the entropy of the entire system. This example of the entropy of mixing illustrates that if a gas is composed of the same kind of molecules, then we cannot divide the molecules into two parts and use (2-39) to calculate the entropies of each part and then add them. For this case of one kind of molecule, then the answer is simply given by (2-38) because for molecules of the same kind there is no problem of mixing. This point will be reconsidered when we discuss the independent property of the various parts of a system.

The Second Law of Thermodynamics

This law is different from the general conservation laws, Newton's laws or the laws in electricity and magnetism. It is a very special law in physics. It states that no engine can continually absorb the internal energy in matter and convert it into work. For example, wouldn't it be nice if an engine could be designed for a ship which extracts energy from the water molecules in the sea to propel the ship? It would then be unnecessary to use steam or oil or nuclear power! But the second law states that this is impossible. In other words, there is no machine whose sole effect is to convert internal energy into work. Of course, engines do convert internal energy into work, such as the steam engine using the energy of the steam to do work, but they all have additional side effects in wasting some energy and increasing the temperature (above that of the environment).

This law may be considered to be empirical. Up to now it has not been known to be violated, but on the other hand, it has really never been proved.

From this law we can immediately draw the following conclusions. If heat is to be moved from low temperature to high temperature, work must be done. If it does no work, then we would be able to transfer the energy from the sea water to the steam engine so that the steam engine can do work. The second law will then be violated. As another example, an isolated system cannot become half cold and half hot.

The most important conclusion is that the entropy change of an adiabatic process cannot decrease. If it decreases, we can supply heat to increase the entropy and the temperature. Then we perform work by an adiabatic process to use up the heat and let the entropy decrease again. Hence we can continually convert heat into work, violating the second law. Every conclusion above can be regarded as a restatement of the second law. The content is the same.

Therefore, a body in equilibrium cannot perform external work and its entropy must be a maximum. Later we shall re-examine this law.

2.7. The Efficiency and Speed of a Heat Engine

Let us review the ideal heat engine, i.e., the Carnot heat engine. Suppose an ideal gas is contained in a heat engine. This heat engine has four processes (Fig. 2-3):

(a) adiabatic expansion, performing work.

(b) isothermal compression, releasing heat and having work done on it while in contact with a low temperature reservoir of temperature T_1.

(c) adiabatic compression with work done on the engine.

(d) isothermal expansion, absorbing heat and performing work while in

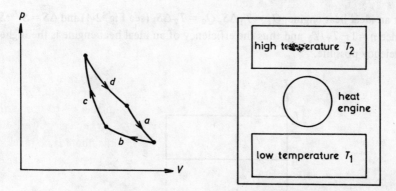

Fig. 2-3 The Carnot cycle: curves (a) and (c) represent adiabatic processes; curves (b) and (d) represent isothermal processes.

contact with a high temperature reservoir of temperature T_2. This engine is said to be ideal because all processes are assumed to be reversible, and heat flow occurs very slowly. When in contact with the upper reservoir, the temperature of the engine is assumed to be T_2, and heat flows very slowly into the engine. Likewise, when in contact with the lower reservoir, the temperature of the engine is assumed to be T_1 and heat flows very slowly out of the engine. For every cycle, the net work is

$$W = Q_2 - Q_1 \ , \tag{2-41}$$

i.e., the absorbed heat less the heat released. The definition of the efficiency of the heat engine is

$$\eta = \frac{W}{Q_2} = 1 - \frac{Q_1}{Q_2} \ , \tag{2-42}$$

i.e., Q_1 is the heat wasted. According to the second law, the above cyclic process must increase the total entropy of the reservoirs. We now regard the high and low temperature reservoirs as a huge adiabatic body (the big box in Fig. 2-3). The increase of total entropy is

$$\frac{Q_1}{T_1} - \frac{Q_2}{T_2} = \frac{Q_2 - W}{T_1} - \frac{Q_2}{T_2} \geqslant 0 \ . \tag{2-43}$$

Hence

$$\eta \leqslant 1 - \frac{T_1}{T_2} \ . \tag{2-44}$$

For an ideal heat engine, $Q_1 = T_1 \Delta S$, $Q_2 = T_2 \Delta S$, (see Fig. 2-4) and $\Delta S = S'' - S'$. Hence $\eta = 1 - T_1/T_2$ and thus the efficiency of an ideal heat engine is the highest efficiency possible.

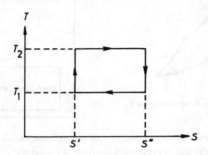

Fig. 2-4 S-T diagram for an ideal heat engine.

If the heat engine has a temperature which is the same as the reservoir when absorbing and releasing heat, then the rate of heat exchange must be zero. Therefore to attain this maximum efficiency, the heat engine must be infinitely slow. This is, however, not practical. To increase the speed we have to sacrifice some efficiency. What is the best compromise? This, of course, depends on the practical considerations. In what follows we shall roughly estimate the relation between the efficiency and the speed.

Let T_1' be the temperature of the heat engine when it is in contact with the low temperature reservoir and T_2' that of the heat engine when it is in contact with the high temperature reservoir. Of course $T_2 > T_2' > T_1' > T_1$ is the necessary condition. Let

$$\alpha_1 \equiv T_1'/T_1 \quad ,$$

$$\alpha_2 \equiv T_2'/T_2 \quad . \tag{2-45}$$

The efficiency of the heat engine is

$$\eta = 1 - \frac{T_1'}{T_2'} = 1 - \frac{T_1}{T_2}\alpha_1\alpha_2 \quad . \tag{2-46}$$

The time spent in each cycle t is equal to the sum of the times spent absorbing and discarding heat:

$$t \propto \frac{Q_2}{T_2 - T_2'} + \frac{Q_1}{T_1 - T_1'} \quad . \tag{2-47}$$

It is assumed in (2-47) that the rate of heat flow is proportional to the temperature difference. Since $Q_2 \propto T_2'$ and $Q_1 \propto T_1'$,

$$t \propto \frac{T_2'}{T_2 - T_2'} + \frac{T_1'}{T_1 - T_1'}$$

$$= \frac{\alpha_1 \alpha_2 - 1}{(\alpha_1 - 1)(\alpha_2 - 1)}. \qquad (2\text{-}48)$$

For simplicity we assume $\alpha_1 = \alpha_2 = \alpha$. Figure 2-5 shows the efficiency η and the cycling frequency $1/t$ versus α^2. Let us suppose we want to maximise the product η/t.

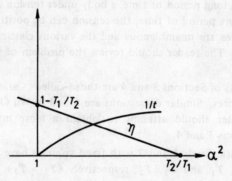

Fig. 2-5 Graphs of efficiency (η) and cycling frequency $1/t$ versus α^2.

If T_1/T_2 is very small, i.e., the original ideal efficiency is very high, then the maximum of η/t is approximately at $\alpha = (T_2/T_1)^{1/3}$. The efficiency at this value of α is then

$$\eta \simeq 1 - \left(\frac{T_1}{T_2}\right)^{1/3}. \qquad (2\text{-}49)$$

Therefore the efficiency is greatly reduced. If T_1/T_2 is close to 1, i.e., the original ideal efficiency is not high, then the maximum of η/t occurs at about $\alpha \approx 1 + \frac{1}{4}(1 - T_1/T_2)$, and the value of η is

$$\eta \simeq \frac{1}{2}\left(1 - \frac{T_1}{T_2}\right). \qquad (2\text{-}50)$$

i.e., half of the ideal value. The above is just a rough estimate. The value of η/t does not necessarily increase with the overall efficiency but this result does show that the efficiency of a practical heat engine is far below that of an ideal one.

The above sections give a brief introduction to several basic concepts in thermodynamics, focusing on the concepts of "coordinate" and "force", and including entropy and temperature into this group of concepts. The discussion of thermodynamics should not stop here. Further discussions are scattered in later chapters, and elaborated upon using a statistical mechanical interpretation.

Problems

1. Tension is a negative pressure. The reader may have learnt the theorem in thermodynamics stating that pressure must be positive. As mentioned in Chapter 1, over a long period of time, a body under tension is not stable. However, during a short period of time, the tension can be positive and the thermodynamic properties are unambiguous and the various elasticity coefficients are also well defined. The reader should review the problem of stability in thermodynamics.

2. The contents of Sections 3 and 4 are the so-called "Caratheodory theorem" and its consequences. Similar discussions are found in Wu (1979), and Wannier (1964). The reader should attempt to obtain a more mathematical understanding of Sections 3 and 4.

3. Two adiabatic fluids 1 and 2 with fixed volumes have heat capacities and temperatures C_1, T_1 and C_2, T_2, respectively ($T_1 > T_2$). Suppose a Carnot engine absorbs heat from 1 and dumps the heat into 2 until 1 and 2 reach the same temperature T_0.

(a) Calculate T_0 and the work done by the Carnot engine.

(b) If heat can flow directly from 1 to 2, what would be the temperature at equilibrium? What is the change of entropy?

(c) If C_1 and C_2 are both positive, prove that entropy must increase. The above assumes C_1 and C_2 do not change with temperature.

4. Derive (2-48). If the heat flow is not proportional to the temperature difference, the result will not be the same. Let the heat flow be $A(\Delta T) + B(\Delta T)^2$ where ΔT is the temperature difference and A, B are constants, with B much smaller than A. Calculate the corrections to (2-48) – (2-50).

5. The equation of state represents a relation among the thermodynamic variables, such as the "coordinates" and the "forces" in thermodynamics, e.g., $pV = NT$ is the equation of state of an ideal gas. The equation of state can be directly determined by experiment. The equation of state of a magnetic body is

$$M = AH/T$$

where M is the magnetic moment, h is the external magnetic field and A is a

constant. Find the relation between entropy and h.
(Hint: from $dE = TdS + hdM$ we can get $\partial S/\partial h = \partial M/\partial T$).

6. The hidden worry of thermodynamics is: we do not know how many coordinates or forces are necessary to completely specify an equilibrium state. Given a set of thermodynamic variable S, there may not be a definite equilibrium state. This has been discussed in Section 2-1. For example the property of steel depends on the arrangement of carbon atoms in the iron. Knowing the number of carbon atoms and that of iron does not specify the equilibrium state. Thermodynamics is not a complete science, and in its applications we must consider many other details. Consider the following example.

Two copper spheres with the same mass and at the same temperature are both under one atmospheric pressure. One is ten times bigger than the other. Why?

Answer: The larger sphere is hollow.

This is not meant merely as a joke. The reader should devote more thought on the coordinates, the time scale, etc.

Chapter 3
THE LAW OF DETAILED BALANCE

Molecular motion can be looked upon as a sequence of reactions and the reactions cause changes. The collision of the molecules causing a change of momentum is a kind of reaction. If the structure of the molecules is changed by the reaction, this is then a chemical reaction. Detailed balance implies that in equilibrium, the number of occurrence of each reaction in the forward direction is the same as that in the reverse direction. That is to say, we have equilibrium not only macroscopically, but also for each microscopic reaction. Detailed balance can be understood from the symmetry of the arrow of time. If time is reversed, the law of the molecular motion is unchanged. Equilibrium is an unchanged state and looks the same whether time flows forwards or backwards. Hence the number of reactions in the forward and backward directions must be the same.

Detailed balance can be applied directly to analyse the ideal gas, including a dense quantum gas. The density in the so-called quantum gas is so high that the wave nature of the particles must be considered. The reaction in an ideal gas is very easy to analyse. If we add some plausible assumptions, we can calculate the density distribution of the particles from detailed balance and it can explain the relationship of temperature and chemical potential with the conserved quantities of the various reactions. Detailed balance can be applied in reverse, i.e. we can determine some properties of the reactions from the known density distribution of the particles. This is the main content of this chapter.

Detailed balance can be used as the starting point of statistical mechanics. In principle it is capable of analysing any equilibrium state. The now commonly used Monte Carlo method is based on detailed balance for calculation. This method utilises numerical calculations on the computer to simulate various reactions. We shall discuss this in Chapter 25.

Detailed balance cannot take the place of the basic assumption in Chapter 5, because we are as yet unable to make careful analysis on more complicated reactions. Even for the ideal gas case, we have to assume the independence of all the reactions. In using detailed balance, we cannot avoid introducing the basic assumption. However, it may appear in different forms.

3.1. The Energy Distribution of the State

Each state is a trajectory, e.g. the state of the electrons in an atomic model is the most familiar. Each state is specified by a group of quantum numbers (n, l, m, σ). The energy of a state in a hydrogen-like atom is

$$-\frac{Z^2 R}{n^2} \; , \qquad n = 1, 2, \ldots ,$$

$$R = 13.6 \text{ eV} \; , \tag{3-1}$$

where Z is the charge of the nucleus (in units of electron charge). The quantum numbers l, m are respectively the angular momentum and its projection in a certain direction, and σ is the projection of the spin of the electron in a certain direction.

Each state can accommodate one electron, because electrons are fermions and must obey the exclusion principle, i.e. two particles cannot occupy the same state. The above atomic model is to put Z electron into the Z lowest energy states which are determined by the electrostatic potential of the nucleus. In most problems, the state of the particles are determined by the potential energy. We review the simplest situation and calculate the energy distribution of the states.

A. The statistics of the energy distribution

Let us first look at the states of a free particle. This particle is confined in a cube of volume L^3 and obeys periodic boundary conditions, i.e. $x = 0$ and $x = L$ are the same side, and similarly in the y, z directions. Hence this is a three dimensional ring-shaped container without boundaries. The results of a quantum mechanical analysis are as follows. The state of the particle is specified by three integers $(n_x, n_y, n_z) \equiv \mathbf{n}$. (For the moment we do not consider the spin of the particle or its other internal states.) The energy for **n** is

$$\epsilon_\mathbf{p} = \frac{p^2}{2m},$$

$$\mathbf{p} = \frac{2\pi}{L}(n_x, n_y, n_z),$$

$$n_x, n_y, n_z = 0, \pm 1, \pm 2, \ldots . \tag{3-2}$$

The wavefunction in the **n** state is a plane wave

$$\phi_\mathbf{n} = \frac{1}{L^{3/2}} e^{i\mathbf{p}\cdot\mathbf{r}} \tag{3-3}$$

(We take $\hbar = 1$ to simplify the formula.) Each configuration is a state. The energy distribution of the states is

$$L^3 g(\epsilon) = \sum_{n_x=-\infty}^{\infty} \sum_{n_y=-\infty}^{\infty} \sum_{n_z=-\infty}^{\infty} \delta\left[\epsilon - \frac{1}{2m}\left(\frac{2\pi}{L}\right)^2 (n_x^2 + n_y^2 + n_z^2)\right]. \tag{3-4}$$

If L is very large, and ϵ not small, the configurations in (3-4) have very large n_x, n_y and n_z, so they can be treated as continuous variables, and Σ is replaced by an integral:

$$\sum_{n_x}\sum_{n_y}\sum_{n_z} = \int dn_x \int dn_y \int dn_z . \tag{3-5}$$

Finally the integration over n can be changed to that over p. (See (3-2).) Hence (3-4) becomes

$$L^3 g(\epsilon) = \frac{L^3}{(2\pi)^3} \int d^3 p \, \delta\left(\epsilon - \frac{p^2}{2m}\right)$$

$$= \frac{L^3}{2\pi^2} m \sqrt{2m\epsilon} . \tag{3-6}$$

Notice that if we use the hard wall boundary condition, i.e. the wavefunction $\phi_\mathbf{n}(\mathbf{r})$ is zero at the boundary, then $\mathbf{n} = (n_x, n_y, n_z)$,

$$\phi_\mathbf{n} = \left(\frac{2}{L}\right)^{3/2} \sin x p_x \sin y p_y \sin z p_z \; ,$$

$$\mathbf{p} = \frac{\pi}{L}(n_x, n_y, n_z) \; ,$$

$$\epsilon = \frac{p^2}{2m} \; , \qquad n_x, n_y, n_z = 1, 2, 3, \ldots \quad . \tag{3-7}$$

Equations (3-7) and (3-2) are not the same.

However, the reader can prove that the $g(\epsilon)$ calculated from (3-7) is the same as (3-6). Provided L is very large, then the boundary condition does not affect $g(\epsilon)$.

B. The statistics of the energy in the short wavelength approximation

In the short wavelength approximation (WKB approximation) the trajectory of the particle can be calculated from classical mechanics, but each trajectory must satisfy a quantum condition. Let us look at the situation of a particle in one dimension. The quantum condition is

$$\oint p(x) \, dx = nh \; , \tag{3-8}$$

$$\epsilon_n = \frac{p(x)^2}{2m} + U(x) \; . \tag{3-9}$$

where $U(x)$ is the potential energy of the particle, $p(x)$ is the momentum, h is Planck's constant and n is an integer. The integration in (3-8) is along a trajectory in phase space. (Fig. 3-1)

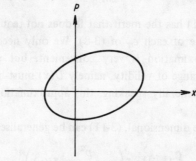

Fig. 3-1 A trajectory in phase space.

34 STATISTICAL MECHANICS

The quantum condition (3-8) determines the various ϵ_n and the trajectories. Each integer specifies a trajectory and each trajectory a state. This method is valid for short wavelength, i.e. over a distance of h/p, $U(x)$ changes very little. Here h/p is the wavelength of the particle and the integral of (3-8) is just the area bounded by the trajectory.

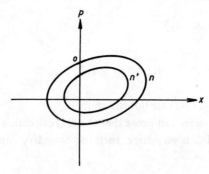

Fig. 3-2 Two trajectories n and n' in phase space.

The annulus region in Fig. 3-2 is specified by two trajectories n and n'. In this region, there are $n-n'$ trajectories. According to (3-8)

$$n - n' = \frac{1}{h} \int dp\, dx \quad . \tag{3-10}$$

That is to say, the area of the region divided by h is the number of trajectories in the region. Therefore, to calculate the states we only need to find the area and then divide it by h. At the same time we can think of the state as points uniformly distributed in phase space with density $1/h$. Therefore,

$$L^3 g(\epsilon) = \frac{1}{h} \int dp\, dx\, \delta\!\left(\epsilon - \frac{p^2}{2m} - U(x)\right) \quad . \tag{3-11}$$

This integral (3-11) has the merit that it does not require solution of the trajectories or the value of each ϵ_n of (3-8). We only need $U(x)$. Hence the short wavelength approximation is very convenient, but we must remember the limitation on its range of validity, namely $U(x)$ must be nearly constant within a distance h/p. Generally speaking, this approximation is only true for states of high energy.

If space is three dimensional, (3-11) can be generalised to

$$L^3 g(\epsilon) = \frac{1}{h^3} \int d^3p\, d^3r\, \delta\!\left(\epsilon - \frac{p^2}{2m} - U(x)\right) \quad . \tag{3-12}$$

If the spin of the particle is not zero, we must add a spin quantum number to specify the state. Bosons have integer spins (in units of $\hbar$) while the spins of fermions are half integers (integer + $\frac{1}{2}$). The most usual fermions are electrons with spin $\frac{1}{2}$, while bosons include photons (spin 1) and ^{4}He (spin 0), etc.

In some solids or liquids, certain motions of small amplitude can be described by special "particles", e.g. phonons (for crystal vibration), spin wave (or magnons, for oscillation of magnetic moment), etc. These particles are usually bosons. (Notice that these particles do not exist in vacuum, but only in matter. Ordinary matter is not homogeneous, but has a crystal structure. Hence the concept of "spin" must be replaced by the symmetry property of the structure of the matter.)

3.2. The Energy Distribution of Fermions

We now use the law of detailed balance to calculate the energy distribution of the particles. This illustrates the content of the law and also emphasises collision as a basic ingredient of the ideal gas; without collisions there would be no thermodynamics.

A gas may be said to be ideal if the collision time (the time the particles are in contact) is very short, and the radius of contact (i.e. the effective interaction distance of the particles) is very small.

Each collision is a reaction. We assume that each reaction is an independent event and there is no relation between successive reactions. This is a rather effective hypothesis, making the analysis much simpler.

Consider a fermionic gas and let f_k be the average number of particles in the k state, where k denotes the momentum and spin of the particles. Because of collisions, each particle continually changes its state, e.g.

$$1 + 2 \longleftrightarrow 3 + 4 \quad , \tag{3-13}$$

i.e. the states 1 and 2 interact to change to states 3 and 4. The rate of this reaction is

$$f_1 f_2 (1 - f_3)(1 - f_4) R \quad , \tag{3-14}$$

where f_1 and f_2 are the probability that there is one particle in the states 1 and 2 respectively and $(1 - f_3)$, $(1 - f_4)$ are the probability that there are no particles in the states 3 and 4. Because the gas is fermionic, if states 3, 4 are occupied, the reaction cannot proceed. This rate must be equal to that in the reverse direction, i.e.

$$f_1 f_2 (1 - f_3)(1 - f_4) R = f_3 f_4 (1 - f_1)(1 - f_2) R' \quad . \tag{3-15}$$

According to quantum mechanics $R = R'$, because the equations of quantum mechanics are invariant under time reversal

$$R = |M|^2 \quad ,$$

$$R' = |M^*|^2 = R \quad . \tag{3-16}$$

Here, M is the matrix element of the scattering operator between states $|1, 2\rangle$ and $|3, 4\rangle$. If time is reversed $|1, 2\rangle$ and $|3, 4\rangle$ are interchanged, and M becomes M^*. From (3-15) we get

$$\frac{f_1}{1-f_1} \cdot \frac{f_2}{1-f_2} = \frac{f_3}{1-f_3} \cdot \frac{f_4}{1-f_4} \quad . \tag{3-17}$$

If there is a reaction of three particles

$$1 + 2 + 3 \to 4 + 5 + 6 \quad ,$$

detailed balance will give results similar to (3-17), namely

$$\frac{f_1}{1-f_1} \cdot \frac{f_2}{1-f_2} \cdot \frac{f_3}{1-f_3} = \frac{f_4}{1-f_4} \cdot \frac{f_5}{1-f_5} \cdot \frac{f_6}{1-f_6} \quad . \tag{3-18}$$

Equations (3-17) and (3-18) are very stringent conditions, to be satisfied for all f_k. We can obtain f_k from (3-17) and (3-18). Of course, if all the f_k satisfy

$$\frac{f_k}{1-f_k} = \text{constant} \quad , \tag{3-19}$$

then (3-17) and (3-18) are certainly satisfied. But this means that each f_k is the same independent of the energy of the states, which is unreasonable. To solve (3-17) and (3-18) we also have to consider the conserved quantities of the reaction. Before and after the reactions (3-17) and (3-18), the energy and the particle number are conserved.

(3-17): particle number = 2, energy $= \epsilon_1 + \epsilon_2 = \epsilon_3 + \epsilon_4$

(3-18): particle number = 3, energy $= \epsilon_1 + \epsilon_2 + \epsilon_3 = \epsilon_4 + \epsilon_5 + \epsilon_6$.

(3-20)

Hence, there is another solution to (3-17) and (3-18), i.e.

$$\frac{f_k}{1-f_k} = e^{-\alpha - \beta \epsilon_k} \quad , \tag{3-21}$$

where α and β are constants. To see that this is a solution, substitute (3-21) into (3-17) and (3-18), and use (3-20) to get

$$e^{-2\alpha-\beta(\epsilon_1+\epsilon_2)} = e^{-2\alpha-\beta(\epsilon_3+\epsilon_4)}$$

$$e^{-3\alpha-\beta(\epsilon_1+\epsilon_2+\epsilon_3)} = e^{-3\alpha-\beta(\epsilon_4+\epsilon_5+\epsilon_6)} \quad . \tag{3-22}$$

Equation (3-21) of course means

$$f_k = \frac{1}{e^{\alpha+\beta\epsilon_k}+1} \quad . \tag{3-23}$$

This is the particle distribution satisfying the condition of detailed balance. The parameters α and β are constants shared by all the particles and should be related to the temperature and chemical potential. The relations of the total energy E and the particle number N with f_k are

$$E = \sum_k \epsilon_k f_k = V\int d\epsilon\, g(\epsilon) \frac{\epsilon}{e^{\alpha+\beta\epsilon}+1} \quad ,$$

$$N = \sum_k f_k = V\int d\epsilon\, g(\epsilon) \frac{1}{e^{\alpha+\beta\epsilon}+1} \quad . \tag{3-24}$$

If the gas is very dilute, i.e., $f_k \ll 1$, then

$$f_k \simeq e^{-\alpha-\beta\epsilon_k} \quad . \tag{3-25}$$

This is the familiar result of the ideal gas, i.e. the Boltzmann distribution, and can be directly calculated from the energy of the states. Because $g(\epsilon) \propto \sqrt{\epsilon}$ (see (3-5)), we get

$$\frac{E}{N} = \frac{3}{2\beta} \quad . \tag{3-26}$$

We discover that $1/\beta$ is the temperature T, because in thermodynamics we know that $E/N = \frac{3}{2}T$. The parameter α can be calculated from (3-24). Using (3-25) for f, we get

$$\frac{N}{V} = e^{-\alpha}\int d\epsilon\, g(\epsilon)\, e^{-\beta\epsilon}$$

$$= e^{-\alpha}(2s+1)(2\pi mT)^{3/2}/h^3 \quad , \tag{3-27}$$

i.e.

$$\alpha = \ln\left[(2s+1)(2\pi mT)^{3/2}\frac{V}{Nh^3}\right] , \qquad (3\text{-}28)$$

where s is the spin of the particle. The chemical potential μ of the ideal gas can be obtained from the results in thermodynamics (2-36)

$$-\frac{\mu}{T} = \left(\frac{\partial S}{\partial N}\right)_E = \ln\frac{V}{N}\left(\frac{CE}{N}\right)^{3/2} - \frac{5}{2}$$

$$= \ln\frac{V}{N} + \ln\left(\frac{3}{2}TC\right)^{3/2} - \frac{5}{2} . \qquad (3\text{-}29)$$

Equations (3-28) and (3-29) are identical except for an undetermined constant C. We can define

$$\frac{\mu}{T} = -\alpha . \qquad (3\text{-}30)$$

This fixes the value of C. Equation (2-36) becomes

$$S = N\ln\left[\frac{V}{N}\left(\frac{4\pi mE}{3Nh^3}\right)^{3/2}(2s+1)\right] + \frac{5}{2}N . \qquad (3\text{-}31)$$

Substituting (3-30) into (3-23) we get the familiar Fermi distribution

$$f_k = \frac{1}{e^{(\epsilon_k - \mu)/T} + 1} . \qquad (3\text{-}32)$$

Notice that in the above we have used the known ideal gas results to determine the meaning of α and β. The ideal gas can be looked upon as a device to measure the temperature and the chemical potential of the gas. To express this idea more clearly: We can make the potential energy $U(\mathbf{r})$ very high in one corner of the container so that the density becomes very low in this corner, so the ideal gas law is applicable. But μ, T are quantities shared by all parts of the body because the various reactions involve all the particles. Hence once we measure μ and T in this corner, then we know the μ and T for the whole gas.

For general densities the entropy can be obtained by integrating the heat capacity:

$$S = \int^T \left(\frac{\partial E}{\partial T}\right)_V \frac{dT}{T} . \qquad (3\text{-}33)$$

The integration constant can be determined by the value at $T \to \infty$, i.e. the ideal gas result (3-31).

Hence, from detailed balance and the results of thermodynamics in Chapter 2, we determine the particle distribution and all the equilibrium properties of a fermionic gas. The constant C in the definition of entropy is determined by (3-30). This is a convenient choice. We shall discuss this at the end of the next section. The most important fermions are the electrons, which we shall discuss in the next chapter.

3.3. Conservation Laws, Chemical Potential and the Constant Term in the Entropy

The above analysis clearly shows the power of the conservation laws. Equation (3-21) is derived using the conservation of particle number and the conservation of energy. Hence T and μ are the forces fixing the total energy and the total particle number.

If there are other conserved quantities, then (3-17) and (3-18) can have more solutions. For example, if momentum is conserved, i.e.

$$\mathbf{p}_1 + \mathbf{p}_2 = \mathbf{p}_3 + \mathbf{p}_4 \quad \text{in (3-17)},$$

$$\mathbf{p}_1 + \mathbf{p}_2 + \mathbf{p}_3 = \mathbf{p}_4 + \mathbf{p}_5 + \mathbf{p}_6 \quad \text{in (3-18)} \quad . \tag{3-34}$$

then this allows a more general solution of (3-17) and (3-18):

$$\frac{f_k}{1 - f_k} = e^{-\alpha - \beta \epsilon_k - \beta \mathbf{v} \cdot \mathbf{p}_k} , \tag{3-35}$$

i.e.

$$f_k = \frac{1}{e^{(\epsilon_k - \mathbf{v} \cdot \mathbf{p}_k - \mu)/T} + 1} , \tag{3-36}$$

where $\mathbf{v}$ is the velocity of the whole gas. But if the container is stationary, collisions of the particles with the wall makes momentum nonconserved, and $\mathbf{v}$ must be 0.

If the spins are unchanged during collisions, then

$$f_k = \frac{1}{e^{(\epsilon_k - \mu - \mathbf{s} \cdot \mathbf{h})/T} + 1} , \tag{3-37}$$

where the parameter $\mathbf{h}$ can be looked upon as the force maintaining the total spin. If $s = \frac{1}{2}$, then $\mathbf{s} \cdot \mathbf{h} = \pm \frac{1}{2} h$.

Each conservation law will introduce a "force". The above "forces" $1/T$, $-\mu/T$, $\mathbf{v}/T$ and $\mathbf{h}/T$ are obvious examples. Conversely, if a conservation law is broken, it will cause a force to become zero, e.g. the walls of a container cause momentum not to be conserved, and $\mathbf{v} = 0$. If the particle number is not conserved, i.e. it can change because of reactions, μ must be zero. More complicated examples are found in chemical reactions. For example, three molecules A, B and C mixing together undergo the reaction

$$A + B \leftrightarrow C \quad . \tag{3-38}$$

The particle numbers of A, B and C are not conserved, but are completely independent. Applying detailed balance to (3-38), then the generalisation of (3-17) is

$$\frac{f_A}{1-f_A} \cdot \frac{f_B}{1-f_B} = \frac{f_C}{1-f_C} \quad . \tag{3-39}$$

Equation (3-21) is still applicable, but the α's must satisfy

$$\alpha_A + \alpha_B = \alpha_C \quad . \tag{3-40}$$

From (3-30), we get

$$\mu_A + \mu_B - \mu_C = 0 \quad . \tag{3-41}$$

Because of the reaction (3-38), the three originally independent chemical potentials are reduced to two, with remaining one determined by (3-41). Each chemical reaction results in a relation between the chemical potentials.

The result (3-41) can be understood from another point of view. In equilibrium the chemical reaction (3-38) cannot perform any work, i.e.

$$\mu_A \, dN_A + \mu_B \, dN_B + \mu_C \, dN_C = 0 \quad . \tag{3-42}$$

But according to (3-38), $dN_A = dN_B = -dN_C$, and we immediately get (3-41). This is a result in thermodynamics and does not require (3-39) for its derivation.

We can see from this that (3-30) is a very ideal choice. If we do not use (3-30), but instead write

$$\frac{\mu}{T} = -\alpha + \eta \quad , \tag{3-43}$$

with η being a constant, then from (3-40) and (3-41) it can be seen that η

cannot be any arbitrary constant but must satisfy

$$\eta_A + \eta_B = \eta_C \ . \tag{3-44}$$

We have to consider all other possible chemical reactions. Each chemical reaction gives a relation similar to (3-44), and each η must satisfy all these relations. Of course, to choose such a η one must start from the conservation laws. All chemical and nuclear reactions must conserve nucleon number A (the total number of protons and neutrons). Hence η_A must be chosen to be

$$\eta_A = \text{constant} \times A \ . \tag{3-45}$$

This constant must be the same for all molecules. If we further extend the definition of chemical potential to the case that the number of nucleons may not be conserved, the only possible choice is to set all the η to zero.

Hence, from the principle of detailed balance and the concept of states in quantum mechanics we obtain α, β and the distribution function (3-23). From the statistics of the states, Planck's constant h enters into α. (See (3-28).) Then from the consideration of conservation laws, h naturally enters into entropy, completing its definition.

3.4. Distribution of Bosons and Black Body Radiation

Bosons are not restricted by the exclusion principle and hence have different reaction rates. The factors $(1 - f_3)$, $(1 - f_4)$ in (3-14) are the results of the exclusion principle and are not valid for bosons. According to quantum mechanics, the reaction of (3-13) should now be

$$f_1 f_2 (1 + f_3)(1 + f_4) R \ , \tag{3-46}$$

i.e. we simply change the negative signs in (3-14) to positive ones. This result is not obvious and requires some calculations to obtain it. We regard this as known. Equation (3-46) says that if states 3 and 4 are already occupied, other states would be more inclined to go to them. Now we only need to change the negative signs in (3-14) − (3-18) to obtain the distribution function for the bosons. Equation (3-32) now becomes

$$f_k = \frac{1}{e^{(\epsilon_k - \mu)/T} - 1} \ . \tag{3-47}$$

The low density case is independent of the statistics and f_k is again the Boltzmann distribution

$$f_k \simeq e^{(\mu - \epsilon_k)/T} \ . \tag{3-48}$$

42 STATISTICAL MECHANICS

This can be called the *classical distribution*, while (3-47) and (3-32) are the quantum distributions. All the results of the last section can be reproduced for the boson case.

Photons are bosons and the photon gas in equilibrium is called the black body radiation. Photons are continually absorbed and emitted by the walls of the cavity. If there are molecules in the cavity, these molecules can also absorb or emit photons. The number of photons is not conserved, and hence its chemical potential is zero. The distribution of the photons is

$$f_k = \frac{1}{e^{\omega_k/T} - 1}, \tag{3-49}$$

where ω_k is the frequency or the energy of the photon ($\hbar = 1$), i.e. $\epsilon_k = \hbar\omega_k$. The energy distribution of the states of the photons can be calculated by the method in Sec. 3.1, and we only have to change $\epsilon_p = p^2/2m$ in (3-6) to $\epsilon_p = cp$, where c is the speed of light, and also include the two polarisations. Thus

$$g(\omega) = \frac{2}{(2\pi)^3} \int d^3p \; \delta(\omega - cp) = \frac{\omega^2}{\pi^2 c^3}. \tag{3-50}$$

Hence the frequency distribution of the black body radiation is

$$\epsilon(\omega) = g(\omega) \frac{\omega}{e^{\omega/T} - 1} = \frac{\omega^3/\pi^2 c^3}{e^{\omega/T} - 1}. \tag{3-51}$$

This is the *Planck distribution*. The discovery of this distribution is a prelude to the birth of quantum mechanics, and precedes the reaction rate considerations in (3-46). Einstein used this together with the principle of detailed balance to derive some properties of the rates of absorption and emission of light as mentioned below.

Let molecule A absorb light of frequency ω to become the excited state A^*. The energy of A^* is higher than that of A by ω. In equilibrium, the ratio of the population N of A to the population N^* of A^* is

$$\frac{N^*}{N} = e^{-\omega/T}. \tag{3-52}$$

This formula can be derived by the principle of detailed balance. (See Prob. 10.) The absorption rate of light is proportional to N and $\epsilon(\omega)$ and its emission rate is proportional to N^*. They must be equal. Therefore,

$$RN\epsilon(\omega) = R'N^* \tag{3-53}$$

where R' is the emission rate of one A^* and R is the absorption rate of the interaction of one A with one photon of energy ω. From (3-52) and (3-53) we get

$$R' = R \frac{N}{N^*} \epsilon(\omega) = R \frac{\omega^3}{\pi^2 c^3} \frac{e^{\omega/T}}{e^{\omega/T} - 1}$$

$$= R \frac{\omega^3}{\pi^2 c^3} [1 + f(\omega)]$$

$$= R \frac{\omega^3}{\pi^2 c^3} + R \epsilon(\omega) . \qquad (3\text{-}54)$$

The factor $1 + f$ is naturally the same as the $1 + f$ in (3-46).

Equation (3-54) shows that:

(1) Even if $\epsilon(\omega) = 0$, i.e. in the case of no photons, A^* can spontaneously emit a photon and become A, with a rate

$$r = R \frac{\omega^3}{\pi^2 c^3} , \qquad (3\text{-}55)$$

i.e. the half life of A^* is $\ln 2/r$.

(2) If $f \neq 0$, the emission by A^* is faster and this is stimulated emission. This is a property special to bosons.

(3) The absorption and emission rates are determined by R, which depends on the structure of the molecule and the details of its interaction with photons.

These results are related to the interaction of molecules with light, and are obtained by detailed balance plus the Planck distribution. They are rather unexpected. Detailed quantum mechanical calculation must involve a complicated analysis of the interaction of A, A^* and the photon, so as to obtain (3-54) (and the value of R).

3.5. Chemical Reactions and the Law of Concentrations

In discussing the conservation laws in Sec. 3.3, we have mentioned chemical reactions. Each chemical reaction involves one conservation law and results in a relation between the chemical potentials like (3-41), which we now use to derive relations between the concentrations of the various molecules.

We start from the reaction (3-38), assuming that A, B and C mix as an ideal gas, each distributed according to the classical formual (3-48). Besides the kinetic energy, the total energy of each molecule also includes the energy of internal motion such as rotation and vibration

$$\epsilon_k = \frac{p_k^2}{2m} + \omega_k \; ,$$

$$f_k = e^{(\mu - \epsilon_k)/T} \; .\tag{3-56}$$

The total population of molecule A is

$$N_A = \sum_k f_{kA}$$

$$= V e^{\mu_A/T} \frac{1}{(2\pi\hbar)^3} \int d^3 p \, e^{-p^2/2m_A T} \sum_\alpha e^{-\omega_{A\alpha}/T}$$

$$\equiv V e^{\mu_A/T} e^{-\phi_A/T} \; ,\tag{3-57}$$

where ϕ_A is a function of the temperature and is related to the mass m and the internal energy $\omega_{A\alpha}$ of the molecules. Let $n_A \equiv N_A/V$ be the concentration of A, then

$$\mu_A = \phi_A + T \ln n_A \; ,$$
$$\mu_B = \phi_B + T \ln n_B \; ,$$
$$\mu_C = \phi_C + T \ln n_C \; .\tag{3-58}$$

From (3-14) we get

$$0 = \phi_A + \phi_B - \phi_C + T \ln \frac{n_A n_B}{n_C} \tag{3-59}$$

or

$$\frac{n_A n_B}{n_C} = e^{-(\phi_A + \phi_B - \phi_C)/T} \; .$$

This is a theorem on equilibrium concentrations. The right hand side of (3-59) can be calculated from the properties of the molecules and is only a function of temperature. In other words, if we know the structure of the molecules we can obtain information on the concentrations. This is very useful for a chemical engineer. If the concentrations of the various molecules can be measured but the structure of the molecules is not clear, then (3-59) can be used to deduce the structure of the molecules. The reader can consult books on chemistry for a deeper understanding of this topic.

THE LAW OF DETAILED BALANCE 45

Problems

1. The concept of a distribution is the most common in statistical data, for example the age distribution, the income distribution or the grade distribution in an examination. A distribution can be represented by a curve. Figure 3-3 is an age distribution $f(x)$ (say in units of ten thousands).

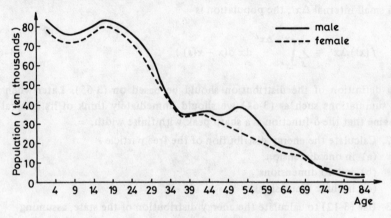

Fig. 3-3 Population distribution plotted against age.

The total area under the curve is the total population.

$$\int_0^\infty f(x)\,dx = \text{total population} \quad . \tag{3-60}$$

The population with age less than x', denoted as $N(x')$, is the area under the curve to the left of $x = x'$.

$$N(x') = \int_0^{x'} f(x)\,dx \quad , \tag{3-61}$$

$$f(x) = \frac{dN(x)}{dx} \quad . \tag{3-62}$$

The meaning of this distribution function is: $f(x)\,dx$ is the population between ages x and $x + dx$. To count the population with age less than x', we can search through the census to collect the names of the people whose age is less than x' and then count the number:

$$N(x') = \sum_s \theta(x' - x(s)) \quad , \tag{3-63}$$

where s is the name of each person and $x(s)$ is the age of the person s. If it is less than x', $\theta(x - x(s))$ is 1 and otherwise it is zero. Equation (3-63) represents this statistical process. The age distribution $f(x')$ is formally the differential of N (see (3-62)):

$$f(x') = \frac{dN(x')}{dx'} = \sum_s \delta(x' - x(s)) \quad . \tag{3-64}$$

In a small internal $\Delta x'$, the population is

$$f(x')\Delta x' = \sum_s \int_{x'}^{x'+\Delta x'} dx\, \delta(x - x(s)) \quad . \tag{3-65}$$

The definition of the distribution should be based on (3-65). Later when we see summations such as (3-64) we should immediately think of its integral or imagine that the δ-function is a sharp peak with finite width.

2. Calculate the energy distribution of the free particle
   (a) in one dimension,
   (b) in two dimensions,
   (c) in three dimensions.

3. Use (3-12) to calculate the energy distribution of the state, assuming

   (a) $U(\mathbf{r}) = \frac{1}{2} K \mathbf{r}^2$
   (b) $U(\mathbf{r}) = -Ze^2/r$.

We can use (b) to estimate the energy of heavy atoms (large Z).

4. Detailed balance comes from the symmetry property of the time direction. This symmetry is a special feature of the equilibrium state. Some states, though apparently steady and unchanging, do not possess this symmetry, and we call them nonequilibrium states. For example, a body with a heat current passing through it cannot be called an equilibrium state, even though the heat current is unchanged in time. But a uniformly moving or rotating body can be regarded as in equilibrium. The reader can think about these situations.

5. Sections 3-2 and 3-3 point out that entropy is intimately related to the concept of a state in quantum mechanics. They also show the importance of conservation laws. The reader may try to derive (3-31) from (3-36) and (3-30).

6. The following reaction occurs inside a star

$$\gamma + \gamma \longleftrightarrow e^+ + e^-$$

where γ is a photon and $e^\pm$ are the positron and the electron respectively. Calculate the population of e^+ and e^- in equilibrium. In a dense gas of protons, electrons and neutrons the following reaction occurs

$$p + e^- \to n + \nu$$

where ν is the neutrino, which escapes after production. The chemical potential of electron satisfies $\mu_e \gg mc^2$, where m is the mass of the electron. Find the population of p, e^- and n in equilibrium.

7. Suppose a star explodes at times $t = 0$, and this star is a mixed gas of protons and electrons. A detector at distance R measures the flux of electrons and protons, with $R \gg$ the radius of the star.

(a) Find the relation between the flux and time. This relation is linked with the distribution of the velocities of the particles.

(b) If the temperature is T after the explosion, calculate the flux.

8. The ideal boson gas can be analysed by (3-47).

(a) Find the relation between μ and N/V. Notice that when the temperature is lower than a certain critical temperature T_c, $\mu = 0$ and f_0 (the population of the state $k = 0$) is proportional to V, i.e. there are macroscopically large numbers of particles occupying the lowest state. This phenomenon is usually called Bose-Einstein condensation.

(b) Calculate the critical temperature T_c.

(c) Calculate the relation of f_0 with T.

(d) Calculate the heat capacity.

(e) Prove that this phenomenon of condensation does not occur in one or two dimensions.

9. The discussion of (3-38) to (3-45) is not limited to fermions. In fact the molecule C in (3-38) must be a boson (two fermions can combine to form a boson). Attempt to generalise the discussion in Section 3-3 to include bosons.

10. From the reaction

$$A^* \longleftrightarrow A + \text{photon} \quad ,$$

derive (3-52). This result is independent of the condition that A is a boson.

Chapter 4
ELECTRONS IN METALS

The topic of electrons in metal is surely a major field of solid state physics. In this chapter we shall pick some simple problems as examples of the theory in the last chapter. We first introduce the free electron model and then discuss the well-known heat capacity of an electron gas. The structure of the crystal is the most important factor affecting the behavior of free electrons. We shall use a one dimensional model to illustrate the interrelation of electrons and the crystal structure. Although the Peierls' instability (i.e. the phenomenon that a one dimensional metal is unstable) is limited to one dimension, it can illustrate some underlying principles. Hence we shall give a rather detailed discussion.

4.1. Dense Ideal Gas

In most metals, the density is high, with the atoms arranged in a crystal structure. The electrons in the inner shells are tightly bound. The electrons in the outermost shell can move freely among the atoms. Many properties of the metal are determined by the motion of these mobile electrons. Now we treat these electrons as an ideal Fermi gas, and this is the free electron model. This simple view produces unexpectedly good results. Let us now see the results of this ideal gas model.

We use k to specify the state of the electron, and ϵ_k its energy. Let μ be the chemical potential of the electron (i.e. the usual electric potential times the charge of the electron). Each state can accommodate at most one electron. Let f_k be the average population of state k (see (3-32)):

$$f_k = f(\epsilon_k - \mu) \quad , \tag{4-1}$$

$$f(\epsilon) \equiv \frac{1}{e^{\epsilon/T} + 1} \quad . \tag{4-2}$$

The energy distribution of the states is an important property. Let

$$g(\epsilon) \equiv \frac{1}{V} \sum_k \delta(\epsilon - (\epsilon_k - \mu_0)) \quad ,$$

$$\mu_0 \equiv \mu(T=0) \quad . \tag{4-3}$$

The function of $g(\epsilon)$ is the energy distribution of the states per unit volume, which we simply call density of states, and ϵ is the energy with respect to μ_0. The calculation of $g(\epsilon)$ has been presented in detail in Section 3-1 of Chapter 3, and gives

$$g(\epsilon) = 2 \int \frac{d^3 p}{(2\pi)^3} \, \delta\left(\epsilon + \mu_0 - \frac{p^2}{2m}\right)$$

$$= \frac{m}{\pi^2} [2m(\epsilon + \mu_0)]^{\frac{1}{2}} \quad . \tag{4-4}$$

The thermodynamic properties of this model can be largely expressed in terms of f and g, e.g. the density N/V of the electrons and the energy density E/V are

$$N/V = \int d\epsilon \, g(\epsilon) \, f(\epsilon - \mu) \quad ,$$

$$E/V = \int d\epsilon \, g(\epsilon) \, f(\epsilon - \mu) (\epsilon + \mu_0) \quad . \tag{4-5}$$

If $T=0$, then all the low energy states are filled up to the Fermi surface. Above this surface all the states are empty. The energy at the Fermi surface is μ_0, i.e. the chemical potential when $T=0$, and is always denoted by ϵ_F:

$$\epsilon_F \equiv \mu(T=0) = \mu_0 \quad . \tag{4-6}$$

The Fermi surface can be thought of as a spherical surface in the momentum space of the electrons. The radius of the sphere p_F is called the Fermi momentum

$$\frac{p_F^2}{2m} = \epsilon_F \quad . \tag{4-7}$$

There are N states with energy less than ϵ_F:

$$\frac{4\pi}{3} p_F^3 \cdot V \cdot \frac{2}{(2\pi)^3} = N \quad , \tag{4-8}$$

i.e. (volume of sphere in momentum space) × (volume) × (spin state (= 2)) ÷ $(2\pi\hbar)^3$, with $\hbar = 1$. Hence

$$p_F = (3\pi^2 n)^{1/3}, \qquad (4\text{-}9)$$

where $n \equiv N/V$, and $a = \hbar/p_F$ is approximately the average distance between the electrons:

$$\epsilon_F \sim \frac{\hbar^2}{ma^2} \qquad (4\text{-}10)$$

is approximately the zero-point energy of each electron. This zero-point energy is a result of the wave nature of the electron or a necessary result of the uncertainty principle. To fix an electron to within a space of size a, its momentum would have to be of order $\hbar/a$.

In most metals, the distance between electrons is about 10^{-8} cm and $\epsilon_F \sim 1$ eV $\sim 10^4$ K (see Table 4-1). Therefore, at ordinary temperatures, $T \ll \epsilon_F$, i.e. the temperature is very low, only electrons very close to the Fermi surface can be excited and most of the electrons remain inside the sphere, experiencing no changes.

Table 4-1. Properties of the electron gas at the Fermi surface

	electron density N/V cm^{-3}	momentum k_F cm^{-1}	velocity v_F cm/sec	energy ϵ_F eV	$T_F = \epsilon_F/k$ K
Li	4.6 × 10^{22}	1.1 × 10^8	1.3 × 10^8	4.7	5.5 × 10^4
Na	2.5	0.90	1.1	3.1	3.7
K	1.34	0.73	0.85	2.1	2.4
Rb	1.08	0.68	0.79	1.8	2.1
Cs	0.86	0.63	0.73	1.5	1.8
Cu	8.50	1.35	1.56	7.0	8.2
Ag	5.76	1.19	1.38	5.5	6.4
Au	5.90	1.20	1.39	5.5	6.4

From Kittel (1966), p. 208

$$\epsilon_F = \frac{\hbar^2}{2m}(3\pi^2 N/V)^{2/3} \quad,$$

$$k_F = (3\pi^2 N/V)^{1/3} = p_F/\hbar \quad,$$

$$v_F = p_F/m \quad,$$

$$T_F = \epsilon_F/k \quad.$$

Notice that ϵ_F is high because the mass of the electron is very small. The high density is also another reason. This electron gas model is quite different from the usual molecular gases. When $T \to 0$, the total energy of this electron gas is

$$E(0) = \frac{3}{5}N\epsilon_F$$

$$= \frac{3N}{10m}\left(\frac{3\pi^2 N}{V}\right)^{2/3} \propto V^{-2/3} \quad, \tag{4-11}$$

producing a pressure, called the zero-point pressure

$$p(0) = -\left(\frac{\partial E}{\partial V}\right)_N = \frac{2}{3}\frac{E}{V} = \frac{2}{5}\frac{N}{V}\epsilon_F \quad. \tag{4-12}$$

If we compare this with the ideal gas pressure $p = NT/V$, we discover that this $p(0)$ corresponds to a temperature of $\frac{2}{5}\epsilon_F$, and N/V is usually much higher than the usual gas. Hence it is an enormous pressure. This pressure makes the metal expand like blowing up a balloon. It will not blow up because of the attractive force between the atoms. This attractive force comes naturally from the negative charge of the electrons and the positive charge of the ions. The calculation of this attractive force is a special topic. The reader can consult books on solid state physics.

4.2. Heat Capacity

Only a very small portion of the electrons is influenced by temperature. Hence the concept of holes appears naturally. The states below the Fermi surface are nearly filled, and empty states are rare. We shall call an empty state a hole. Now this model becomes a new mixed gas of holes together with electrons above the Fermi surface. (We shall call these outer electrons.) The momentum of a hole is less than p_F, while that of the outer electrons is larger than p_F. The lower the temperature T, the more dilute the gas is. At $T = 0$, this gas disappears. This is a very convenient way of thinking.

At $T = 0$ the total energy is zero, i.e. there are no holes or outer electrons. The holes are also fermions because each state has at most one hole. Hence, a state of energy $-\epsilon'$ can produce a hole of energy ϵ'. The average population of the hole is (for each state):

$$1 - f(-\epsilon' - \mu) = f(\epsilon' + \mu) \quad . \tag{4-13}$$

Now the origin of the energy is shifted to the energy at the Fermi surface, i.e. $\mu_0 = 0$. The energy of a hole is the energy required to take an electron from inside the Fermi surface to the outside (Fig. 4-1). As $T \ll \epsilon_F$, the energy of

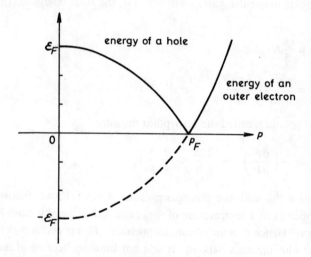

Fig. 4-1 Electron (hole) energy versus momentum; p_F and ϵ_F are the Fermi momentum and energy.

the holes or the outer electrons cannot exceed T by too much. In this interval of energy, $g(\epsilon)$ is essentially unchanged, i.e.

$$g(\epsilon) \simeq g(0) = m p_F / \pi^2 \quad . \tag{4-14}$$

Hence the energy distribution is the same for the holes or the outer electrons. Therefore, $\mu \cong 0$. All the calculations can now be considerably simplified, e.g. the total energy is

$$\frac{1}{V}[E(T) - E(0)] = \int_0^\infty d\epsilon \, 2g(0) f(\epsilon) \epsilon \quad , \tag{4-15}$$

where $2g(0)$ is the density of states of the holes plus the outer electrons. The energy of a hole cannot exceed ϵ_F, but $\epsilon_F \gg T$ and so the upper limit of the integral in (4-15) can be taken to be ∞. This integration is easy:

$$\int_0^\infty d\epsilon\, f(\epsilon)\epsilon = T^2 \int_0^\infty dx\, \frac{x}{e^x + 1} = \frac{\pi^2}{12} T^2 \qquad (4\text{-}16)$$

Substituting in (4-15), and differentiating once, we get the heat capacity

$$\frac{1}{V} C = \frac{1}{V} \frac{\partial E}{\partial T} = \frac{\pi^2}{3} g(0) T \qquad (4\text{-}17)$$

This result is completely different from that of the ideal gas in which $C = \frac{3}{2}N$. In that case each gas molecule contribute a heat capacity of $\frac{3}{2}$. Now only a small portion of the electrons is involved in motion and the number of active electrons is about NT/ϵ_F,

$$Vg(0) \int_0^\infty d\epsilon\, f(\epsilon) \sim \frac{NT}{\epsilon_F} \qquad (4\text{-}18)$$

Each active electron contribute approximately 1 to the heat capacity C. Hence $C \sim N(T/\epsilon_F)$. From (4-17) we get

$$C = N \frac{\pi^2}{2} \left(\frac{T}{\epsilon_F} \right) \qquad (4\text{-}19)$$

In the initial stage of studying metals it is usual to treat the electrons as a dilute ideal gas. But experiments show that C is in fact much less than $\frac{3}{2}N$. Only when the fermionic nature was discovered did this problem of heat capacity have a satisfactory explanation.

Equation (4-19) is the first term of an expansion in T. What is the next one? If $g(\epsilon) = g(0)$, we do not get the next term. Hence we expand $g(\epsilon)$ to $O(\epsilon)$ to get the next term:

$$g_+(\epsilon) \simeq g(0) + \epsilon g'(0) \quad ,$$

$$g_-(\epsilon) \simeq g(0) - \epsilon g'(0) \quad , \qquad (4\text{-}20)$$

where g_+ is the density of states of the outer electrons and g_- that of the holes. As $g_+ \neq g_-$, at $T = 0$ the chemical potential is not zero. The population in each state of the outer electron and the hole is

$$f(\epsilon - \mu)$$

and

$$1 - f(-\epsilon - \mu) = f(\epsilon + \mu) \quad . \tag{4-21}$$

The value of μ can be easily calculated because the population of the holes and the outer electrons should be equal:

$$\int_0^\infty d\epsilon \, g_+(\epsilon) f(\epsilon - \mu) = \int_0^\infty d\epsilon \, g_-(\epsilon) f(\epsilon + \mu) \quad , \tag{4-22}$$

where both μ and $g'(0)$ are small numbers. Expanding (4-22), we can solve for μ:

$$\mu = -\frac{T^2 \pi^2 g'(0)}{6 g(0)} \quad . \tag{4-23}$$

Because $g'(0)/g(0) = \tfrac{1}{2} \epsilon_F$, hence $\mu \sim -T^2/\epsilon_F$. Besides μ, $g'(0)$ may cause other direct corrections to the energy (4-15). Equation (4-15) should be

$$\frac{1}{V}[E(T) - E(0)] = \int_0^\infty d\epsilon \, g_+(\epsilon) f(\epsilon - \mu)\epsilon + \int_0^\infty d\epsilon \, g_-(\epsilon) f(\epsilon + \mu)\epsilon \quad .$$

Upon expanding, all terms of $g'(0)$ are cancelled. Hence the results of (4-17) or (4-19) do not contain the T^2 terms. Therefore,

$$C = N \frac{\pi^2}{2} \frac{T}{\epsilon_F} \left[1 + O\left(\frac{T^2}{\epsilon_F^2}\right) \right] \quad . \tag{4-24}$$

The correction terms come from $g''(0)$ and $g'(0)^2$.

Besides the electrons, the vibration of the atoms in the metal also contributes to the heat capacity. Above room temperatures, its contribution is about $3NT$, a dominant term. The Debye temperature θ_D of a metal is around 100 K to 400 K. The heat capacity due to lattice vibration at $T > \theta_D$ is about N. (This will be discussed fully in Chapter 6, we just use the results here.) At very low temperatures, i.e. $T \ll \theta_D$ it is $N(T/\theta_D)^3$. Because $\epsilon_F \gg \theta_D$, hence this T^3 term is much larger than that correction term in (4-24). Therefore, at low temperatures, the total energy capacity is

$$\frac{C}{N} = \frac{\pi^2}{2} \left(\frac{T}{\epsilon_F}\right) + \frac{12\pi^4}{5} \left(\frac{T}{\theta_D}\right)^3 \quad . \tag{4-25}$$

The reader should notice that the analysis from (4-20) to (4-24) is very important, in that it demonstrates that the heat capacity of the electron does not have a T^2 term, but only T and T^3 terms.

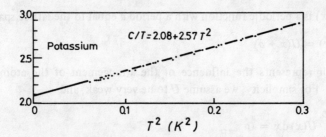

Fig. 4-2 The low temperature heat capacity of potassium/T (10^4 erg K^{-2} mole^{-1}). From W. H. Lien and N. E. Phillips, *Phys. Rev.* **133A** 1370 (1964).

Figure 4-2 is a plot of C/T versus T^2. According to (4-25) it should be a straight line. This experiment is a crucial test of this free electron model. From the experimental data we can determine the coefficients of T and T^3 in (4-25). The coefficient of T is

$$r \equiv \frac{\pi^2 m}{p_F^2} = \frac{\pi^2 k^2 m}{\hbar^2 (3\pi^2 n)^{2/3}}. \tag{4-26}$$

We have put in k and $\hbar$ explicitly, and n is the electron density. The experimental value of r is larger than that calculated by (4-26). The reason for this is not clear. Of course, the free electron model is oversimplified, and it is surprising that the results are so good. The interaction of the electrons with the impurities and the lattice vibrations plus the periodic structure of the lattice have not been taken into account here.

The metallic properties of a material denote those properties similar to the free electron gas, and depend crucially on the near constancy of $g(\epsilon)$ at the Fermi surface. If $g(\epsilon)$ is affected, then its metallic property is affected too.

4.3. Influence of Periodic Structure

The above discussion points out the importance of $g(\epsilon)$, the density of states. In solids there are many factors affecting $g(\epsilon)$. The crystal structure is naturally an important factor. In the following we shall use the simplest one dimensional model to discuss the influence of a periodic arrangement of the atoms.

The energy ϵ of the electron states can be solved from the wave equation

$$-\frac{1}{2m}\frac{d^2}{dx^2}\psi(x) + U(x)\psi(x) = \epsilon\psi(x) \quad , \tag{4-27}$$

where $U(x)$ is a periodic function with a period a equal to the lattice spacing:

$$U(x) = U(x + a) \tag{4-28}$$

and which represents the influence of the arrangement of the atoms on the electrons. For simplicity, we assume U to be very weak, and

$$\int_0^L U(x)\,dx = 0 \quad , \tag{4-29}$$

where L is the total length of the body. If we neglect U, it is simple to obtain the solution of (4-27)

$$\psi_p = \frac{1}{\sqrt{L}} e^{ipx} \quad ,$$

$$\epsilon_p = p^2/2m \quad . \tag{4-30}$$

Now we calculate the correction due to U on ϵ_p. The perturbation theory in quantum mechanics tells us that the first order correction to ϵ_p is

$$\langle \psi | U | \psi \rangle = \frac{1}{L} \int_0^L U(x)\,dx = 0 \quad . \tag{4-31}$$

But this is not valid, because this formula is appropriate only for nondegenerate energies. In (4-30), there is degeneracy, i.e. the energies of ψ_p and ψ_{-p} are the same, with value $p^2/2m$. To deal with this, we note that $U(x)$ can be written as

$$U(x) = \sum_{n=-\infty}^{\infty} U_n e^{inqx} \quad ,$$

$$q \equiv 2\pi/a \quad . \tag{4-32}$$

The potential $U(x)$ has interaction between $|p\rangle$ and $|p + nq\rangle$, given by the matrix element

$$\langle p | U | p' \rangle = \sum_{n=-\infty}^{\infty} U_n \delta_{p, p+nq} \quad . \tag{4-33}$$

Therefore, if
$$p \simeq -p + nq,$$
i.e.
$$p \simeq \frac{n}{2}q, \qquad n = \pm 1, \pm 2, \ldots \qquad (4\text{-}34)$$

then the result of (4-31) is not valid. Let us now look at the situation $n = \pm 1$. Let

$$p_1 \equiv \frac{1}{2}q + k,$$

$$p_2 = -\frac{1}{2}q + k, \qquad k \ll q. \qquad (4\text{-}35)$$

Quantum mechanics tells us to consider these two states together in order to obtain the first order correction term. The energy operator in the space of these two states is represented by a matrix H:

$$H = \begin{pmatrix} \epsilon_1 & H_{12}^* \\ H_{12} & \epsilon_2 \end{pmatrix}, \qquad H_{12} = U_1, \qquad (4\text{-}36)$$

$$\epsilon_1 = \frac{1}{2m}\left(\frac{q^2}{4} + k^2 + qk\right),$$

$$\epsilon_2 = \frac{1}{2m}\left(\frac{q^2}{4} + k^2 - qk\right). \qquad (4\text{-}37)$$

The eigenvalues of H are

$$\epsilon_\pm = \frac{1}{2m}\left(\frac{q^2}{4} + k^2\right) \pm \sqrt{\frac{q^2 k^2}{4m^2} + \Delta^2},$$

$$\Delta \equiv |U_1|. \qquad (4\text{-}38)$$

These results are plotted in Fig. 4-3. When $k = 0$, i.e. when $p = \pm q/2$, $\epsilon_+ - \epsilon_- = 2\Delta$. The situation is similar for $n = \pm 2, \pm 3, \ldots$, etc. The density of states therefore has a gap (Fig. 4-4) when the energy is

$$\frac{1}{2m}\left(\frac{n\pi}{a}\right)^2, \qquad n = \pm 1, \pm 2, \pm 3, \ldots \qquad (4\text{-}39)$$

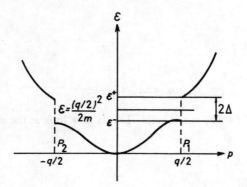

Fig. 4-3

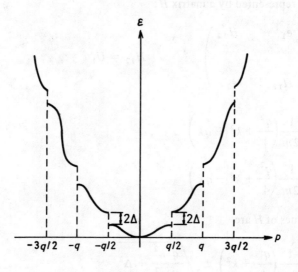

Fig. 4-4

This is the most important effect of the periodic potential.

Let the total number of electrons be N. Suppose

$$N = L/a \quad , \qquad (4\text{-}40)$$

i.e. there is one electron per lattice site on the average, then what is p_F? Each p

has two states (the two directions of the spin), therefore,

$$N = \frac{2L}{2\pi} \int_{-p_F}^{p_F} dp = \frac{2p_F L}{\pi} \quad ,$$

$$p_F = \frac{\pi}{2a} = \frac{q}{4} \quad , \tag{4-41}$$

i.e. the Fermi surface consists of the two points $p = p_F$ and $p = -p_F$. As the gap at $p = \pm q/2$ is far from p_F, if $T/\epsilon_F \ll 1$ it would exhibit the metallic property discussed in the last section.

If each lattice site has two electrons, then

$$p_F = \frac{\pi}{a} = \frac{q}{2} \quad . \tag{4-42}$$

The states between $q/2$ and $-q/2$ are fully occupied. If the electron is to be excited, it must overcome the gap and must have an energy larger than 2Δ. With this in mind, we can calculate $g(\epsilon)$ and the heat capacity.

We choose the origin of the energy to be in the middle of the gap, and we get

$$g(\epsilon) = \frac{2}{2\pi} \int dp \, \delta(\epsilon - \epsilon_p)$$

$$\simeq \left(\frac{2ma}{\pi^2}\right) \frac{|\epsilon|}{\sqrt{\epsilon^2 - \Delta^2}} \quad , \quad \epsilon^2 > \Delta^2 \quad . \tag{4-43}$$

The above assumes $\epsilon \ll \epsilon_F$. We use the method of holes to calculate the total energy. Holes are empty states below the gap (see Fig. 4-5). The number of holes and that of outer electrons are equal and in (4-43) $g(\epsilon) \approx g(-\epsilon)$. Therefore $\mu = 0$. The total energy and heat capacity are

$$\frac{1}{N}[E(T) - E(0)] \simeq \frac{L}{N} \int_0^\infty d\epsilon \, 2g(\epsilon) \frac{1}{e^{\epsilon/T} + 1}$$

$$\simeq 4ma^2 \, \pi^{-3/2} \, \Delta^{3/2} \, \sqrt{2T} \, e^{-\Delta/T} \quad ,$$

$$\frac{C}{N} \simeq 4\sqrt{2} \, ma^2 \, \pi^{-3/2} \, \Delta^{5/2} \, T^{-3/2} \, e^{-\Delta/T} \tag{4-44}$$

The above assumes $T \ll \Delta$.

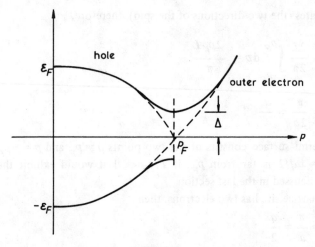

Fig. 4-5

This result is obviously different from the situation in (4-19). The energy gap makes the inner electron difficult to excite and $e^{-\Delta/T}$ is the most important factor in (4-44). The population of holes and outer electrons is directly related to the conducting property of the body. If Δ is not very large, this population can be effectively controlled by the temperature. This body becomes a semiconductor. If Δ is very large, this population is small and the body is now an insulator.

4.4. The Structural Change Caused by Electrons

The crystal structure produces a gap in the energy of the electron. In the above example, if $N = 2L/a$, then the influence of this gap is very important. We have already seen the influence of Δ on C. Now let us see how it affects the zero-point energy:

$$\frac{1}{N}[E(\Delta) - E(\Delta = 0)] \simeq a \int_{-\epsilon_F}^{\Delta} d\epsilon\, [g(\epsilon) - g_0(\epsilon)]\, \epsilon$$

$$\simeq -\pi^{-2}\, ma^2\, \Delta^2\, \ln(\epsilon_F/\Delta) < 0 \quad,$$

$$\epsilon_F = \frac{p_F^2}{2m} = \frac{\pi^2}{2ma^2} \quad. \tag{4-45}$$

This result is calculated from the approximation (4-43). More precise calculation gives similar results. The main conclusion here is that $E(0)$ is lowered because of the gap.

Now we return to the situation $N = L/a$, i.e. $p_F = \frac{1}{4}q$. In this case ϵ_F is still far away from the gap. We have mentioned that the result is not much different from the case without $U(x)$. However, the atoms in the crystal can move. Suppose every other atom moves by b (Fig. 4-6), then the period changes from

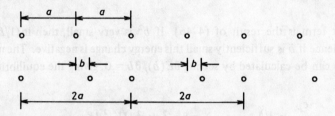

Fig. 4-6 The vibration of atoms in a crystal.

a to $2a$. If b is not very large, then the increase in energy of the lattice is proportional to b^2, since $b = 0$ is the equilibrium condition:

$$\frac{1}{N}[E(b) - E(b=0)] = \frac{1}{2}Kb^2 , \qquad (4\text{-}46)$$

where K is the elastic constant. After displacement, the potential $U(x)$ seen by the electrons is also changed. The new potential energy has a period of $2a$. Therefore,

$$U(x, b) = U(x, b=0) + b \sum_{n=-\infty}^{\infty} e^{inq'x} U_n' + 0(b^2) ,$$

$$q' = \frac{2\pi}{2a} = \frac{q}{2} . \qquad (4\text{-}47)$$

We have a new bU_1':

$$|bU_1'| \equiv \Delta' . \qquad (4\text{-}48)$$

This bU_1' has the same effect as the above U_1, producing an energy gap, but the energy gap occurs at

$$p = \pm \frac{q'}{2} = \pm \frac{q}{4} , \qquad (4\text{-}49)$$

which is the present $\pm p_F$. This new gap can lower the energy of the electrons, by an amount obtained by replacing Δ' for Δ and $2a$ for a in (4-45). We then obtain

$$\frac{1}{N}[E(b) - E(b=0)] = -4ma^2 \pi^{-2} b^2 |U_1'|^2 \ln\left(\frac{\epsilon_F}{b|U_1'|}\right) + \frac{1}{2} Kb^2 \quad . \tag{4-50}$$

The last term is the result of (4-46). If b^2 is very small, then $\ln(1/b)$ is very large. Hence if b is sufficiently small this energy change is negative. The minimum of $E(b)$ can be calculated by setting $\partial E(b)/\partial b = 0$, giving the equilibrium value of b as

$$b \sim \frac{\epsilon_F}{|U_1'|} e^{-1/\lambda} \quad , \qquad \lambda = 8\pi^{-2} ma^2 |U_1'|^2/K \quad . \tag{4-51}$$

The conclusion is that the structure of period a is unstable. Because of the interaction of the electron with the lattice, a structure with period $2a$ is lower in energy, and at the same time there is an energy gap at $p = \pm p_F$. The original metallic property is lost. This was instability first pointed out by Peierls.[a]

The most important conclusion is that the structure of the lattice and the motion of the electrons are mutually related, and must be considered together.

This analysis is very simple, but it clearly demonstrates that a one dimensional crystal cannot be a metal, i.e. there must be an energy gap at $p = \pm p_F$. Notice that the above calculation uses $p_F = q/4, q/2$. The reader can easily see that p_F can be any value, and even if p_F is not a rational multiple of q, the above conclusion remains unchanged: the period of the lattice structure must be $2\pi/(2p_F)$.

Of course, this model has not accounted for the vibration of the lattice. One dimensional vibration has a very large amplitude. (See the last section of Chapter 13). In addition, the interaction between the electrons has not been considered. Therefore, the above conclusion may be modified. A one-dimensional substance in the laboratory must be part of a whole physical body. For example the electrons of some crystals can conduct easily in a certain x direction, but not in the y or z direction. This substance can be thought of as composed of a bundle of one dimensional crystals. But this kind of substance is more complicated than the above model and the above conclusion can only be a reference but cannot be directly applied. The motion of one dimensional electrons involve many problems, both theoretical and experimental, and these are awaiting a solution.

[a] Peierls (1955).

Problems

1. This chapter is mainly calculational. The reader should derive every result, especially those in the last two sections, otherwise he will not be able to fully understand the implications.

2. Prove that at low temperatures the pressure of the electron gas is

$$p(T) = \frac{2}{3}\frac{E}{V} = \frac{2}{5}\frac{N}{V}E_F\left[1 + \frac{5\pi^2}{12}\left(\frac{T}{\epsilon_F}\right)^2 + \ldots\right].$$

This is the equation of state at low temperatures.

3. Dense electron gas appears in astrophysics, e.g. in the theory of the white dwarf. Roughly speaking, this is a cluster of helium nuclei and electrons at high temperatures and high pressures.

Express the pressure of the electron in terms of the radius of the star. The velocity of the electrons is close to the speed of light, and we must use a relativistic treatment.

This problem is fully discussed in Huang (1963), p. 230.

4. Let ϕ be the difference of the chemical potential μ of the electron with the potential outside of the metal. (Fig. 4-7). Assume that T is rather low, $T \ll \phi$, ϵ_F.

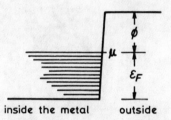

Fig. 4-7 Idealization of a metal; the difference ϕ must be overcome to eject electrons.

Light irradiated on the metal surface can eject electrons out of the metal. This is the photo-electric effect. Let I be the average number of ejected electrons produced by one photon.

(a) At $T = 0$, what is the relationship of I with the frequency ω of the light?

(b) When $T \neq 0$, find $I(\omega)$. Notice that when $T \neq 0$ there are ejected electrons even when there is no light irradiated on the surface. This is the effect of thermal electron radiation.

5. If there is a crack in a metal, what is the density of electrons in this crack? Notice that Fig. 4-7 is a highly simplified model surface, and is far from realistic.

64 STATISTICAL MECHANICS

6. Section 4 discusses the one dimensional model. The situation is much more complicated in two or three dimensions, although there is a formal similarity. The interaction of the lattice with the electrons produces a new periodic density. This can be called the "density wave", or "charge density wave".

PART II
HYPOTHESIS

Statistical mechanics is established on a daring assumption: the Boltzmann formula for the entropy

$$S = \ln \Gamma ,$$

where Γ is the volume of the region of motion. This region of motion represents the region within which the states of the molecules of a body are constrained to vary and Γ essentially measures the product of the amplitudes of the various motions.

This part is divided into five chapters. We first introduce the physical content and motivation of this basic assumption and use a few examples to illustrate its use (Chapters 5 and 6). Chapters 7 and 8 contain further deductions and more examples where the discussion focuses on the implications of the assumption, and its relation with the equilibrium observation time, the reaction rate and the region of motion. Chapter 9 discusses the mathematical problem of the macroscopic limit or the thermodynamic limit. We use the Yang-Lee theorem as the centre of our discussion.

The aim of this part is to point out the applications of the basic assumption of statistical mechanics, as well as certain points of ambiguity, thus enabling the reader to apply the assumption properly without being led blindly by the principle.

All the results of statistical mechanics can be derived from this basic assumption together with the law of large numbers, without any other new concepts. The reader should notice that this assumption has never been proved and our present understanding of it is still incomplete.

Chapter 5

THE BASIC ASSUMPTION OF ENTROPY AND MOLECULAR MOTION

In thermodynamics, entropy is a measure of the small scale molecular motion. In the preceding chapter we have emphasised this point and treated entropy as a coordinate, on the same level as volume, particle number, and magnetic moment, etc. Indeed how does entropy measure the molecular motion? Statistical mechanics provides the answer to this question. This is its most important achievement. This answer is not derived step by step from any basic principles, but it is based on a daring assumption. Up to now the origin and definition of this assumption are still not completely clear. This chapter discusses the various concepts and calculational problems involved in this assumption, including its motivation, the statistics of the states, the law of large numbers and the special feature of high dimensions, etc.

5.1. Region of Motion

The discussion of the ideal gas in Chapter 2 has provided a clue about the relation of entropy to the molecular motion. We examine (2-36):

$$S = N \ln \left[\left(\frac{CE}{N} \right)^{3/2} \frac{V}{N} \right] , \qquad (5\text{-}1)$$

where N is the total number of molecules, E/N is the average energy of each molecule and V/N is the average volume occupied by each molecule. Let

$$\frac{(\Delta p)^2}{2m} \equiv \frac{E}{N} , \qquad (\Delta x)^3 \equiv \frac{V}{N} , \qquad (5\text{-}2)$$

where m is the mass of a molecule. Because of the continual collisions, the momentum and position of each molecule is changing incessantly. Roughly speaking, Δp is the magnitude of the change of momentum of the molecules and Δx is that of the change of position. The reader may ask the following question. As each molecule can reach any point in the container, should not $(\Delta x)^3$ be V? Why do we use V/N instead? The reason is: each molecule is identical and indistinguishable. Hence, each molecule essentially does not move out of a range of V/N. If it did move out and another molecule moves in, that would be as if no such motion had taken place.

Now we can write (5-1) as

$$S = N \ln \left(\frac{\Delta x \, \Delta p}{w} \right)^3 = \ln \Gamma ,$$

$$\Gamma \equiv \left(\frac{\Delta x \, \Delta p}{w} \right)^{3N} . \qquad (5\text{-}3)$$

where $w = 2m/C$ is a constant and Γ is the volume of a $6N$ dimensional region (in units of w^{3N}). This region has $3N$ directions of length Δx, and $3N$ directions of Δp.

This region can be said to represent the extent of the motion of the N molecules, and each direction has a length representing the magnitude of change of one variable of motion. There are $6N$ variables in all, i.e. $3N$ coordinates and $3N$ momenta $(x_k, y_k, z_k, p_{xk}, p_{yk}, p_{zk}, k = 1, 2, \ldots N)$. Hence the relation of entropy and the region of motion is very clear in this model of the ideal gas.

5.2. Assumption for the Calculation of the Entropy

Of course, the ideal gas is a very simple and special case. Its entropy is easily calculated in the last section, but most of the other problems are not simple. Nevertheless, the result of the ideal gas has clearly indicated that to understand the meaning of entropy we must start from the magnitude of the molecular motion. Now we generalise (5-3).

At any instant, the positions and momenta of N molecules can be represented by a point in the above $6N$ dimensional space. This point is called the state of the system at that instant, and the space is called phase space. The state changes with time, tracing out a trajectory in this space. Therefore the trajectory is a set of actual states. In a sufficiently long period of observation, this trajectory traces out a very long and meandering curve scattered in a region of the phase space. The lengths of this region in the $6N$ directions are the amplitude of change of the $6N$ variables of motion. We call this region the distribution region of the trajectory or the region of motion of the states. Let Γ be the volume of this region, and assume that entropy is the logarithm of Γ:

$$S = \ln \Gamma \quad . \tag{5-4}$$

This is the basic assumption of statistical mechanics.[a] It is not derivable from any known principle. Here we regard it as a proposal motivated by (5-3). Whether the S in (5-4) is indeed the entropy in thermodynamics should be validated by its applications in thermodynamics. The above definition of Γ is not complete. We have only given some names, and uttered some words. Indeed, how can we determine this region of motion from the molecular motion? This is a difficult question. However, we defer the discussion to later chapters. At the moment we use the following assumption to patch up some explanation.

A supplement to the basic assumption is: The region of motion of the states is equal to the region allowed by the invariant quantities.

This indeed is a daring assumption, but it is not so intangible. We consider a simple example in order to understand the implication of this assumption. A fly moving inside a room has a very complicated trajectory of motion and its position is changing continually. If the observation time is much longer than that required by the fly to go from one side of the room to the other, we can ask: what is the magnitude of the change of position of the fly? The position of the fly is specified by three coordinates (x, y, z). We need not analyse the trajectory, but we can simply say that the magnitude of the changes of x, y, z is determined by the floor, the ceiling and the four walls. These are invariant quantities. The region of motion is the room and the volume of the room is the product of all the magnitude of the changes in the three directions.

We now generalise the (x, y, z) space to the phase space determined by the six variables of motion. These positions of the walls and ceiling are generalised to the various invariant quantities. No matter how complicated the molecular motion is, the region of motion in phase space is determined by these invariant quantities. This is the implication of this assumption.

We must emphasise the following — we must have motion and change so as to define the region of motion. This indeed is the implication of the word "motion". If the fly does not move, then the room cannot represent its region of motion. Each molecule must be moving continually, producing collisions and changes of momentum. The validity of this assumption should be built upon the continual change.

[a] This assumption is a masterpiece of Boltzmann (1844-1906).

For a gas, the invariant quantities include the total energy, the volume and the total number of molecules. The situation for solids is more complicated. Besides the energy and the total number of molecules, there can be other invariant quantities due to its structure, like the positions of the frozen impurity atoms. All states allowed by the invariants must be included in the region of motion. This supplementary assumption is very nice. It reduces the complicated problem of the trajectory to a relatively simple problem of calculating the volume of the region. Of course whether this assumption is correct should be determined by the result of analysing the trajectory. This supplementary assumption should be regarded as part of the basic assumption.

Now we must clarify an important point. The question is — what variables should we use to define the phase space? In the above discussion we have used the positions and momenta of all the particles as the variables. Why do we not use the square of the positions and the energy as the variables? This is an important question because if we use different variables, we have a different volume. Hence, if the variables are not specified, Γ will be meaningless. We use an example to illustrate this point. We first look at the momentum p and energy $E = p^2/2m$ of a particle. Let a certain region be defined as $0 < E < E_0$, or $|p| < \sqrt{2mE_0}$. If we use p as the variable, the volume of this region is

$$\Gamma = 2\sqrt{2m}\sqrt{E_0} \quad . \tag{5-5}$$

If we use E as the variable, the volume is

$$\Gamma' = E_0 \quad . \tag{5-6}$$

The unit or the constant $2\sqrt{2m}$ is not important. The important observation is $\Gamma \propto \sqrt{E_0}$ but $\Gamma' \propto E_0$. In calculating the volume, the high energy parts are not so important in Γ as in Γ'. Hence, this problem of the right variables is a problem of evaluating the weighting of each state in the region of motion. We now add a definite rule to this assumption to avoid any confusion later. This rule states that: the problem of statistical weighting is very obscure in continuous variables. The only way out is to forego the continuous variables, and instead use discrete variables. Quantum mechanics provides the solution. The phase space can be regarded as a collection of discrete points and each point represents a state. We now define the rule:

Γ is the total number of states in the region of motion, i.e. each state has a statistical weight of 1.

Now the rules for calculating entropy are ready. We start from the entropy of an ideal gas in (5-3) to point out its relation with the magnitude of the motion, and then we introduce the concept of the region of motion. Finally we use an assumption to change the problem of calculating the entropy to a statistical problem of calculating the number of states in the region of motion. We must remember that this is just an assumption which has not yet been proved. Its implication and validity await further analysis.

Now we first discuss some problems of the techniques of calculation as a preparation for the actual calculation.

5.3. Fermions

The implications of the basic assumption cannot be clearly explained in a few words or by a few examples. In fact, the main aim of this book is to clarify the implications of this assumption. We start from the simplest example.

In Chapter 3 we used the law of detailed balance to discuss the ideal gas including the quantum gas. The collisions or the reactions between the particles are the crucial point of analysis. Now we shall start from the basic assumption to tackle the same problem.

Let L states have the same energy, of which N are occupied by the fermions. If there is some cause making these particles change their states continually, then what is the entropy?

To use the basic assumption in computing the entropy, we must first find Γ, i.e. the total number of states allowed by the invariant quantities. Here the invariants are N, L and ϵ. Hence Γ is the total number of ways of putting N particles into L states and each state accommodates just one particle. This total number is

$$\Gamma = \frac{L!}{N!(L-N)!} \,. \tag{5-7}$$

According to the basic assumption, $S = \ln \Gamma$.

But (5-7) involve factorials, which are of little use unless evaluated more explicitly. In fact, the basic assumption is true only for system of many particles i.e. $N, L \gg 1$. Entropy is an extensive quantity. If L, N are very large, (5-7) can be simplified by the law of large numbers. The derivation of this law is given in Sec. 5.6. Now we just apply the result. If $N \gg 1$ then

$$N! \simeq N^N e^{-N} (2\pi N)^{1/2} \left[1 + O\!\left(\frac{1}{N}\right)\right] . \tag{5-8}$$

Substituting this into (5-7), we obtain after some simplifications

$$S = L \left[\frac{N}{L} \ln \frac{L}{N} + \frac{L-N}{L} \ln \frac{L}{L-N} \right]$$
$$+ \frac{1}{2} \ln \frac{L}{2\pi(L-N)N} . \qquad (5\text{-}9)$$

Suppose L and N are of the same order, i.e. N/L is $O(1)$ and $S = O(L)$. The last term of (5-9) is $O(\ln L)$ and can be neglected. Because each state has the same energy ϵ, the total energy is

$$E = \epsilon N . \qquad (5\text{-}10)$$

We now differentiate the result (5-9) and get

$$dS = \ln \left(\frac{L-N}{N} \right) dN . \qquad (5\text{-}11)$$

According to thermodynamics, the differential of the entropy is related to the temperature T and chemical potential μ, i.e.

$$dS = \frac{1}{T} dE - \frac{\mu}{T} dN$$
$$= \frac{\epsilon - \mu}{T} dN . \qquad (5\text{-}12)$$

The above dE is $\epsilon\, dN$ (see (5-10)). Equations (5-12) and (5-11) are equivalent. Therefore

$$\ln \frac{L-N}{N} = \frac{\epsilon - \mu}{T} ,$$

hence

$$f \equiv \frac{N}{L} = \frac{1}{e^{(\epsilon-\mu)/T} + 1} . \qquad (5\text{-}13)$$

This result is, of course, that of Chapter 3 where f is the average population of the states.

The reader may ask: is this system too simple? The energy of all states are ϵ, but the states in an actual system have different energies. The answer is that this system can be regarded as part of a large system, i.e. we can group the set of states with the same energy ϵ together to form the system of the above example. Now the important requirement is that each part of the large system should have the same temperature T and chemical potential μ. This requirement is indeed a most important achievement of the basic assumption. If the basic assumption is true, we can then prove that T and μ are uniform throughout the system. The proof is left to the next chapter. We first carefully consider a case where the particles have different energies.

5.4. The Ideal Gas

Let us now look at a more complicated example, i.e. to consider all the particles of an ideal gas and use the basic assumption to calculate its entropy.

Suppose the ideal gas consists of N molecules, whose coordinates and momenta are $\mathbf{r}_k$, $\mathbf{p}_k$, $k = 1, 2, \ldots, N$. In Sec. 3.1 we obtained the statistical formula of the state of one particle:

$$\int d\Gamma = \frac{1}{h^3} \int d^3r \, d^3p \tag{5-14}$$

i.e. in a volume of $d^3p \, d^3r$ there are $d\Gamma$ states. We now generalise (5-14) to the statistical formula of N particle states

$$\int d\Gamma = \int d\Gamma_1 \, d\Gamma_2 \ldots d\Gamma_N$$

$$= \frac{1}{h^{3N}} \int d^3r_1 \ldots d^3r_N \, d^3p_1 \ldots d^3p_N \ , \tag{5-15}$$

i.e. the number of states in a volume $d^3r_1 \ldots d^3r_N \, d^3p_1 \ldots d^3p_N$ in $6N$ dimensional space.

To calculate Γ we must consider the restriction of the various invariants. Here the constraints are: (1) the volume is invariant, i.e. the positions of all the molecules must be inside this volume and (2) the total energy is invariant with a definite value E.

(1) Let the integration of r_k be restricted as

$$0 < x_k, y_k, z_k < L, \qquad k = 1, 2, \ldots, N \quad . \tag{5-16}$$

The volume of the container is $V = L^3$.

(2) Energy invariance can be achieved by imposing a delta function $\delta(E - H)$ where H is the total energy function

$$H = \sum_{k=1}^{N} \frac{p_k^2}{2m} \quad . \tag{5-17}$$

The restrictions (5-16) and (5-17) are not enough and we have to divide the statistical result by $N!$, because each particle is identical and interchanging the labels of a pair of particles will not change the state. As (5-16) and (5-17) put no restriction on the labels so the number of states has been overcounted $N!$ times. This point has been mentioned at the beginning of this chapter. If the particles are not the same, should we again divide the statistical result by $N!$ times? This will be discussed at the end of this section.

Now we use (5-15) to calculate Γ with the constraints (5-16) and (5-17). We get

$$\Gamma(E) = \frac{V^N}{N! h^{3N}} W(E) \ ,$$

$$W(E) = \int d^3 p_1 \ldots d^3 p_N \ \delta(E - H) \ , \tag{5-18}$$

where $W(E)$ can be regarded as the surface area of a sphere in $3N$ dimensions with a radius determined by $H = E$:

$$\left[\sum_{k=1}^{N} p_k^2 \right]^{1/2} = 2mE \ . \tag{5-19}$$

Points on the surface of this sphere satisfy (5-19). The calculation of the surface area of a sphere is an interesting mathematical problem, and we defer its discussion to the next section. For the results see (5-39) and (5-31). Now we write down the answer first:

$$S = \ln \Gamma$$
$$= N \left[\frac{3}{2} \ln \left(\frac{4 \pi m E}{3 N h^3} \right) + \frac{5}{2} + \ln \frac{V}{N} + O \left(\frac{1}{N} \ln N \right) \right] . \tag{5-20}$$

With this result we can find the differentials of all the quantities; because $dE = -p \, dV + T \, dS$, so

$$\frac{1}{T} = \left(\frac{\partial S}{\partial E} \right)_V = \frac{3}{2} \frac{N}{E} \ ,$$

$$p = T \left(\frac{\partial S}{\partial V} \right)_E = T \frac{N}{V} \ . \tag{5-21}$$

The other properties, such as the heat capacity or the expansion coefficient can all be calculated.

We have already emphasised that if the basic assumption is to be true, the various variables must be changing continually. How do the variables in this ideal gas change? The change of position is because each molecule has nonzero velocity while the change of momenta of the molecules is due to collisions either with the wall or with another molecule. If the molecules collide with the wall of the container, there is energy exchange. Therefore even though there is no collisions between the molecules, the magnitude and direction of the momentum of a particle is always changing. If there is no energy exchange at all, then the directions of the momenta are changing but their magnitudes are not and this basic assumption cannot be established. If the molecules collide with each other, then the change of momenta is always occurring. Let λ be the mean free path of the molecules, i.e. the distance between two successive collisions of each molecule. Then

$$\tau = \frac{\lambda}{v} , \qquad (5\text{-}22)$$

is the mean free time, where v is the average velocity. If

$$L \gg \lambda , \qquad (5\text{-}23)$$

then the collisions between atoms are frequent and the observation time t must satisfy the condition

$$t \gg \tau , \qquad (5\text{-}24)$$

in order for the basic assumption to be established. If $L \ll \lambda$, i.e. collisions are infrequent, then

$$t \gg \frac{L}{v} , \qquad (5\text{-}25)$$

is the necessary condition. At room temperatures and normal pressure the mean free distance of air molecules is $\lambda \sim 10^{-5}$ cm $= 1\,000$ Å, $v = 5 \times 10^4$ cm/sec and $\tau \sim 10^{-9}$ sec. Notice that

$$\lambda \sim \frac{1}{n\sigma} ,$$

where n is the density and $\sigma \sim 10$ Å^2 is the effective cross-section area of the collisions of the molecules. The average distance between air molecules is $n^{-1/3} \sim 30$ Å. At other temperatures and pressures λ and τ can be calculated by the formula $v \propto \sqrt{T}$ and with T/p replacing n.

Of course, the collisions between the molecules are caused by the interaction, but (5-17) does not include the interaction energy. Hence, we have utilised the interaction causing collisions in order to satisfy the condition of the basic assumption, but on the other hand we have neglected the interaction energy in calculating the entropy. This does not lead to a contradiction, because the time of contact of molecule with molecule is extremely short, and the interaction only acts at the time of contact. The main effect of collision is to change the momenta after the collision. From the definitions (5-16) and (5-17) of the region of motion to the calculation of the various properties is a sequence of mathematical manipulation which in principle does not involve any new concept. The usual application of this basic assumption is essentially like this. However, the details of calculations may be different and more complicated.

Once the region of motion is fixed, the subsequent mathematical calculations are usually very complicated. These problems are very interesting and each has its own method of solution. We shall discuss these step-by-step and simultaneously introduce new concepts to help us understand the problem.

We must emphasize again that the determination of the region of motion is the first and the most important step. This is also often the most unclear step and the most difficult one.

The above result (5-20) is the entropy calculated from the basic assumption and it is the same as (3-31) in Chapter 3. In Chapter 3 we pointed out that once the chemical potential μ is defined, then the entropy is defined also. There is little choice for the definition of the chemical potential. It can be defined by the conservation laws and the analysis of the chemical reactions. If we do not use the definition in Chapter 3 then the μ in each equation must be rewritten as $\mu + c$ where c is a constant. But each substance will then have a different c and this is rather inconvenient. The entropy determined by the basic assumption does not involve the structure or property of any substance in the definition. Therefore, if the entropy calculated by this assumption is correct, the constant c in the μ must be zero. It is not a coincidence that (5-20) is exactly the same as (3-31).

Now we return to discuss the $N!$ factor in (5-15). If all the molecules are different, can we just discard this $N!$ factor? Of course, this is a guess. It would not be easy to put N different molecules together for $N \gg 1$. Even if this were possible, the gas then may not be ideal. There is a reason for discussing this problem. If the molecules are all different we should ask: in the observation period how far can a molecule go? If the density is very dilute and the container is very small then each molecule can travel between the walls many times. Therefore, the region of motion is still defined by (5-16) and (5-17) and we only need to drop $N!$ in (5-15) and (5-18). If the density is high, and collisions are frequent, in the period of observation a molecule cannot go very far away. Hence the restriction (5-16) is not enough and we have to add restrictions due to the finite time. If the restriction of this time scale cannot be definitely defined, then the basic assumption is useless. We shall frequently discuss the time restriction and the problems involved in applying the basic assumption.

Of course, if $N!$ is dropped, S will not be proportional to N and is not additive. This is the case of a small container with no collision between the molecules. There is nothing wrong that S is not proportional to N. As the molecules are different and travelling between the walls, it is impossible to divide the body into parts with the same property. If we partition the container into two parts, then molecules at one side cannot go to the other side, and the situation is different from the unpartitioned case. The entropy will be different then.

Equations (5-15) and (5-18) are not only appropriate for gases, but they can be applied to the solute molecules in a dilute solution as well. However the following condition must be satisfied: the solute molecules must be moving fast, i.e. during the observation time the distance travelled must be much larger than the average distance between the solute molecules. These formulas are not valid for frozen impurity atoms which are dispersed in a solid. The positions of the frozen atoms are unchanged. These shall be analysed when we discuss impurities in detail in Chapter 18. Notice that if the atoms are frozen, then it does not matter whether the atoms are identical or not. Each integration of $\mathbf{r}_k$ in (5-15) will be limited to the neighbourhood of a fixed frozen position and $N!$ should be omitted, whether or not the atoms are the same.

To sum up, the region of motion must be determined by the molecular motion. Each factor has its own meaning and cannot be inserted or dropped at will.

5.5. Some Special Features of High Dimensional Space

In the simple example of Sec. 5.3, the phase space has 2^L states, that is to say, there are L positions and each has 2 states. The region of motion is part of this space and is determined by the total number of particles.

In the last section the phase space is defined by points in a $6N$ dimensional space. Each point is a state. The density of the points is $1/h^{3N}$ uniformly distributed in this $6N$ dimensional space and the region of motion is determined by $E = H$, i.e. the $\delta(E - H)$ in (5-18). (The total number n is fixed already.) The region of motion can be regarded as the surface of a sphere. In the example of Sec. 5.3 the phase space can be considered as the 2^N vertices of a cube in the L dimensional space, but the geometrical meaning of the region of motion is not so clear.

Strictly speaking, a collection of a finite number of discrete points has no geometrical meaning. The phase space in the above section can also be represented by the population of the particle states. The representation by the $6N$ dimensional space is not necessary. However, if the points are numerous, geometrical concepts are always very useful. The use of the $6N$ dimensional space in the last section is a very good example. The important property of a macroscopic physical system is the large number of variables of motion, hence the phase space is related to some properties of a high dimensional space.

In this section we introduce some geometrical properties of high dimensional space and discuss the law of large numbers.

Let $x = (x_1, x_2, \ldots, x_n)$ be a point in an N dimensional vector space. The length of x is defined as

$$|x| = \left[\sum_{i=1}^{N} x_i^2 \right]^{1/2} . \tag{5-26}$$

Suppose N is a very large number, such as 10^{20}. If at least some x_i are not zero, then

$$|x| \sim N^{1/2} . \tag{5-27}$$

This is a large number. If each x_i changes by a bit, $\Delta x_i \sim \epsilon$, then

$$|\Delta x| \sim \epsilon N^{1/2} . \tag{5-28}$$

Even if ϵ is not large, $|\Delta x|$ can be very large. Conversely, even though x may traverse a large distance in this space, the change of each x_i can be very small.

The most special feature is the volume in this high dimensional space. Now let us look at the volume of a sphere. Let the radius of the sphere be R, then its volume is $Q(R) = CR^N$, and the surface area of the sphere is $Q'(R) = NCR^{N-1}$. The constant C can be determined as follows. Let

$$I = \int_{-\infty}^{\infty} dx_1 \, dx_2 \ldots dx_N \, e^{-\frac{1}{2}(x_1^2 + x_2^2 + \ldots + x_N^2)}$$

$$= \left[\int_{-\infty}^{\infty} dx \, e^{-\frac{1}{2}x^2} \right]^N$$

$$= (2\pi)^{N/2} . \tag{5-29}$$

78 STATISTICAL MECHANICS

If we use spherical coordinates, we get

$$I = \int_0^\infty dR\, Q'(R) e^{-\frac{1}{2}R^2}$$

$$= NC \int_0^\infty dR\, R^{N-1} e^{-\frac{1}{2}R^2}$$

$$= 2^{N/2} C \left(\frac{N}{2}\right)! \quad . \tag{5-30}$$

Comparing (5-29) and (5-30) we get C. Hence the volume of a sphere is

$$Q(R) = \frac{\pi^{N/2} R^N}{\left(\dfrac{N}{2}\right)!} \quad . \tag{5-31}$$

If the radius is increased from R to $(1 + \epsilon)R$, then $Q(R)$ will increase to

$$Q(R)(1 + \epsilon)^N \quad . \tag{5-32}$$

Even though ϵ is not large, $(1 + \epsilon)^N$ can be an enormous number. Hence the volume of a spherical shell of width ϵ is

$$Q(R)[(1 + \epsilon)^N - 1] \quad . \tag{5-33}$$

This is not much different from the volume of the sphere in (5-32). Hence the volume of a sphere is concentrated on its surface shell.

Essentially, the length in a high dimensional space is a large number, but the volume is an "enormously large" number

$$\text{Length} \sim a\sqrt{N} \quad ,$$

$$\text{Volume} \sim a^N \quad , \tag{5-34}$$

where a is a length scale. The special properties above are due to the fact that this space has many directions.

5.6. Peak Integration and the Law of Large Numbers

The following uses the derivation of the law of large numbers to illustrate some common techniques in mathematics. We write $N!$ as an integral:

$$N! = \int_0^\infty dX\, e^{-X} X^N$$

$$= \int_0^\infty dX\, e^{S(X)} \quad,$$

$$S(X) \equiv N \ln X - X \quad. \tag{5-35}$$

If N is a large number, X^N increases rapidly with X while e^{-X} decreases rapidly and $e^{S(X)}$ is very small except at the neighbourhood of its maximum. The maximum is easily obtained:

$$\frac{\partial S}{\partial X} = \frac{N}{X} - 1 = 0 \quad. \tag{5-36}$$

Its solution is $X_c = N$. Near X_c,

$$S(X) = S(X_c) + \frac{1}{2}\left(\frac{\partial^2 S}{\partial X^2}\right)_c (X - X_c)^2 + \ldots$$

$$\simeq N \ln N - N - \frac{1}{2N}(X - X_c)^2 \quad. \tag{5-37}$$

Hence

$$N! \simeq e^{N \ln N - N} \int_{-N}^{\infty} dx'\, e^{-x'^2/2N} \quad,$$

$$= N^N e^{-N} (2\pi N)^{1/2} \left[1 + O\!\left(\frac{1}{N}\right)\right] . \tag{5-38}$$

Roughly speaking, $X^N e^{-X}$ is a peak with centre at $X_c = N$, and its width is $\sqrt{2\pi}$. This width is much smaller than N. Hence the lower limit of the x' integration can be taken as $-\infty$. The $O(1/N)$ term in the equation can be determined by a more carefully calculation. Equation (5-38) is the so-called law of large numbers or Stirling's formula. The above method of integration is applicable to peak-like integrands.

80 STATISTICAL MECHANICS

Using (5-38), the volume of a sphere $Q(R)$ and the surface area of the sphere $Q'(R)$ can be simplified as

$$Q(R) \simeq \exp\left[\frac{N}{2}\left(\ln\frac{2\pi R^2}{N} + 1 - \frac{1}{N}\ln \pi N\right)\right],$$

$$Q'(R) \simeq \exp\left[\frac{N}{2}\left(\ln\frac{2\pi R^2}{N} + 1 - \frac{1}{N}\ln \pi N + \frac{2}{N}\ln\frac{N}{R}\right)\right]. \quad (5\text{-}39)$$

Hence, when $N \gg 1$,

$$\frac{1}{N}\ln Q(R) \simeq \frac{1}{N}\ln Q'(R)$$

$$\simeq \frac{1}{2}\left[\ln\frac{2\pi R^2}{N} + 1\right] + O\left(\frac{\ln N}{N}\right). \quad (5\text{-}40)$$

5.7. The Law of Equipartition of Energy

We next consider another example for demonstration, i.e. the law of equipartition of energy. This law states that if the total energy function is the sum of M square terms:

$$H = \sum_{i=1}^{M} A_i q_i^2, \quad (5\text{-}41)$$

where q_i are coordinates or momenta and A_i are positive constants, then the relation of the total energy E and the temperature is

$$\frac{E}{M} = \frac{1}{2}T, \quad (5\text{-}42)$$

i.e. each variable q_i acquires an average energy $T/2$.

Now we go back to result (5-21) for the ideal gas, i.e.

$$\frac{E}{3N} = \frac{1}{2}T. \quad (5\text{-}43)$$

The total energy H (see (5-17)) is the sum of $3N$ terms (each p_k^2 is in fact three terms $p_{kx}^2 + p_{ky}^2 + p_{kz}^2$). Hence (5-41) shows that the average energy of each term is $T/2$. This is an example of the law of equipartition of energy.

Another example is a system of simple harmonic oscillators with total energy

$$H = \sum_{i=1}^{N} \left(\frac{1}{2} p_i^2 + \frac{1}{2} \omega_i^2 x_i^2 \right) . \tag{5-44}$$

Here each p_i is a variable, so H has $2N$ terms, and each term, according to the above law, contributes an energy $T/2$. Hence the total energy is

$$E = \left(\frac{1}{2}T\right) 2N = NT . \tag{5-45}$$

The deduction of this law is very simple. The calculation of $\Gamma(E)$ is similar to that of the ideal gas case:

$$\Gamma(E) = \frac{1}{h^m} \int dq_1 \, dq_2 \ldots dq_m \, \delta(E - H)$$

$$= \frac{1}{h^m} (A_1 A_2 \ldots A_m)^{-1/2} \int dx_1 \, dx_2 \ldots dx_m$$

$$\delta(E - x_1^2 - x_2^2 - \ldots - x_m^2) ,$$

$$x_i \equiv \sqrt{A_i} \, q_i . \tag{5-46}$$

The integration of (5-46) is also the volume of a sphere. Using (5-40), we get

$$S(E) = \ln \Gamma(E)$$

$$= \frac{M}{2} \left(\ln \frac{2\pi E}{Mh^2} + 1 \right) - \frac{1}{2} \sum_{i=1}^{M} \ln A_i ,$$

$$\frac{\partial S}{\partial E} = \frac{1}{T} = \frac{M}{2} \cdot \frac{1}{E} . \tag{5-47}$$

T is only related to the $(M/2) \ln E$ term.

5.8. Perspectives of the Region of Motion

From the beginning of this chapter, we have emphasised that the meaning of the region of motion is the magnitude of change of the various variables. The above discussion of the complicated techniques may make the reader forget this meaning. In this section we use the point of view of projection in geometry to make this meaning more concrete.

We first digress on some basic facts about projection. Projection is quite familiar to us in our daily life. What we see are the projection of the three dimensions on our retina. Engineering plans are also the projections of the three dimensions on a two-dimensional plane, and are perspectives in various directions. For the same reason, points in an n-dimensional vector space can be projected onto an n'-dimensional base space (of course $n' < n$). It is a crosspoint. We have to examine the projection of a collection of points in the various directions and its density distribution.

We use an example to illustrate the method of calculating the density distribution in the projected space. Suppose points are distributed uniformly on a three-dimensional spherical surface

$$x^2 + y^2 + z^2 = R^2 \ . \tag{5-48}$$

What is its projected density on the xy plane? The answer is

$$\gamma(x,y) = \frac{\Gamma}{4\pi R^2} \int dz \, \delta(\sqrt{x^2 + y^2 + z^2} - R)$$

$$= \frac{\Gamma}{2\pi R} \frac{\theta(R^2 - x^2 - y^2)}{\sqrt{R^2 - x^2 - y^2}} \ . \tag{5-49}$$

In this equation Γ is the total number of points and the projection is a disc with high density of points near the boundary. The reader can find out the reason for this himself. To get the projected density on the x axis, we need to integrate once more.

$$\gamma(x) = \frac{\Gamma}{4\pi R^2} \int dz \, dy \, \delta(\sqrt{x^2 + y^2 + z^2} - R)$$

$$= \frac{\Gamma}{2R} \theta(R - |x|) \ . \tag{5-50}$$

The meaning of this integration is to sum all the points perpendicular to the projected surface.

Generally speaking, a uniform distribution on a n-sphere gives in n' dimensions a projected density

$$\gamma(x_1, x_2, \ldots, x_{n'}) \propto \theta(R - r)(R^2 - r^2)^{(n-n')/2 - 1}$$

$$r^2 = x_1^2 + x_2^2 + \ldots + x_{n'}^2 \ . \tag{5-51}$$

If $n \gg n'$ the density concentrates at the centre, because

$$(R^2 - r^2)^{M/2} \simeq R^M e^{-Mr^2/2R^2} \ , \tag{5-52}$$

and if M is a large number, this distribution is essentially concentrated within a radius of

$$r \sim R/\sqrt{M} \ .$$

This is very different from the situations in (5-49) and (5-50). Equation (5-52) is a normal or Gaussian distribution. This distribution has been discussed in the peak integration in Sec. 5.6 and will often reappear.

The region of motion of an ideal gas is inside the surface of a sphere in $6N$ dimensions and the states are represented by discrete points with uniform distribution. The magnitudes in the various directions of this region of motion are the magnitudes of change of the various variables of motion. How can we measure these magnitudes? The most direct method is to project the region of motion in the various directions. From the distribution of points on the projected plane we can get the answer.

Now let us project the points in the region of motion onto the $\mathbf{p}_1$ space. The density distribution on the projected line is

$$\gamma(\mathbf{p}_1) = \frac{V^N}{N! h^{3N}} \int d^3 p_2 \, d^3 p_3 \ldots d^3 p_N \, \delta\left(E - \frac{p_1^2}{2m} - H'\right) \ , \tag{5-53}$$

$$H' = \sum_{i=2}^{N} \frac{p_i^2}{2m} \ , \quad H = \frac{p_1^2}{2m} + H' \ . \tag{5-54}$$

In (5-53) $\mathbf{p}_1$ is not integrated but is fixed, $\gamma(\mathbf{p}_1) d^3 p_1$ is the total number of states in $d^3 p_1$ for the first particle and H' is the energy after taking out the first particle. Hence according to the definition of $\Gamma(N, E)$ in (5-18),

$$\begin{aligned}
\gamma(\mathbf{p}) &= \frac{V}{Nh^3} \Gamma\left(N-1, E - \frac{p^2}{2m}\right) \\
&= \frac{V}{Nh^3} \exp\left\{S\left(N-1, E - \frac{p^2}{2m}\right)\right\} \\
&= \frac{V}{Nh^3} \exp\left\{S(N, E) + \frac{\mu}{T} - \frac{1}{T}\left(\frac{p^2}{2m}\right) + O\left(\frac{1}{N}\right)\right\} \\
&\propto \exp\left\{\frac{\mu}{T} - \frac{p^2}{2mT}\right\} \ .
\end{aligned} \tag{5-55}$$

This result is in fact (5-52) where $2mE = R^2$, $p^2 = r^2$ and $T = R^2/N$. The meaning of the various symbols in (5-55) is very clear:

$$\frac{1}{T} = \left(\frac{\partial S}{\partial E}\right)_N ,$$

$$\frac{\mu}{T} = -\left(\frac{\partial S}{\partial N}\right)_E . \tag{5-56}$$

The relation between the chemical potential μ and S can be obtained from the discussion of work and heat in Chapter 2:

$$dE = \mu dN + T dS . \tag{5-57}$$

(We do not consider the other differentials.) The second differentiations of S with respect to N and E are quantities of order $1/N$. Hence the projection of the region of motion on the $\mathbf{p}_1$ direction or any $\mathbf{p}_i$ direction is a normal distribution. The magnitude of this region of distribution is about

$$\Delta p = \sqrt{mT} . \tag{5-58}$$

This can be regarded as an estimate of the magnitude of change of the momentum of the molecule.

We can use a bigger space as the projection plane, e.g., the ($\mathbf{p}_1 \, \mathbf{p}_2$) space or an even larger space. The above analysis can be slightly generalised to give the density distribution of the projected points:

$$\gamma(\mathbf{p}_1, \mathbf{p}_2, \ldots, \mathbf{p}_n) \propto \exp\left\{-\frac{1}{T}\sum_{k=1}^{n}\frac{p_k^2}{2m}\right\} \exp\left\{\frac{\mu n}{T}\right\} . \tag{5-59}$$

If $n \ll N$, this result is established.

Generally speaking, projection is statistics by classification. We group together states of the same class and count their number.

The above analysis of projection tells us what the region of motion looks like in different directions. These, of course, are results of the basic assumption. From here we can return to Sec. 5.1 and compare with the discussion of this section.

5.9. Trajectory and the Region of Motion

Now we have some idea of the geometrical property of the region of motion. In Sec. 5.2 we used a fly moving in a room to illustrate the meaning of the region of motion. The trajectory of the fly fills the room.

But the phase space of N molecules is not quite like the room because N is a large number, not just 3. The volume of the phase space is a large number of order e^N. The trajectory is the set of states during the observation time. How many states are visited during this time? The number is about

$$n \sim N(T/\tau) \ . \tag{5-60}$$

where τ is the mean free time and T/τ is the number of collisions in the time T. Each collision changes the state once, and n is a large number of order N. (Of course, in between collisions, the molecules move also, and n is larger than (5-60) but is still a large number of order N.) A number of order N is far less than e^N, i.e. the trajectory in the region of motion is a very tiny thread. The region of motion must be defined by the trajectory. At first sight, this tiny thread can remain in an extremely small region. How can this tiny thread fill up this enormous volume? This question is hard to answer. We shall make a more detailed analysis in Chapter 25. Here, from the point of view of geometry, we give a crude treatment.

Let $(\mathbf{p}_1, \mathbf{p}_2, \ldots \mathbf{p}_N) \equiv \mathbf{p}$ be the momenta of N molecules. The space of $\mathbf{p}$ is a $3N$ dimensional space. For simplicity we omit the position space. If two molecules collide (e.g. molecules 1 and 2), then the momenta of these two molecules are changed from $\mathbf{p}_1, \mathbf{p}_2$ to $\mathbf{p}'_1, \mathbf{p}'_2$. Let

$$\mathbf{q}_{12} = (\mathbf{p}'_1 - \mathbf{p}_1, \mathbf{p}'_2 - \mathbf{p}_2, 0, 0, \ldots, 0) \ . \tag{5-61}$$

That is to say, state $\mathbf{p}$ moves to $\mathbf{p} + \mathbf{q}_{12}$. Let the next collision be between i and j, and the state moves to $\mathbf{p} + \mathbf{q}_{12} + \mathbf{q}_{ij}$. After n collisions there are n $\mathbf{q}$ vectors. The length of each $\mathbf{q}$ is different. Their average is the magnitude of the change of momentum of each molecule, that is the length of the sides of the region of motion.

Notice that if 12 and ij in $\mathbf{q}_{12}$ and $\mathbf{q}_{ij}$ are different molecules, then $\mathbf{q}_{12}$ and $\mathbf{q}_{ij}$ are orthogonal vectors, i.e.

$$\mathbf{q}_{12} \cdot \mathbf{q}_{ij} = \sum_{\alpha=1}^{N} q_{12\alpha} q_{ij\alpha} = 0 \ , \tag{5-62}$$

because $\mathbf{q}_{12}$ has only the nonzero components $\alpha = 1$ and 2 and $\mathbf{q}_{ij}$ only those $\alpha = i$ and j. In the n collisions, $n \sim N$, the n **q** vectors are almost all mutually orthogonal because the molecules in the different parts are colliding.

We now need a bit of imagination. If we have several rods and we want to make up a frame of a space, how can we do this? The answer is that the rod should be mutually perpendicular so as to make up a frame. In building a house, the pillars and rafts must be perpendicular. Now we are in a $3N$ dimensional space with $n \sim 3N$ **q** vectors as the rods. To make up a frame for the $3N$ dimensional space, the **q**'s should be mutually orthogonal. If the rods are distributed evenly in all the direction, they cannot make up a frame for the space because there are e^N directions requiring e^N rods.

The trajectory is linked by the various **q** vectors. Because the **q** vectors are mutually perpendicular, the trajectory can make up a frame of the volume of a $3N$ dimensional space. The states in the trajectory is very scarce compared to the total volume, but it can, like the pillars and rafts of a house, make up a frame for a finite volume.

The above is a crude explanation for the relation between the trajectory and the region of motion. The orthogonality of the various **q** vectors will be explained later as the independence property. In Chapter 6 we shall analyse these problems further.

Problems

1. Apply the analysis of Sec. 5.3 to bosons.
    (a) Calculate the entropy.
    (b) Find the relation of the population with T and μ.

2. The Hamiltonian of a paramagnetic substance is

$$H = - \sum_{i=1}^{N} s_i h , \qquad (5\text{-}63)$$

where s_i can be $+1$ or -1.
The change of the spin s_i is determined by a certain rule of motion. We disregard this here. The magnetic field h is a fixed constant.
    (a) Using the basic assumption, calculate the entropy.
    (b) Find the relation between the energy E and the temperature T. The total magnetic moment $M = \sum_{i=1}^{N} s_i$ is just $-E/h$.
    (c) This model has many points of similarity with the fermion model in Sec. 5.3. Discuss the relation between them.

3. Let a be the length of the side of an N-dimensional cube. Its volume is a^N. Calculate the volumes of the circumscribed and inscribed spheres. Compare their magnitudes. (N is a large number.)

4. A uniform string of mass M and length L is fixed at the two ends with tension F.

(a) From the following equation calculate the frequencies of vibration.

$$\frac{\partial^2 y}{\partial t^2} = c^2 \frac{\partial^2 y}{\partial x^2} , \quad c^2 = FL/M , \quad 0 < x < L . \quad (5\text{-}64)$$

(b) Calculate the energy, the entropy and the heat capacity as a function of temperature, assuming that the highest frequency is ω_D.

5. Assume that the string in the above problem is in a room of temperature T. Let d be the diameter of the string and η the coefficient of viscosity of air.

(a) Considering the air resistance, calculate the correction term to the wave equation.

(b) The observation time should be longer than the fluctuation period τ so that equilibrium can be defined. If we include air resistance, find τ.

(c) Is the result of (b) in the last problem influenced by τ? If the string is immersed in water, will the result be different?

6. Derive (5-51) and (5-52) and clarify the relation between (5-54) and (5-52). Equation (5-55) is, of course, the Boltzmann distribution already discussed in Chapter 3. Here it is regarded as a projected distribution from a high dimensional sphere.

7. Calculate the projection in the s_1 direction of the region of motion in problem 2.

8. Projection of a cube.

The vertices of an n-dimensional cube can be represented by the coordinates $(1, 1, \ldots, 1), (-1, 1, 1, \ldots, 1), (1, -1, 1, \ldots,), \ldots, (-1, -1, \ldots, -1)$ and there are 2^n such vectors. Call these vectors $\mathbf{R}_i$, $i = 1, 2, \ldots, 2^n$. Take any pair of mutually perpendicular vectors $\mathbf{e}_x$ and $\mathbf{e}_y$,

$$\mathbf{e}_x \cdot \mathbf{e}_x = 1, \quad \mathbf{e}_y \cdot \mathbf{e}_y = 1, \quad \mathbf{e}_x \cdot \mathbf{e}_y = 0 . \quad (5\text{-}65)$$

Then the projection of $\mathbf{R}_i$ in the xy plane is

$$R_{ix} = \mathbf{R}_i \cdot \mathbf{e}_x \quad \text{and} \quad R_{iy} = \mathbf{R}_i \cdot \mathbf{e}_y \quad (5\text{-}66)$$

(a) Draw the projection of the three-dimensional cube, i.e. $n = 3$. Avoid the overlapping projections of the vertices.

(b) Draw the projection of the 4- and 5-dimensional cubes. These pictures can give the reader some impression of higher dimensional spaces.

Chapter 6

SOME ELEMENTARY APPLICATIONS OF THE BASIC ASSUMPTION

In this chapter we start from the basic assumption and derive some properties of the equilibrium states, i.e. the fact that the temperature, pressure and chemical potentials, etc. are the same throughout the system. We use these properties to analyse some examples, such as the problem of the vibration of the lattice, including the Debye model and second sound.

6.1. Systems and Subsystems

A system can be regarded as being composed of its subsystems. Now we start from this basic assumption to see the relation of the entropy of a system with those of its subsystems. Divide the system into two parts 1 and 2. After fixing the states s_1 and s_2 of the two parts, the state of the system is fixed also. Therefore

$$s = (s_1, s_2) \ . \tag{6-1}$$

The total energy H is the sum of $H_1 + H_2$ plus the interaction energy H_{12} across the boundary,

$$H(s) = H_1(s_1) + H_2(s_2) + H_{12}(s_1, s_2) \ . \tag{6-2}$$

Now we neglect H_{12} because it is only effective at the boundary and is very small. This is not to say that it is unimportant, since the exchange of energy between the two parts is mediated by H_{12}. But its main function is to maintain energy exchange between the two parts, and its contribution to the total energy is very small. So we drop it for the moment. The entropy of the system is then

$$S(E) = \ln \Gamma(E) , \qquad (6\text{-}3)$$

$$\Gamma(E) = \sum_{s_1} \sum_{s_2} \delta(E - H_1(s_1) - H_2(s_2))$$

$$= \int dE_1 \, dE_2 \, \delta(E - E_1 - E_2) \, \Gamma_1(E_1) \, \Gamma_2(E_2)$$

$$= \int dE_1 \, dE_2 \, \delta(E - E_1 - E_2) \, e^{S_1(E_1) + S_2(E_2)} , \qquad (6\text{-}4)$$

where

$$\Gamma_i(E_i) = \sum_{s_i} \delta(E_i - H_i(s_i))$$

$$\equiv e^{S_i(E_i)} , \qquad i = 1, 2 . \qquad (6\text{-}4')$$

We wish to apply peak integration on (6-4), hence we must find the maximum of $S_1(E_1) + S_2(E_2)$. However, E_1 and E_2 are not independent variables and are constrained by $E_1 + E_2 = E$. We can use the method of Lagrange multipliers for this problem and we now introduce the principle of the method.

Consider a particle whose potential energy is $V(\mathbf{r})$ but whose position is constrained on a curved surface $\phi(\mathbf{r}) = 0$. Where is the equilibrium position? That is, where is the minimum of $V(\mathbf{r})$ on this surface? The particle is not only acted upon by a force $-\nabla V(\mathbf{r})$ but also by the constraint force of the surface. This constraint force is normal to the surface (neglecting friction) and hence is proportional to $-\nabla \phi(\mathbf{r})$ and can be written as $-\lambda \nabla \phi$. At equilibrium this force should cancel $-\nabla(\mathbf{r})$:

$$-\nabla V - \lambda \nabla \phi = -\nabla(V + \lambda \phi) = 0 ,$$

$$\phi(\mathbf{r}) = 0 . \qquad (6\text{-}5)$$

The parameter (or Lagrange multiplier) λ measures the magnitude of the constraint force. Equation (6-5) shows that we should find the minimum of $V + \lambda \phi$; the equilibrium point and the value of λ can then be solved from the two equations in (6-5). For the particle, the restriction to the surface $\phi(\mathbf{r}) = 0$

gives a constraint force with magnitude λ and this force keeps the particle on the surface.

We now return to (6-4). We use the method of Lagrange multipliers to find the maximum of $S_1(E_1) + S_2(E_2)$. According to (6-5) we write

$$\frac{\partial}{\partial E_1}[S_1(E_1) + \beta(E - E_1 - E_2)] = 0 ,$$

$$\frac{\partial}{\partial E_2}[S_2(E_2) + \beta(E - E_1 - E_2)] = 0 ,$$

$$E - E_1 - E_2 = 0 , \qquad (6\text{-}6)$$

where β is the strength of the constraint force. The meaning of (6-6) is obvious: $\partial S_1/\partial E_1 = T_1$, $\partial S_2/\partial E_2 = T_2$ are the temperatures of the two parts. Equation (6-6) shows that

$$T_1 = T_2 = \frac{1}{\beta} \equiv T . \qquad (6\text{-}7)$$

When the temperatures are the same, the integral of (6-4) is maximum.

Let $E_1 = \bar{E}_1$, $E_2 = \bar{E}_2$ be the solution of (6-6); then the expansion of $S_1 + S_2$ around the neighbourhood of this maximum is

$$S_1(E_1) + S_2(E_2)$$
$$= S_1(\bar{E}_1) + S_2(\bar{E}_2) + \frac{1}{T}(E_1 - \bar{E}_1 + E_2 - \bar{E}_2)$$
$$- \frac{1}{2}\left[\frac{1}{K_1}(E_1 - \bar{E}_1)^2 + \frac{1}{K_2}(E_2 - \bar{E}_2)^2\right] + \ldots , \qquad (6\text{-}8)$$

$$\frac{1}{K_i} \equiv -\left(\frac{\partial^2 S_i}{\partial E_i^2}\right)_{E_i = \bar{E}_i} = (T^2 C_i)^{-1} , \qquad i = 1, 2, ,$$

$$C_i = \frac{\partial E_i}{\partial T} . \qquad (6\text{-}9)$$

SOME ELEMENTARY APPLICATIONS OF THE BASIC ASSUMPTION 91

The first order term in (6-8) is zero under the condition $(E_1 + E_2) - (\bar{E}_1 + \bar{E}_2) = 0$. Substituting (6-8) into (6-4) we get

$$\Gamma(E) \simeq \exp\{S_1(\bar{E}_1) + S_2(\bar{E}_2)\} \cdot \int dE_1' \, dE_2' \, \delta(E_1' + E_2')$$

$$\cdot \exp\left\{-\frac{E_1'^2}{2K_1} - \frac{E_2'^2}{2K_2}\right\},$$

$$= \sqrt{2\pi K} \, \exp\, S_1(\bar{E}_1) + S_2(\bar{E}_2) \quad ,$$

$$K = K_1 K_2 / (K_1 + K_2) \quad , \qquad \text{assuming } K_1 K_2 > 0 \quad . \tag{6-10}$$

Hence the entropy of the system is

$$S(E) = S_1(\bar{E}_1) + S_2(\bar{E}_2) + \ln(\sqrt{2\pi K}) \quad . \tag{6-11}$$

The last term can be neglected, as it is of order $\ln N$, and S_1, S_2 are of order N, N being the total number of particles.

The above derivation seems very simple, but we want to further explain some obscure points. $\Gamma(E)$ is the integral of $\Gamma_1(E_1) \Gamma_2(E_2)$ (see (6-4)). The product form $\Gamma_1(E_1) \Gamma_2(E_2)$ is the result of statistics by classification. This statistics by classification groups together states of a special kind, i.e. the energy of the first part is E_1 and that of the second is E_2. The word "class" refers to (E_1, E_2). The value of each (E_1, E_2) is a class. Of course, $E_1 + E_2 = E$, because only these kinds of states are in the region of motion. Equation (6-4) sums up the results of this statistics by classification, adding up all the kinds of states, and obtains the volume $\Gamma(E)$ of the region of motion. The integration of (6-4) is termed statistics by summation.

The above calculation points out to us that $\Gamma_1(E_1)$ and $\Gamma_2(E_2)$ are maximum at $(\bar{E}_1, \bar{E}_2)$, and that the number of states in the neighbourhood of $(\bar{E}_1, \bar{E}_2)$ comprises nearly all the states. "Neighbourhood" here means

$$E_1' = E_1 - \bar{E}_1 \sim \sqrt{K_1} \quad ,$$

$$E_2' = E_2 - \bar{E}_2 \sim \sqrt{K_2} \quad , \tag{6-12}$$

This is because

$$\Gamma_1(E_1)\,\Gamma_2(E_2) \propto \exp\left\{-\frac{E_1'^2}{2K_1} - \frac{E_2'^2}{2K_2}\right\}, \qquad (6\text{-}12')$$

(see Eq. (6-10)). Hence states far from $(\bar{E}_1, \bar{E}_2)$ are few in number, i.e. the integration of (6-4) gives

$$\Gamma(E) \simeq \Gamma_1(\bar{E}_1)\,\Gamma_2(\bar{E}_2)\,\Delta E\,, \qquad \Delta E = \sqrt{2\pi K}\,,$$

$$S(E) = \ln \Gamma(E) \simeq S_1(\bar{E}_1) + S_2(\bar{E}_2)\,. \qquad (6\text{-}13)$$

(Although $\Delta E \sim \sqrt{N}$ but $S_1, S_2 \sim N$, hence $\ln \Delta E$ can be ignored.) That is to say, the entropy obtained after summation is the maximum of the entropies of the classes. This summation is in fact the step required to find the maximum and yields the class with the most states. The entropy of this class is very nearly the sum of the entropies of the parts. That is to say — "the whole region of motion is nearly occupied by the states in this class." This result is peculiar. It can be said to be the property of a superlarge number of order e^N. The functions $\Gamma_1(E_1)$ and $\Gamma_2(E_2)$ are all superlarge numbers. Once a variable is changed the changes of these superlarge functions are enormous. Hence the maxima of the functions are quite disparate, and they are far larger than the non-maximum values. Consequently only the maximum values are important.

This class with the largest number of states is just the class with $T_1 = T_2$, i.e. the two parts with the same temperature. The other classes have a minute effect on $S(E)$. This conclusion constitutes a bridge linking equilibrium phenomena with the basic assumption of statistical mechanics.

The above discussion can be generalised as follows. We divide the system into m parts, each specified by the state s_i with entropy $S_i(E_i)$, energy $H_i(s_i)$ and $S_i(E_i) = \ln \Gamma_i(E_i)$,

$$\Gamma_i(E_i) = \sum_{s_i} \delta(E - H_i(s_i))\,, \qquad i = 1, 2, \ldots, m\,. \qquad (6\text{-}14)$$

The total entropy of the system is

$$S(E) = \ln \Gamma(E)\,,$$

$$\Gamma(E) = \sum_{s_1} \sum_{s_2} \ldots \sum_{s_m} \delta(E - H_1(s_1) - H_2(s_2) - \ldots - H_m(s_m))\,.$$

$$(6\text{-}15)$$

SOME ELEMENTARY APPLICATIONS OF THE BASIC ASSUMPTION 93

We can repeat the calculations of Eqs. (6-4) to (6-11) and obtain

$$S(E) = \sum_{i=1}^{m} S_i(\bar{E}_i) \ . \tag{6-16}$$

T and $\bar{E}_i$ are the solutions of the following equations

$$\frac{\partial S_i(E_i)}{\partial E_i} = \frac{1}{T} \ , \qquad i = 1, 2, \ldots, m \ ,$$

$$E = E_1 + E_2 + \ldots + E_m \ . \tag{6-17}$$

The derivation is left as an exercise for the reader.

Of course, statistics by classification is not limited to energy classification. The above result can be generalised to any variable of motion. We only require this variable to be a number of order N. We now discuss an important example.

In the above, we regarded the particle number as unchanged in each subsystem. If there is exchange of particles between the subsystems, then the particle numbers $N_1(s_1), N_2(s_2), \ldots, N_m(s_m)$ of each subsystem become the variables of motion. Assume that the total particle number is a constant:

$$N = \sum_{i=1}^{m} N_i(s_i) \ . \tag{6-18}$$

Let

$$S_i(E_i, N_i) = \ln \Gamma_i(E_i, N_i) \ ,$$

$$\Gamma_i(E_i, N_i) = \sum_{s_i} \delta(E_i - H_i(s_i)) \, \delta(N_i - N_i(s_i)) \ , \tag{6-19}$$

i.e. besides classifying by energy we need also to make a classification according to particle number. This is a generalisation of (6-14) and the generalisation of the conclusions (6-16) and (6-17) is

$$S(E, N) = \sum_{i=1}^{m} S_i(\bar{E}_i, \bar{N}_i) \ . \tag{6-20}$$

$\bar{E}_i, \bar{N}_i$ and T, μ are the solutions of the following set of equations:

$$\frac{\partial S_i}{\partial E_i} = \frac{1}{T}, \qquad i = 1, 2, \ldots, m \ ,$$

$$\frac{\partial S_i}{\partial N_i} = -\frac{\mu}{T},$$

$$E = E_1 + E_2 + \ldots + E_m,$$

$$N = N_1 + N_2 + \ldots + N_m. \tag{6-21}$$

Hence each subsystem has the same temperature T and the same chemical potential μ. Further consideration leads to the conclusion that each part has the same pressure if we assume that the volumes of the different parts are also the variables of motion.

The above discussions show that from the basic assumption we can directly derive some properties of the equilibrium states, giving us confidence about this assumption. The reader should notice some incorrect motions such as "statistical mechanics proves the existence of equilibrium", or "statistical mechanics proves that entropy is a maximum". In fact we assume that entropy is a maximum (statistics by classification is smaller than the summation, and the total entropy is the result of the statistics by summation). But the assumption is still an assumption and not a proof.

We now look at some applications.

6.2. Vibrations

Quantum mechanics can be applied to many problems, but there are four cases giving very simple results. They are (1) the free particle, (2) the simple harmonic oscillator, (3) the hydrogen atom and (4) spin $\frac{1}{2}$ particles. These are the basic applications. We had dwelt with the statistics of states of the free particle. Later we shall encounter more of this. Now let us look at simple harmonic oscillators.

The small amplitude vibration of a solid can be regarded as a collection of simple harmonic oscillators. Each fundamental vibration has its own frequency. According to quantum mechanics, a vibration of frequency ω has energy $\epsilon_n (\hbar = 1)$:

$$\epsilon_n = (n + \tfrac{1}{2})\omega, \qquad n = 0, 1, 2, \ldots. \tag{6-22}$$

That is to say, the state of vibration can be specified by the integer n. If there are N fundamental vibrating elements, then the state is specified by N integers

SOME ELEMENTARY APPLICATIONS OF THE BASIC ASSUMPTION

$(n_1, n_2, \ldots, n_N) \equiv s$, with total energy

$$H(s) = \sum_{i=1}^{N} (n_i + \tfrac{1}{2}) \omega_i \; . \tag{6-23}$$

Now let us calculate the entropy:

$$S(E) = \ln \Gamma(E) \; ,$$

$$\Gamma(E) = \sum_{s} \delta(E - H(s))$$

$$= \sum_{n_1 = 0}^{\infty} \sum_{n_2 = 0}^{\infty} \cdots \sum_{n_N = 0}^{\infty} \delta\!\left(E - \sum_{i=1}^{N} (n_i + \tfrac{1}{2}) \omega_i\right) . \tag{6-24}$$

These sums are not easy to evaluate. Let us divide these N elements into groups so that frequency is nearly the same in each group. We calculate the entropy for each group and use the results of the last section to sum up the entropies.

Let the α group have N_α elements with frequency ω_α. Then the entropy of this group is

$$S_\alpha(E_\alpha) = \ln \Gamma_\alpha(E_\alpha) \; ,$$

$$\Gamma_\alpha(E_\alpha) = \sum_{n_1 = 0}^{\infty} \cdots \sum_{n_{N_\alpha} = 0}^{\infty} \delta\!\left(E_\alpha - \sum_{i=1}^{N_\alpha} (n_i + \tfrac{1}{2}) \omega_\alpha\right) . \tag{6-25}$$

The summations in (6-25) are easier to do because there is only *one* frequency ω_α. Let

$$M_\alpha \equiv \frac{E_\alpha}{\omega_\alpha} - \frac{1}{2} N_\alpha \; , \tag{6-26}$$

then the δ-function in (6-25) is[a]

$$\frac{1}{\omega_\alpha} \delta\!\left(M_\alpha - \sum_{i=1}^{N_\alpha} n_i\right) . \tag{6-27}$$

---

[a] M_α is regarded as an integer. The variable in the δ-function (6-27) is also regarded as an integer. We define $\delta(M) = 1$ if $M = 0$, $\delta(M) = 0$ if $M \neq 0$ and M is likewise an integer.

96 STATISTICAL MECHANICS

Suppose we have N_α boxes, and take n_i as the number of balls in the i-th box, the total number of balls must be M_α so as to satisfy (6-25). Hence, Γ_α is the total number of ways of putting M_α balls into N_α boxes (divided by ω_α):

$$\Gamma_\alpha(E_\alpha) = \frac{1}{\omega_\alpha} \frac{(M_\alpha + N_\alpha - 1)!}{M_\alpha!(N_\alpha - 1)!} . \qquad (6\text{-}28)$$

The reader may derive (6-28) himself. This is not the same as the calculation of the surface area of a sphere in the last chapter, but the aim and the principle are the same.

According to the result of the last section, the total entropy is the maximum of the sum (see (6-16) and (6-17)):

$$S(E) = \sum_\alpha S_\alpha(E_\alpha) , \qquad (6\text{-}29)$$

$$\frac{\partial S_\alpha}{\partial E_\alpha} = \frac{1}{T} , \quad E = \sum_\alpha \bar{E}_\alpha . \qquad (6\text{-}30)$$

$\bar{E}_\alpha$ and T must be solved from (6-30). Using the law of large numbers $\ln N! \approx N \ln N - N$, (6-28) can be simplified and we get:

$$S_\alpha(E_\alpha) = \ln \Gamma_\alpha(E_\alpha)$$
$$= (M_\alpha + N_\alpha) \ln(M_\alpha + N_\alpha) - M_\alpha \ln M_\alpha - N_\alpha \ln N_\alpha , \qquad (6\text{-}31)$$

$$\frac{\partial S_\alpha}{\partial E_\alpha} = \frac{1}{\omega_\alpha} \frac{\partial S_\alpha}{\partial M_\alpha}$$

$$= \frac{1}{\omega_\alpha} \ln \frac{M_\alpha + N_\alpha}{M_\alpha}$$

$$= \frac{1}{T} . \qquad (6\text{-}32)$$

From the definition of M_α, (6-26) and (6-32), we obtain

$$\bar{E}_\alpha = N_\alpha \left(\frac{1}{e^{\omega_\alpha/T} - 1} + \frac{1}{2} \right) \omega_\alpha . \qquad (6\text{-}33)$$

SOME ELEMENTARY APPLICATIONS OF THE BASIC ASSUMPTION

The relation between the total energy and T is determined by (6-30), i.e.

$$E = \sum_\alpha N_\alpha \left(\frac{1}{e^{\omega_\alpha/T} - 1} + \frac{1}{2} \right) \omega_\alpha . \tag{6-34}$$

Similarly from (6-31), (6-32) and (6-29) the entropy is found to be

$$S = \sum_\alpha N_\alpha \left(\frac{\omega_\alpha/T}{e^{\omega_\alpha/T} - 1} + \ln\left(\frac{1}{1 - e^{\omega_\alpha/T}} \right) \right) . \tag{6-35}$$

In many applications, the frequency distribution is very dense. Then the sums in the above two equations can be replaced by integrations:

$$\sum_\alpha N_\alpha f(\omega_\alpha) = \int d\omega\, g(\omega) f(\omega) V ,$$

$$V g(\omega) = \sum_i \delta(\omega - \omega_i) = \sum_\alpha N_\alpha \delta(\omega - \omega_\alpha) , \tag{6-36}$$

where $g(\omega)$ is the frequency distribution of the vibrating elements and depends on the characteristics of the physical system. Before giving an example of $g(\omega)$, we first point out the significance of the zero-point energy.

As $T \to 0$, the energy of each element is $E_\alpha/N_\alpha = \frac{1}{2}\omega_\alpha$. (See Eq. (6-33).) This is the so-called zero-point energy. This is, of course, a very important term. At high temperatures, i.e. when $\omega_\alpha/T \ll 1$

$$e^{\omega/T} - 1 \simeq \frac{\omega}{T} + \frac{1}{2}\left(\frac{\omega}{T}\right)^2 + O\left(\frac{\omega^3}{T^3}\right)$$

$$= \frac{\omega}{T}\left(1 + \frac{\omega}{2T}\right) + O\left(\frac{\omega^3}{T^3}\right) . \tag{6-37}$$

Substituting into (6-33) we get

$$\frac{\overline{E}_\alpha}{N_\alpha} = T - \frac{\omega_\alpha}{2} + O\left(\frac{\omega_\alpha^2}{T}\right) + \frac{\omega_\alpha}{2}$$

$$= T + O\left(\frac{\omega_\alpha^2}{T}\right) . \tag{6-38}$$

The first term T is, of course, the result of the law of equipartition of energy discussed in the last chapter such that each element has an energy T. The term $O(\omega_\alpha^2/T)$ can be neglected when T is very large. Notice that this zero-point energy is still very important, because it cancels the term $-\omega_\alpha/2$. Were it not for the zero point energy, the energy at high temperature would have a constant term.

6.3. The Debye Model

Now we apply the above results to the lattice vibrations. It will be necessary to find the frequency distribution $g(\omega)$. The structure of a crystal is very complicated and the analysis of its vibrating modes is a special topic in itself. Here we discuss a very crude estimation, i.e. the Debye model.

The vibration at low frequencies is the elastic wave, whose frequency is inversely proportional to its wavelength. Let k be the wave number, i.e. $2\pi/$(wavelength), then

$$\omega_\parallel = c_\parallel k \; ,$$
$$\omega_\perp = c_\perp k \; . \tag{6-39}$$

where $\parallel$ denotes longitudinal wave, while $\perp$ denotes transverse wave, and $c_\parallel$ and $c_\perp$ are their respective speeds of propagation. Equation (6-39) is the law of elasticity and is valid for long wavelength vibrations ($1/k \gg$ lattice spacing).

The calculation of $g(\omega)$ is similar to calculating the energy distribution of the states of a particle in Chapter 3. We first solve the elastic wave equation for the plane wave (or stationary wave, depending on the boundary condition). If we use cyclic boundary conditions, then $\mathbf{k}$ must satisfy

$$\mathbf{k} = \frac{2\pi}{L}(n_x, n_y, n_z) \; ,$$
$$n_x, n_y, n_z = 0, \pm 1, \pm 2, \ldots \; , \tag{6-40}$$

assuming the crystal to be a cube of length L. Hence the normal mode is specified by three integers. The frequency distribution (according to the definition (6-36)) is

$$V g(\omega) = \sum_{n_x} \sum_{n_y} \sum_{n_z} [\delta(\omega - \omega_\parallel) + 2\delta(\omega - \omega_\perp)] \; . \tag{6-41}$$

SOME ELEMENTARY APPLICATIONS OF THE BASIC ASSUMPTION 99

Note that there are two kinds of transverse waves (left polarised, and right polarised), and one kind of longitudinal wave. We can change the summation in (6-41) into an integration to obtain

$$g(\omega) = \frac{1}{(2\pi)^3} \int d^3k \ [\delta(\omega - c_\| k) + 2\delta(\omega - c_\perp k)]$$

$$= \frac{1}{(2\pi)^3} 4\pi \omega^2 \left(\frac{1}{c_\|^3} + \frac{2}{c_\perp^3} \right) . \qquad (6\text{-}42)$$

This is valid for low frequency vibrations. When ω is large, (6-42) is obviously not correct. However, we are just making a rough estimate, so (6-42) can be used throughout. We also have to notice that

$$\int g(\omega) \, d\omega = \frac{N}{V} , \qquad (6\text{-}43)$$

is a necessary condition (see Eq. (6-36)). If the crystal has N_A atoms, then $N = 3N_A$ because each atom has three degrees of freedom. To satisfy (6-43), let us set ω_D as the maximum frequency, i.e.

$$g(\omega) = 0 , \qquad \omega > \omega_D . \qquad (6\text{-}44)$$

If $\omega < \omega_D$ we use (6-42) for $g(\omega)$. From (6-43) we get

$$\omega_D^3 = \frac{2\pi^2 N}{V} c^3 ,$$

$$\frac{3}{c^3} \equiv \frac{1}{c_\|^3} + \frac{2}{c_\perp^3} . \qquad (6\text{-}45)$$

Here ω_D can be determined from the elastic wave, and this $g(\omega)$ fixes the Debye model. The frequency ω_D is called the Debye frequency as determined by $g(\omega)$. The various equilibrium properties can be derived by differentiating (6-34) to (6-36). The most important example is the heat capacity (Fig. 6-1):

$$C_V = \frac{\partial E}{\partial T} = 3N_A \ f\!\left(\frac{\omega_D}{T}\right) , \qquad (6\text{-}46)$$

$$f(y) = \frac{3}{y^3} \int_0^y dx \ \frac{x^4 e^x}{(e^x - 1)^2} . \qquad (6\text{-}47)$$

100 STATISTICAL MECHANICS

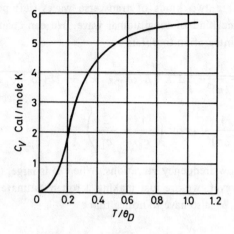

Fig. 6-1 Plot of the heat capacity C_V as a function of y (see (6-46), (6-47)). This is a universal curve for all solids. From Kittel (1966) p. 179.

The reader can derive these for himself. Notice that we have used units with $\hbar = k = 1$. If we use the usual units, then ω_D/T should be replaced by $\hbar\omega_D/kT \equiv \theta_D/T$ where θ_D is the Debye temperature. According to (6-46), the ratio C_V/N_A of all solids should be the same function of θ_D/T. This conclusion agrees quite well with experiments, as can be seen from Table 6-1. (Notice that the various Debye temperatures are close to room temperature.)

Table 6-1 Debye temperature [b]

Solids	θ_D (from C_V) K	θ_D (from elastic coefficients) K
NaCl	308	320
KCl	230	246
Ag	225	216
Zn	308	305

[b] See Kittel (1966), p. 132.

The right-hand column in the table is the value of θ_D obtained from the elastic coefficients (see Eq. (6-45)). The middle column is obtained by fitting (6-46) with the experimentally determined C_V. If the Debye model is completely correct, the two columns should be identical. The agreement is actually not bad.

Now the question arises: Why is this crude model so successful? The main reasons are: (1) the low frequency part of $g(\omega)$ is precise, hence when $T \ll \theta_D$ (6-46) should be very precise; (2) at high temperature, i.e. for $T \gg \theta_D$, only the integral of $g(\omega)$ (the total number of modes) appears in C_V, hence the result of (6-46) should not be bad for high temperatures. As the high temperature and low temperature cases are good approximations, the medium temperature case is not expected to be too bad.

Figures 6-2 and 6-3 show more precise calculations of $g(\omega)$.

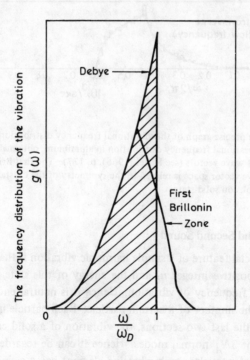

Fig. 6-2 A more precise graph of the vibrational frequency distribution of a lattice. (Again see Kittel (1966), p. 177.) The curve here is calculated from (6-42), but the integration of k is taken in the first Brillouin zone, with a cubic crystal. (Diagram by Fang.)

102 STATISTICAL MECHANICS

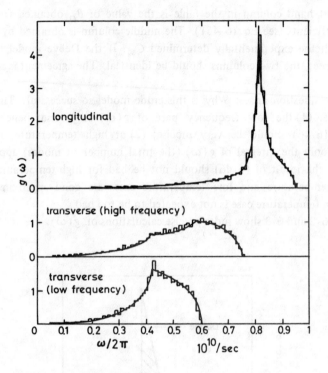

Fig. 6-3 A more precise graph of the vibrational frequency distribution of a lattice. The vibrational frequency distribution in aluminium, obtained by the statistics of 2791 wave vectors (see Kittel (1966), p. 177). The first Brillouin zone of the wave vector space is related to the symmetry of the crystal structure. See books on solid state.

6.4. Phonons and Second Sound

The most special feature of a simple harmonic vibration is that its state can be specified by a positive integer n_i and the energy of this state is $n_i \omega_i + \frac{1}{2} \omega_i$, where ω_i is the frequency of vibration. Hence it is natural enough to think of n_i as denoting the number of a certain particle. Each particle has an energy ω_i. As discussed in the last two sections, the vibration of a solid can be thought of as the motion of $3N_A$ normal modes. Hence it can be regarded as the motion of a collection of particles, each vibrating element being a single particle state and n_i is the number of particles in this state. These particles are bosons because n_i can be larger than 1 (see Chapter 3 for the discussion of the particle state and its boson property). If the solid is a crystal then the vibrating element is a plane wave, described by the wave vector **k**. Hence n_i is the number of particles with momentum **k** ($=\hbar$**k**, $\hbar = 1$). The vibration of the lattice can then be analysed from the point of view of a Bose gas, and these particles are called phonons.

From the analysis of this phonon gas, we can obtain some properties of a solid related to vibrations. Because we are familiar with the analysis of a gas, if a collection of particles can be represented by a gas we can immediately apply all the known techniques of analysing a gas to deduce the properties of this collection of particles.

There is a very common phenomenon related to vibrations, namely the sound wave. Phonons are elastic waves and can be regarded as sound waves. But now we have a phonon gas and there should be a sound wave in this gas and this sound wave is propagated in this phonon gas and is different from the phonons themselves. We call it second sound. Let us look at this interesting problem.

The sound wave in common gases can be explained by the equations of fluid mechanics (see Eq. (1-1) in Chapter 1). Let us review the theory. Assuming that the amplitude of the sound wave is very small, then (1-1) can be simplified as

$$\frac{\partial \rho'}{\partial t} + \rho \nabla \cdot \mathbf{v} = 0 \; , \tag{6-47}$$

$$\rho \frac{\partial \mathbf{v}}{\partial t} = -\nabla p = -\frac{dp}{d\rho} \nabla \rho' \; . \tag{6-48}$$

Here ρ' represents the change of the mass density, and ρ is the density when there is no sound wave. From these two equations, we can immediately obtain the wave equation

$$\frac{\partial^2 \rho'}{\partial t^2} = c^2 \nabla^2 \rho' \; , \tag{6-49}$$

$$c^2 = \frac{dp}{d\rho} \; , \tag{6-50}$$

where c is the speed of propagation. Because the diffusion of heat is slower than the periodic change of the energy, the density and pressure changes in the sound wave are adiabatic processes. If we assume that the period of vibration is much longer than the mean free time of the molecules and that the wavelength is much longer than the mean free path (see the discussion of local equilibrium after (1-1)) then c^2 can be measured from the equilibrium quantities:

$$c^2 = \left(\frac{\partial p}{\partial \rho}\right)_S . \tag{6-51}$$

In the adiabatic process of an ideal gas (see Eqs. (2-24) and (2-27)),

$$p \propto V^{-5/3} \propto \rho^{5/3} .$$

Hence the speed of sound in an ideal gas is

$$c = \sqrt{\frac{5p}{3\rho}} = \sqrt{\frac{5T}{3m}} . \tag{6-52}$$

We expect to be able to derive equations like (6-47) and (6-48) for the phonon gas. However, our phonon is massless and, in addition, the total number of phonons is not a constant but changes continually. We must remember that the number of phonons is just a state of the normal mode. As the state changes, the number of phonons changes also. Hence a phonon gas has no continuity equation like (6-47). But since each phonon has a definite energy and we can write down an equation for the conservation of energy:

$$\frac{\partial \epsilon}{\partial t} + \nabla \cdot \mathbf{J} = 0 , \tag{6-53}$$

where $\epsilon(\mathbf{r}, t)$ is the energy density and $\mathbf{J}(\mathbf{r}, t)$ is the energy flux. Equation (6-48) is just Newton's second law, i.e. the change of momentum is equal to the applied force. Each phonon has its momentum. We let $\mathbf{q}(\mathbf{r}, t)$ be the density of momentum; then

$$\frac{\partial \mathbf{q}}{\partial t} = -\nabla p . \tag{6-54}$$

From the viewpoint of local equilibrium, ϵ, $\mathbf{q}$ and $\mathbf{J}$ are functions of the temperature T and the flow velocity $\mathbf{v}$. To find these functions, we have to do some preparatory work.

We now digress from the present problem and consider a body with momentum $\mathbf{P}$ and assume that this body is in free motion, so that $\mathbf{P}$ does not change.

According to our assumption, the entropy of this body is

$$S(E, \mathbf{P}) = \ln \Gamma(E, \mathbf{P}) ,$$

$$\Gamma(E, \mathbf{P}) = \sum_s \delta(E - H(s)) \, \delta(\mathbf{P} - \mathbf{P}(s)) , \qquad (6\text{-}55)$$

where $\mathbf{P}(s)$ is the momentum of the state s. Following our previous discussions we regard the body as the sum of m parts with momenta $\mathbf{P}_1, \mathbf{P}_2, \ldots, \mathbf{P}_m$. The generalisation of (6-16), (6-17) and (6-20), (6-21) is

$$S(E, \mathbf{P}) = \sum_{i=1}^{m} S_i(\bar{E}_i, \bar{\mathbf{P}}_i) . \qquad (6\text{-}56)$$

$\bar{E}_i, \bar{\mathbf{P}}_i$ and T, $\mathbf{v}$ are the solutions of the following set of equations:

$$\frac{\partial S_i}{\partial E_i} = \frac{1}{T} ,$$

$$\frac{\partial S_i}{\partial \mathbf{P}_i} = -\frac{\mathbf{v}}{T} , \qquad i = 1, 2, \ldots, m , \qquad (6\text{-}57)$$

$$E = E_1 + E_2 + \ldots + E_m ,$$

$$\mathbf{P} = \mathbf{P}_1 + \mathbf{P}_2 + \ldots + \mathbf{P}_m . \qquad (6\text{-}58)$$

Thus, besides having the same temperature, all the parts of the body should have the same value of $\mathbf{v}$, which can be interpreted as the velocity of the body. The reason is as follows. If we use the centre-of-mass coordinate system the total energy is $E - \mathbf{P}^2/2M$, where M is the total mass, and the total momentum is zero. The entropy remains the same because entropy is a statistical quantity and is independent of the coordinate system:

$$S(E, \mathbf{P}) = S\left(E - \frac{\mathbf{P}^2}{2M}, 0\right) . \qquad (6\text{-}59)$$

Therefore

$$\frac{\partial S}{\partial \mathbf{P}} = \frac{\partial S}{\partial E}\left(-\frac{\mathbf{P}}{M}\right) = -\frac{\mathbf{v}}{T} . \qquad (6\text{-}60)$$

This shows that $\mathbf{v}$ is just $\mathbf{P}/M$, i.e. the velocity of the body.

In (6-57), $\mathbf{v}/T$ is a "force", like $1/T$. Now we apply these results to the phonon gas. Let us take a phonon gas with total momentum $\mathbf{P}$. We divide these phonons into classes, with the α class consisting of phonons with nearly the same momentum $\mathbf{k}_\alpha$ and the same polarisation. (Each class has the same frequency because frequency is determined by the wave vector and polarisation.) The entropy of this class is

$$S_\alpha(\mathbf{P}_\alpha) = \ln \Gamma_\alpha(\mathbf{P}_\alpha) \; ,$$

$$\Gamma_\alpha(\mathbf{P}_\alpha) = \sum_{n_1 = 0}^{\infty} \cdots \sum_{n_{N_\alpha} = 0}^{\infty} \delta\left(\mathbf{P}_\alpha - \sum_{i=1}^{N_\alpha} n_i \mathbf{k}_\alpha\right) , \qquad (6\text{-}61)$$

where N_α is the total number of states of this class of phonons, n_i is the number of phonons in the i-th state and $\mathbf{P}_\alpha$ is the total momentum of this class of phonons. Notice that the momentum of each phonon is $\mathbf{k}_\alpha$ and its polarisation is the same. As $\mathbf{k}_\alpha$ is fixed, ω_α is also fixed and once $\mathbf{P}_\alpha$ is fixed, the energy is fixed likewise. We need not use a δ-function such as

$$\delta\left(E_\alpha - \sum_{i=1}^{N_\alpha} (n_i + \tfrac{1}{2})\omega_\alpha\right) \; ,$$

to restrict the energy. We introduce the definition

$$\mathbf{P}_\alpha = M_\alpha \mathbf{k}_\alpha \; . \qquad (6\text{-}62)$$

The δ-function in (6-61) may be replaced by[c]

$$a\, \delta\left(M_\alpha - \sum_{i=1}^{N_\alpha} n_i\right) , \qquad (6\text{-}63)$$

$S_\alpha(\mathbf{P}_\alpha)$ will be the same as (6-31)

$$S_\alpha(\mathbf{P}_\alpha) = (M_\alpha + N_\alpha) \ln(M_\alpha + N_\alpha) - M_\alpha \ln M_\alpha - N_\alpha \ln N_\alpha \; . \qquad (6\text{-}64)$$

---

[c] Notice that a is an unimportant constant and can be neglected after taking the logarithm:

$$\delta(\mathbf{P}_\alpha - \sum_{i=1}^{N_\alpha} n_i \mathbf{k}_\alpha) = (k_{\alpha x}, k_{\alpha y}, k_{\alpha z})^{-1}\, \delta(M_\alpha - \sum_{i=1}^{N_\alpha} n_i) \; ,$$

$$\delta(E_\alpha - \sum_{i=1}^{N_\alpha} (n_i + \tfrac{1}{2})\omega_\alpha) = \frac{1}{\omega_\alpha} \delta(M_\alpha - \sum_{i=1}^{N_\alpha} n_i) \; .$$

SOME ELEMENTARY APPLICATIONS OF THE BASIC ASSUMPTION 107

Note that $\ln a$ can be neglected, because M_α, and N_α are large numbers. M_α is of course the total number of phonons in the α class.

We now sum over these classes using (6-56) and (6-57). Because

$$E_\alpha = \left(M_\alpha + \frac{1}{2}N_\alpha\right)\omega_\alpha ,$$

$$\mathbf{P}_\alpha = M_\alpha \mathbf{k}_\alpha , \qquad (6\text{-}65)$$

we can rewrite (6-57) as

$$dS_\alpha = \frac{1}{T}dE_\alpha - \frac{\mathbf{v}}{T} \cdot d\mathbf{P}_\alpha$$

$$= \frac{1}{T}(\omega_\alpha - \mathbf{k}_\alpha \cdot \mathbf{v})dM_\alpha ,$$

i.e.

$$\frac{dS_\alpha}{dE_\alpha} = \frac{1}{T}(\omega_\alpha - \mathbf{k}_\alpha \cdot \mathbf{v})$$

$$= \ln \frac{M_\alpha + N_\alpha}{M_\alpha} . \qquad (6\text{-}66)$$

This is nearly the same as (6-32) except that ω_α is replaced by $\omega_\alpha - \mathbf{k}_\alpha \cdot \mathbf{v}$. Let

$$\bar{n}(\omega) \equiv \frac{1}{e^{\omega/T} - 1} . \qquad (6\text{-}67)$$

Then from (6-66) we have for M_α

$$\frac{M_\alpha}{N_\alpha} = \bar{n}(\omega_\alpha - \mathbf{k}_\alpha \cdot \mathbf{v}) . \qquad (6\text{-}68)$$

The last two equations of (6-57) are

$$E = \sum_\alpha N_\alpha \bar{n}(\omega_\alpha - \mathbf{k}_\alpha \cdot \mathbf{v})\omega_\alpha + \sum_\alpha \frac{1}{2}N_\alpha \omega_\alpha ,$$

$$\mathbf{P} = \sum_\alpha N_\alpha \bar{n}(\omega_\alpha - \mathbf{k}_\alpha \cdot \mathbf{v})\mathbf{k}_\alpha . \qquad (6\text{-}69)$$

Each phonon has an energy flux of $\omega_\alpha c\hat{\mathbf{k}}_\alpha/V$ where $c\hat{\mathbf{k}}_\alpha$ is the velocity and $\hat{\mathbf{k}}$

is the direction of **k**. Hence the total energy flux is

$$J = \sum_\alpha N_\alpha \, \bar{n} \, (\omega_\alpha - \mathbf{k}_\alpha \cdot \mathbf{v}) \, \omega_\alpha \frac{c \hat{\mathbf{k}}_\alpha}{V} \quad . \tag{6-70}$$

Now replace $\sum_\alpha N_\alpha$ by $\dfrac{3V}{(2\pi)^3} \int d^3k$ and from (6-69) and (6-70) we obtain the energy density E/V, the momentum density $\bar{\mathbf{P}}/V$ and the energy flux **J**:

$$\epsilon = \frac{E}{V} = \frac{3}{(2\pi)^3} \int d^3k \, \bar{n} \, (\omega_k - \mathbf{k} \cdot \mathbf{v}) \, \omega_k \quad ,$$

$$\mathbf{q} = \frac{\mathbf{P}}{V} = \frac{3}{(2\pi)^3} \int d^3k \, \bar{n} \, (\omega_k - \mathbf{k} \cdot \mathbf{v}) \, \mathbf{k} \quad ,$$

$$\mathbf{J} = \frac{3}{(2\pi)^3} \int d^3k \, \bar{n} \, (\omega_k - \mathbf{k} \cdot \mathbf{v}) \, \omega_k \, c \hat{\mathbf{k}} \quad ,$$

$$\bar{n}(\omega) = \frac{1}{e^{\omega/T} - 1} \quad . \tag{6-71}$$

The factor of 3 in the equations above comes from the three polarisations. We did not take into account the difference between the frequencies of the longitudinal and the transverse waves in order to avoid unnecessary complications. The zero-point energy has also been dropped because it does not affect the following discussion.

Because $\bar{n}$ is a function of T, ϵ, $\mathbf{J}$ and $\mathbf{q}$ become functions of T and $\mathbf{v}$. Now we regard T and $\mathbf{v}$ as functions of position $\mathbf{r}$ and time t, $T = T(\mathbf{r}, t)$ and $\mathbf{v} = \mathbf{v}(\mathbf{r}, t)$. Therefore, ϵ, $\mathbf{q}$ and $\mathbf{J}$ become functions of $\mathbf{r}$ and t. We use these in (6-53) and (6-54) to discuss the motion of the phonon gas. Notice that we must not forget that $\mathbf{r}$ and t here are variables on macroscopic scales and dt and $d\mathbf{v}$ are quantities much larger than the scales of the motion of the phonons. These scales have to be discussed further.

Using (6-71), we can recast (6-53) and (6-54) into equations involving T and $\mathbf{v}$. Since $\mathbf{v}$ is a small quantity,

$$\bar{n}(\omega - \mathbf{k} \cdot \mathbf{v}) \simeq \bar{n}(\omega) - \mathbf{k} \cdot \mathbf{v} \, \frac{\partial \bar{n}}{\partial \omega} \quad . \tag{6-72}$$

SOME ELEMENTARY APPLICATIONS OF THE BASIC ASSUMPTION 109

Substituting this into (6-71) we obtain

$$\mathbf{J} = \frac{3}{(2\pi)^3} \int d^3k \, \omega c \hat{\mathbf{k}} \left(-\mathbf{k} \cdot \mathbf{v} \frac{\partial \bar{n}}{\partial \omega} \right)$$

$$\simeq c^2 \rho \mathbf{v} ,$$

$$\mathbf{q} \simeq \rho \mathbf{v} ,$$

$$\rho \equiv \frac{3}{(2\pi)^3} \int d^3k \, \frac{k^2}{3} \left(-\frac{\partial \bar{n}}{\partial \omega} \right) . \qquad (6\text{-}73)$$

In the above we replaced ω with ck and interpreted ρ as the "mass density" of the phonon gas. In addition we have

$$\frac{\partial \epsilon}{\partial t} = \frac{\partial \epsilon}{\partial T} \frac{\partial T}{\partial t} ,$$

$$\nabla p = \frac{1}{3} \nabla \epsilon = \frac{1}{3} \frac{\partial \epsilon}{\partial T} \nabla T ,$$

$$\frac{\partial \epsilon}{\partial T} = \frac{3}{(2\pi)^3} \int d^3k \, \frac{\omega^2}{T} \left(-\frac{\partial \bar{n}}{\partial \omega} \right) = \frac{3c^2}{T} \rho , \qquad (6\text{-}74)$$

where we used $p = \frac{1}{3}\epsilon$, $\frac{\partial \bar{n}}{\partial \omega} = -\frac{\omega}{T}\frac{\partial \bar{n}}{\partial \omega}$. The reader can verify that after substituting (6-73), (6-74) into (6-53) and (6-54) we obtain

$$\frac{\partial T}{\partial t} = -\frac{1}{3} T \nabla \cdot \mathbf{v} ,$$

$$\frac{\partial \mathbf{v}}{\partial t} = -c^2 \nabla T . \qquad (6\text{-}75)$$

It follows that

$$\frac{\partial^2 T}{\partial t^2} = c_2^2 \nabla^2 T , \qquad c_2 = \frac{c}{\sqrt{3}} . \qquad (6\text{-}76)$$

We have finally obtained the equation for the sound wave which is called the second sound and its speed is $1/\sqrt{3}$ of that of the phonons. However, this result is the simple part of the analysis. The difficult part has not started yet, i.e. the analysis of the validity of the various assumptions. If some assumption is invalid then (6-76) will not be true. Let us now see what assumptions we have made in the above discussion:

(1) The momenta of the phonons must change continually. If there are no such changes, our basic assumption would not be applicable. The cause of these changes are:

(i) The vibration is not perfectly harmonic, resulting in an interaction between the various modes and causing the number of phonons in each mode to change continually;

(ii) The defect or impurity atoms in a crystal can produce similar effects.

We will assume that these interaction are strong enough to maintain equilibrium but weak enough so as not to effect the statistical calculations.

(2) Although the momenta of the phonons are changing, the total momenta must be conserved (if there is no pressure gradient). From the point of view of the wave motion the conservation of momentum is due to the homogeneity of the medium. Vacuum is homogeneous. Hence we have the law of conservation of momentum. The crystal is not a homogeneous medium and hence the total momentum of the phonons is not conserved. But the arrangement of atoms in a crystal is periodic with a fixed lattice spacing. Therefore, for the long wave phonons the medium is effectively homogeneous and this short-distance periodic change does not have any effect. Hence the change due to (i) does not affect the total momentum of a collection of long wave phonons. But if the number of short wave phonons is large, momentum will not be conserved. There is then a problem with Eq. (6-54). Short wave phonons have high energy. Therefore the temperature has to be very low

$$T \ll \theta_D \,, \tag{6-77}$$

so that short wave phonons will be few in number and (6-54) can then be valid. However if T is low, then the long wave phonons will be scarce too (the total number of phonons is αT^3). When the gas is too dilute, sound waves cannot propagate. The change due to (ii), i.e. the change caused by the defects and impurities, totally disobeys the law of conservation of momentum. Thus only in perfect crystals is (6-54) valid.

Hence the second sound can only exist in a very perfect crystal at intermediate temperatures. Now the speed of sound is large (around 10^5 cm/sec) and for wavelengths of 1 cm the frequency will be 10^5. But if the frequency is too high, local equilibrium can only be maintained by a quick response of the phonons. Thus we also must consider the speed of response. This problem is rather complicated and is beyond the scope of our discussion here. We will content ourselves by mentioning that, although big perfect crystals are here to grow, there are reports that a crystal like CdS can maintain second sound at about 10 K.[d]

Because this second sound does not occur easily, it is not generally regarded as important. But the analysis of it has high pedagogical value. Our discussion started from the very abstract basic assumption, leading ultimately to wave motion in a phonon gas. This is indeed a useful illustrative example.

6.5. General Conclusions, Average Values and the Normal Distribution

The main point in this chapter is on the relation between a system and its subsystems. In equilibrium, the forces in the various parts of a body have constant values, i.e. the parts have the same temperature, pressure and chemical potential. If the whole body is moving, then every part should have the same velocity. (If it is rotating, then each part has the same angular velocity. The reader can deduce this from the basic assumption.) This is the result of the basic assumption. The uniformity of the forces is a property of equilibrium.

We should notice that each force is connected with an invariant quantity: temperature with energy, pressure with volume, chemical potential with particle number, flow velocity with momentum. When this force varies then there will be changes in the invariant quantities. The analysis of the sound wave is one such example. Slow- and large-scale changes are usually related to these invariant quantities.

The various parts of a body are not necessarily divided according to the volume occupied. Our discussion of vibration shows that we may classify according to the energy or momentum of the modes. Each class can be regarded as a part of the body. Nevertheless, the method of calculating the entropy of the classes is meaningful only when the parts are essentially independent. The independence of the parts means that the motions of the various parts are not correlated, and that entropy can be calculated separately. This property of independence is a very complicated problem and later we shall study this aspect more carefully.

From the viewpoint of statistics, this chapter discusses only the steps of classification and summation and this is the characteristics of functions of

---

[d] See Prohofsky and Krumhansl (1964) and references therein.

large value. The results obtained above are not restricted to invariant quantities like energy, particle number, etc. We can make classifications according to any macroscopic quantities. We now explain this point as follows.

Let us suppose that $\hat{A}(s)$ is a variable of motion of order N. Let

$$\Gamma(E, A) = \sum_s \delta(E - H(s)) \, \delta(A - \hat{A}(s))$$

$$\equiv e^{S(E, A)} \; . \tag{6-78}$$

This is the result of statistics by classification. Each value of A is a class. In the states of energy E, we collect those states with $\hat{A}$ equal to A and count their number. The function $\Gamma(E, A)$, of course, is the distribution of states in (E, A) space, and $\Gamma(E)$ can be obtained by a summation of $\Gamma(E, A)$

$$\Gamma(E) = \int dA \; \Gamma(E, A)$$

$$= \int dA \; e^{S(E, A)} \; . \tag{6-79}$$

Peak integration will be used to evaluate (6-79). We first find the maximum of $S(E, A)$:

$$\frac{\partial S(E, A)}{\partial A} = 0 \; . \tag{6-80}$$

Let its solution be $\bar{A}(E)$. If $\bar{A}$ is a maximum, then a necessary condition is

$$(A - \bar{A})^2 \left(\frac{\partial^2 S}{\partial A^2} \right)_E \equiv -\frac{(A - \bar{A})^2}{K_A} < 0 \; . \tag{6-81}$$

From (6-15) we get

$$\Gamma(E) \simeq e^{S(E, \bar{A})} \int dA' \, e^{-A'^2/2K_A}$$

$$= e^{S(E, \bar{A})} \sqrt{2\pi K_A} \; , \tag{6-82}$$

$$S(E) = \ln \Gamma(E)$$

$$= S(E, \bar{A}(E)) \; . \tag{6-83}$$

Example: $A(s) = H_1(s_1) - H_2(s_2)$.

Here H_1, and H_2 are the energies of the first and second parts of the system, (see (6-1) and (6-2)), as discussed in the beginning of this chapter. Then

$$\Gamma(E,A) = \sum_s \delta(E-H(s))\,\delta(A-\hat{A}(s))$$

$$= \sum_{s_1,s_2} \delta(E - H_1(s_1) - H_2(s_2))\,\delta(A - H_1(s_1) + H_2(s_2))$$

$$= \sum_{s_1,s_2} \delta\left(\frac{E+A}{2} - H_1(s_1)\right) \delta\left(\frac{E-A}{2} - H_2(s_2)\right)$$

$$= \Gamma_1\left(\frac{E+A}{2}\right)\Gamma_2\left(\frac{E-A}{2}\right) \qquad (6\text{-}84)$$

For the definition of $\Gamma_1(E_1)$ and $\Gamma_2(E_2)$ see Eq. (6-4'). Therefore

$$S(E,A) = \ln \Gamma(E,A)$$

$$= S_1\left(\frac{E+A}{2}\right) + S_2\left(\frac{E-A}{2}\right) .$$

The result of (6-80) is now cast in the form

$$\frac{\partial S}{\partial A} = \frac{1}{2}\left(\frac{\partial S_1}{\partial E_1} - \frac{\partial S_2}{\partial E_2}\right) = 0 ,$$

where

$$E_1 = \frac{1}{2}(E+A) ,$$

$$E_2 = \frac{1}{2}(E-A) . \qquad (6\text{-}85)$$

Equation (6-85), of course, is the same as (6-6) and (6-7) and $\bar{A} = \bar{E}_1 - \bar{E}_2$ is the solution of (6-85), while $\bar{E}_1$, $\bar{E}_2$ are the solutions of (6-6) and (6-7).

The analysis just given indicates that from the basic assumption we can calculate the value $\bar{A}$ of any macroscopic $\hat{A}(s)$. This involves the calculation of the entropy of the classes and then finding the maximum (see (6-80)).

We see that the distribution of states in A is a normal one

$$\frac{\Gamma(E,A)}{\Gamma(E)} = \frac{1}{\sqrt{2\pi K_A}} e^{-(A-\bar{A})^2/2K_A} . \tag{6-86}$$

Hence $\bar{A}$ can also be written as

$$\bar{A} = \int dA \, \frac{\Gamma(E,A)}{\Gamma(E)} A$$

$$= \frac{1}{\Gamma(E)} \sum_s \delta(E - H(s)) \, \hat{A}(s) , \tag{6-87}$$

i.e. $\bar{A}$ is the average value of $\hat{A}(s)$ with respect to all the states in the region of motion. The results derived above follow from the observation that in the region of motion $\hat{A}(s)$ is very nearly $\bar{A}(1 + O(K_A^{-1/2}))$ and $K_A \sim N$ is a very large number. The use of arguments of this type is typical in statistical mechanics.

Let us now review the meaning of "region of motion". It is a region in phase space bounded by the trajectory such that the trajectory winds inside it. The conclusion we arrived at is that in the region of motion, the value of $\hat{A}(s)$ is $\bar{A}$ nearly everywhere. As the trajectory winds inside this region, it is not difficult to imagine that along the trajectory the values of $\hat{A}(s)$ are all nearly equal to $\bar{A}$.

The usual experimentally determined quantities are time averages within an observation period:

$$\langle \hat{A} \rangle = \frac{1}{\mathcal{T}} \int_0^{\mathcal{T}} dt \, \hat{A}(s(t)) , \tag{6-88}$$

where $s(t)$ is the trajectory in phase space representing the detailed molecular movement. The observation time is from $t = 0$ to $t = \mathcal{T}$. As $\hat{A}(s(t))$ is nearly always equal to $\bar{A}$, our conclusion is that

$$\langle \hat{A} \rangle = \bar{A} . \tag{6-89}$$

Hence, besides calculating the entropy, the basic assumption provides a definite rule for computing $\langle \hat{A} \rangle$. We only need to know the region of motion to calculate $\bar{A}$.

The reader may ask: since the value of $\hat{A}(s)$ in nearly every state is $\bar{A}$, why do we not just choose a state and calculate $\bar{A}$? Why do we have to do the average? This problem involves many aspects. In principle this is permissible, but in practice, choosing a state at random is not a simple matter. The calculation by averaging is still necessary. We shall discuss this later.

Notice that we have not yet discussed a way of determining the region of motion from the trajectory. This is a hard problem and will also be discussed later. Now the determination of the region of motion is part of the assumption, i.e., any state allowable by the invariant quantities is included in the region of motion. Hence the above conclusion is just an extension of the assumption.

The statement $\langle \hat{A} \rangle = \bar{A}$ can be expressed as follows: The time average of a macroscopic variable of motion is equal to its average in the region of motion.

The reader should note that: (1) This conclusion is applicable for a macroscopic variable only, i.e., for variables of order N; (2) This conclusion does not tell us that the trajectory spends equal time in each point of the region of motion. Because $\hat{A}$ is almost $\bar{A}$ nearly everywhere in the region of motion, if the trajectory is not too exceptional, i.e. not staying in those exceptional points for too long, then $\langle \hat{A} \rangle = \bar{A}$ can be established. Hence $\langle \hat{A} \rangle = \bar{A}$ is not a surprising conclusion. It does not impose a stringent condition on the trajectory.

Suppose we divide the region of motion into M large regions, where M is not too large a number, e.g. $M \sim \sqrt{N}$. Let each region have the same volume. If we require the trajectory to spend essentially the same time in each large region, $\langle \hat{A} \rangle = \bar{A}$ can be comfortably established. This requirement is completely different from that of requiring the trajectory to pass each point.

Although this requirement is not stringent, the question of whether the trajectory actually satisfies this condition is still unanswered. If the trajectory prefers to stay in places where $\hat{A}(s) \neq \bar{A}$, we can only admit that we misidentified the region of motion or that the basic assumption is wrong.

Finally we want to mention the meaning of the normal distribution. Why do distributions like (6-86) repeatedly appear? From our derivation, we know that this is because A is a large number of order N. A large number of the order N is a sum of many variables of molecular motion and these molecular variables are essentially independent because the system can be divided into many independent parts. Later we shall discuss the central limit theorem, which states that if A is the sum of many essentially independent variables, then the distribution of A is normal. Hence the normal distribution of macroscopic variables can be said to be the result of the parts of the system being essentially independent.

Problems

1. Use the method of calculating entropy by classification and summation to find the relations of T, μ with N, E for a system of (a) N bosons and (b) N fermions.

2. Derive (6-46) and (6-47) and prove that at low and high temperatures

$$C_V \simeq \frac{12\pi^4}{5} N_A \left(\frac{T}{\theta_D}\right)^3 , \qquad T \ll \theta_D ,$$

$$C_V \simeq 3N_A , \qquad T \gg \theta_D , \qquad (6\text{-}90)$$

i.e., the low temperature T^3 law (Einstein's law). Notice that the electronic heat capacity is proportional to T; see Chapter 4.

3. The diagram below shows the heat capacity of a certain crystal. Prove that the area of the shaded part is the zero-point energy. For a one-dimensional crystal made up of N atoms, calculate the frequency distribution $g(\omega)$.

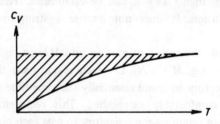

Fig. 6-4 Heat capacity C_V versus T.

4. The theory of black body radiation is similar to the theory of lattice vibrations. (See Chapter 3, Sec. 3.4.) Make a comparison.

5. Black body radiation is a photon gas. Can sound wave phenomena exist in this gas? Make a detailed analysis.

6. The calculation in Sec. 6.4 is rather complicated, but it is quite helpful for understanding the basic principle and technique of the calculation. The reader should do the derivation himself, paying attention to every small point. The second sound was first found in liquid helium; the reader may consult Woo (1972), Wilks (1967).

7. In Sec. 6.5 we emphasised the fact that $S = O(N)$, $\Gamma = O(e^N)$. Were this not to be so the basic assumption might not have applied. This "order of magnitude" requirement is not simple. If there is a large range interaction it may not be established.

SOME ELEMENTARY APPLICATIONS OF THE BASIC ASSUMPTION 117

Solve the following model:

$$H = -\frac{J}{2} \sum_{i,j} s_i\, s_j \;, \qquad i,j = 1, 2, \ldots, N \;,$$

$$s_i = \pm 1 \;. \tag{6-91}$$

This is the so-called Ising model with long range interaction. Each spin s_i has interaction with all the other variables. The general discussion of the Ising model is in Chapter 17.

Hint: Write H as $-\frac{1}{2} JM^2$, where M is the sum of all s_i and $\Gamma(E)$ can be immediately calculated out.

Analyse the order of magnitude of its various quantities. Compare Prob. (5.2) with the present one.

Chapter 7

RULES OF CALCULATION

The basic assumption is a rule for calculating entropy. This rule can be written as many different rules for calculation. These new rules of calculation in turn have the same implications and consequences as the basic assumption, except that the method of calculation is different. One uses these rules to calculate some thermodynamic potentials and then derive the entropy and other equilibrium properties from these potentials. In many situations this method of calculation is more convenient than that of calculating the entropy directly. We first discuss the definitions of the various thermodynamic potentials and their implications, and then derive the various rules for calculation from the basic assumption.

7.1. Various Thermodynamic Potentials

The energy in the above discussion refers to the energy of the molecular motion in the system and its molecular interactions. It may be called the internal energy. Now we can add to it the energy related to the interaction of the system with external forces and define various thermodynamic potentials; the free energy is an example. The inclusion of different external forces will result in different thermodynamic potentials. Let us consider a simple example.

Example 1:

The iron rod in Fig. 7-1 is elongated by an external force f, supplied by a weight (of mass M), and hence $f = Mg$, where g is the acceleration due to gravity. Let X_0 be the unstretched length of the iron rod.

RULES OF CALCULATION 119

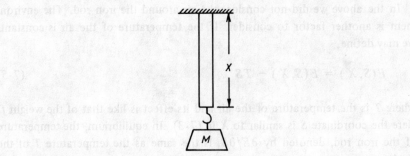

Fig. 7-1 An iron rod under an external force.

The internal energy of the iron rod is

$$E(S, X) = E(S, X_0) + \frac{1}{2} K(S)(X - X_0)^2 \quad , \tag{7-1}$$

where S is the entropy and we have omitted the other coordinates. $K(S)$ is the adiabatic elastic coefficient of this iron rod. In equilibrium, the external force f is balanced by the internal tension:

$$f = K(S)(X - X_0) = \left(\frac{\partial E}{\partial X}\right)_S . \tag{7-2}$$

We may include the potential energy of the weight and we obtain

$$\begin{aligned} H(S, X) &= E(S, X) - MgX \\ &= E(S, X) - fX . \end{aligned} \tag{7-3}$$

That is, we have included the coordinate X and its interaction energy with the external force $-fX$. In equilibrium

$$\left(\frac{\partial H}{\partial X}\right)_S = 0 \quad , \tag{7-4}$$

i.e. H is a minimum. This, of course, is the same as (7-2). In mechanics, we are accustomed to identifying the equilibrium state with the minimum energy. Hence (7-4) looks nicer than (7-2). If we do not hang the weight, then $E(S, X)$ is more convenient for the discussion of the other properties. If we have the weight, then $H(S, X)$ is more appropriate.

In the above we did not consider the air around the iron rod. The environment is another factor to consider. If the temperature of the air is constant, we may define

$$F(S,X) = E(S,X) - TS \, , \qquad (7\text{-}5)$$

where T is the temperature of the air and its effect is like that of the weight f. Here the coordinate S is similar to X in (7-3). In equilibrium, the temperature of the iron rod, denoted by $\partial E/\partial S$, is the same as the temperature T of the air, i.e.

$$\left(\frac{\partial E}{\partial S}\right)_X = T \, , \quad \text{or}$$

$$\left(\frac{\partial F}{\partial S}\right)_X = 0 \, , \qquad (7\text{-}6)$$

so F, like H, is a minimum. Of course we can define still another thermodynamic function

$$G(S,X) = E(S,X) - TS - fX \, , \qquad (7\text{-}7)$$

which includes the energy due to the two external forces T and f acting on the coordinates S and X respectively. At a fixed T and f, G must be a minimum.

We call H, F and G the thermodynamic potentials. H can also be called the thermodynamic potential at constant tension; F is the thermodynamic potential at constant temperature; and G is the thermodynamic potential at constant tension and temperature.

Example 2:

The gas in Fig. 7-2 is confined in a vessel with cross-sectional area A and height y. The pressure comes from a weight on the piston: $p = Mg/A$ where Mg is the weight. Following the above steps the reader can define the thermodynamic potential

$$H = E + Mgy$$

$$= E + pV \, , \qquad (7\text{-}8)$$

Fig. 7-2 A gas confined under constant pressure.

where $V = Ay$. The other thermodynamic potentials are:

$$F = E - TS \;,$$

$$G = E + pV - TS \;,$$

$$\Omega = E - \mu N - TS \;. \qquad (7\text{-}9)$$

The differentials of these thermodynamic potentials have special forms. For instance, the differential of H in (7-3) is

$$\begin{aligned} dH &= dE - f\,dX - X\,df \\ &= T\,dS - X\,df \;, \end{aligned} \qquad (7\text{-}10)$$

since $dE = T\,dS + f\,dX$. The differential of F in (7-5) is

$$dF = -S\,dT + f\,dX \;. \qquad (7\text{-}11)$$

We see from (7-10) that H can be regarded as a function of S and f: $H = H(S, f)$. Likewise F is a function of T and X: $F = F(T, X)$. Hence if we regard T as a coordinate, then $-S$ appears as the force. If we regard f as the coordinate, then $-X$ becomes the force. This interpretation is followed by the above results.

We list the differentials of the various thermodynamic potentials in Table 7-1.

Table 7-1 The differentials of the thermodynamic potentials

	Definition	Differential
E	E	$dE = T\,dS - p\,dV + \mu\,dN$
H	$E + pV$	$dH = T\,dS + V\,dp + \mu\,dN$
F	$E - TS$	$dF = -S\,dT - p\,dV + \mu\,dN$
G	$E - TS + pV$	$dG = -S\,dT + V\,dp + \mu\,dN$
Ω	$E - TS - \mu N$	$d\Omega = -S\,dT - p\,dV - N\,d\mu$

The partial derivatives can be directly read from the above table, e.g.

$$S = -\left(\frac{\partial F}{\partial T}\right)_{V,N} , \qquad N = -\left(\frac{\partial \Omega}{\partial \mu}\right)_{T,V} ,$$

$$V = \left(\frac{\partial G}{\partial p}\right)_{T,N} , \qquad T = p\left(\frac{\partial V}{\partial S}\right)_{E,N} ,$$

$$\mu = \left(\frac{\partial G}{\partial N}\right)_{T,p} , \qquad p = -\left(\frac{\partial \Omega}{\partial V}\right)_{T,\mu} . \qquad (7\text{-}12)$$

These differentials will be very useful when we do calculations.

7.2. The Various Rules of Calculation

A. Rule for constant energy and for constant temperature

We assume that the motion is confined to a region in phase space with a definite energy, i.e., the energy at each point of the region is E. The calculation of entropy can be reduced to

$$\Gamma(E) = \sum_s \delta(E - H(s))$$

$$S(E) = \ln \Gamma(E) , \qquad (7\text{-}13)$$

where s denotes the various states, and $H(s)$ is the energy of the s state. This summation is a statistical process. The factor $\delta(E-H)$ in the equation restricts the energy and we only collect those states with energy E, and count their total number. This statistical step is called the rule of constant energy. In practice this rule is not very convenient and we have to modify it. In this section we discuss the various ways of modifying it.

Now we replace the factor $\delta(E-H)$ in (7-13) by

$$\Delta(E-H) \equiv e^{-\beta(H-E)}, \qquad (7\text{-}14)$$

and define a new quantity

$$\overline{\Gamma}(E) = \sum_s \Delta(E-H), \qquad (7\text{-}15)$$

where $\beta > 0$; its value is to be determined later. Notice that in (7-13) we collect all those states with energy E and assign to them the same statistical weight. In (7-15), all the states are collected whatever their energies but are assigned different statistical weights, proportional to $e^{-\beta H(s)}$. At first sight, the determination of the region of motion in (7-15) is somewhat ambiguous, and may seem to have no specific meaning.

The function $\Delta(E-H)$ is, of course, not the same as $\delta(E-H)$. Hence $\overline{\Gamma}$ and Γ are not the same function; they are related by

$$\begin{aligned}\overline{\Gamma}(E) &= \sum_s \int dE' \; \Delta(E-E') \; \delta(E'-H(s)) \\ &= \int dE' \; \Gamma(E') \; e^{-\beta(E'-E)} \\ &= \int dE' \; \exp\left\{S(E') - \beta(E'-E)\right\} \quad . \end{aligned} \qquad (7\text{-}16)$$

Assuming that $\partial S(E')/\partial E' > 0$ we see that $\Gamma(E')$ increases rapidly with E'. As E' is a large number, $e^{-\beta E'}$ decreases rapidly with E'. Thus the integrand of (7-16) has a sharp peak; its maximum is determined from

$$\frac{\partial}{\partial E'}[S(E') - \beta(E'-E)] = 0, \qquad (7\text{-}17)$$

i.e. $\partial S(E')/\partial E' = \beta$. We put

$$\beta = \frac{1}{T} \equiv \frac{\partial S(E)}{\partial E} . \tag{7-18}$$

The maximum of the integrand of (7-16) occurs at $E' = E$. As $S(E')$ and E' are large numbers we can use the law of large numbers of the last chapter to calculate this integral. (See Eqs. (6-8) to (6-13).) Expanding $S(E')$ and retaining terms up to second order only, we have

$$\overline{\Gamma}(E) \simeq e^{S(E)} \int dE'' \, e^{-(E'')^2/2K} ,$$

$$E'' \equiv E' - E , \tag{7-19}$$

$$\frac{1}{K} \equiv -\frac{\partial^2 S}{\partial E^2} = \left(T^2 \frac{\partial E}{\partial T}\right)^{-1} . \tag{7-20}$$

Because S, E and E' are all proportional to N, K is too. The integration with respect to E' gives $\sqrt{2\pi K} \propto \sqrt{N}$. Therefore,

$$S(E) \cong \ln \overline{\Gamma}(E) - \frac{1}{2}\ln(2\pi K) , \tag{7-21}$$

$$\ln \overline{\Gamma} \propto N , \qquad \ln K \sim \ln N . \tag{7-22}$$

Thus if N is very large, $\ln N$ may be neglected, we have then

$$S(E) = \ln \overline{\Gamma}(E) ,$$

$$\overline{\Gamma}(E) = \sum_s e^{-(H(s)-E)/T} . \tag{7-23}$$

Although $\Delta(E-H)$ and $\delta(E-H)$ are not the same, if N is very large, they give completely identical results for S because $\Delta(E-H)$ cuts-off the high energy states. It does not suppress the low energy states (instead it magnifies their statistical weight), but the number of low energy states is too few. Hence only the states of energy around E count in $\overline{\Gamma}(E)$. In (7-23) the relation between T and E is through (7-18); this is not quite convenient. Because the temperature T is easier to measure and energy is not, it is better to eliminate E in favour of T. Let

$$F = -T \ln Z ,$$

$$Z = \sum_s e^{-H(s)/T} . \qquad (7\text{-}24)$$

Then from (7-23) we obtain

$$Z \simeq e^{S(E)} e^{-E/T} ,$$

$$F = E - TS . \qquad (7\text{-}25)$$

This of course, is one of the thermodynamic potential discussed in the previous section, and S can be obtained by direct differentiation (see (7-12)). The function Z is called the partition function at constant temperature.

The calculation in (7-24) takes T as a constant. This way of calculating F and then S is logically called the rule for constant temperature. It gives the same result as the above rule for constant energy. In calculations it is more convenient than the rule for constant energy. F is the thermodynamic potential at constant temperature.

B. *Rule for restricted energy*

As we can change $\delta(E - H)$ in (7-13) to $\Delta(E - H)$ we can also change it to $\theta(E - H)$, i.e. we do not count the states with energy larger than E. The proof is very simple:

$$\sum_s \theta(E - H(s)) = \int dE' \; \theta(E - E') \; \Gamma(E')$$

$$= e^{S(E)} \int_0^E dE' \; e^{S(E') - S(E)}$$

$$\simeq e^{S(E)} \int_0^\infty dX \; \exp\left\{ -\frac{X}{T} - \frac{X^2}{2K} \right\} , \qquad (7\text{-}26)$$

where $X \equiv E - E'$ and K are defined by (7-20). Hence, if N is very large,

$$S(E) = \ln \sum_s \theta(E - H(s)) . \qquad (7\text{-}27)$$

We call this the rule for restricted energy and it is quite useful in some cases as well as in general discussions.

C. Rule for constant pressure and temperature

In the rules of calculation given above we had assumed that the system had a fixed volume. If the phase space includes states with variable volume, then $\hat{V}(s)$ becomes a function of the state s, just like $H(s)$. We have to add the word "for constant volume" to the different rules above, e.g. the rule for constant volume and energy. The equations of calculation will be changed. For example, (7-13) becomes

$$\Gamma(E, V) = \sum_s \delta(E - H(s)) \, \delta(V - \hat{V}(s)) \, ,$$

$$S(E, V) = \ln \Gamma(E, V) \, . \tag{7-28}$$

Similarly we can replace $\delta(V - \hat{V})$ by

$$\Delta(V - \hat{V}) \equiv e^{-p(V - \hat{V})/T} \, ,$$

$$\frac{p}{T} = \frac{\partial S(E, V)}{\partial V} \, . \tag{7-29}$$

Here p is just the pressure of the system and T is the temperature. Hence we obtain the rule for constant energy and pressure. We can also write $\delta(E - H)$ as $\Delta(E - H)$ at the same time, and then we get the rule for constant temperature and pressure. The steps of the calculation are similar to the above rules and the reader may derive them as an exercise. Here we only mention the thermodynamic potential G at constant temperature and pressure. The definition of G is

$$G = - T \ln \bar{Z} \, ,$$

$$\bar{Z} = \sum_s e^{-(H(s) + pV(s))/T} \, . \tag{7-30}$$

This is a generalisation of (7-24) and $\bar{Z}$ is the partition function at constant temperature and pressure. The generalisation of (7-25) is

$$G = E - TS + pV \, ,$$

$$S = -\left(\frac{\partial G}{\partial T}\right)_p ,$$

$$V = \left(\frac{\partial G}{\partial p}\right)_T . \qquad (7\text{-}31)$$

D. The rule for open systems

In the above rules, the total number of molecules in each state is assumed to be constant. Now if states with different numbers of molecules are included in the phase space, then the number of molecules $\hat{N}(s)$ becomes a function of the state. The above rules use $\delta(N - \hat{N}(s))$ to collect states with definite number N. We may now change $\delta(N - \hat{N}(s))$ to

$$\Delta(N - \hat{N}(s)) = e^{(\hat{N}(s) - N)\mu/T} ,$$

$$\frac{\mu}{T} = \left(\frac{\partial S(E, V)}{\partial N}\right)_E , \qquad (7\text{-}32)$$

where μ is the chemical potential of the molecules. We can similarly write down a collection of rules. We mention here the commonly used rules for constant temperature for an open system. The entropy can be calculated by the following equation

$$S(E, N) = \ln \tilde{\Gamma}(E, N) ,$$

$$\tilde{\Gamma}(E, N) = \sum_s e^{-(H(s) - \mu \hat{N}(s) - E + \mu N)/T} . \qquad (7\text{-}33)$$

This is a generalisation of (7-23). The rule can directly use μ, T as fixed values to calculate Ω, the thermodynamic potential at constant temperature for an open system:

$$\Omega = -T \ln \xi ,$$

$$\xi = \sum_s e^{-(H(s) - \mu \hat{N}(s))/T} , \qquad (7\text{-}34)$$

where ξ is the partition function at constant temperature for an open system.

128 STATISTICAL MECHANICS

The entropy and particle number may be obtained by differentiating Ω:

$$\Omega = E - TS - \mu N ,$$

$$S = -\left(\frac{\partial \Omega}{\partial T}\right)_\mu ,$$

$$N = -\left(\frac{\partial \Omega}{\partial \mu}\right)_T . \tag{7-35}$$

These are the generalisations of (7-24) and (7-25). "Open" implies that the molecules are not strictly confined in the system, but the entropy of a closed system can still be calculated by this rule.

E. The rule of maximum value

All the states allowed by the invariant quantities are included in the region of motion. Hence the volume of this region of motion is a maximum. (If it were not the maximum, then we could put more states into the region of motion.) Hence the basic assumption is that the entropy is a maximum. In many calculations, this particular property is very useful. The above rules can be regarded as the result of finding the maximum. We shall use the rule for constant temperature to illustrate this.

Let $\hat{A}(s)$ be a variable of large magnitude. We may divide phase space into regions according to its value, so that $\hat{A}(s)$ in each region is fixed. Let

$$\Gamma(A) \equiv \sum_s \delta(A - \hat{A}(s)) \equiv e^{S(A)} . \tag{7-36}$$

This is the volume of a particular region of phase space. Using the familiar method of peak integration and we obtain

$$S(A) = \ln \overline{\Gamma}(A) ,$$

$$\overline{\Gamma}(A) = \sum_s e^{-\alpha(\hat{A}(s) - A)} \qquad \alpha = \frac{\partial S(A)}{\partial A} ,$$

$$A = \frac{\sum_s \hat{A}(s) e^{-\alpha \hat{A}(s)}}{\sum_s e^{-\alpha \hat{A}(s)}} = \langle \hat{A} \rangle . \tag{7-37}$$

We may now ask the following question. Why is the value of $\alpha\hat{A}(s)$ such that S is maximum and $\langle H \rangle = E$? The definition of $\langle H \rangle$ is

$$\langle H \rangle = \frac{\sum_s H(s)\, e^{-\alpha\hat{A}(s)}}{\sum_s e^{-\alpha\hat{A}(s)}} . \tag{7-38}$$

This $\langle H \rangle$ is the average value of the energy in this region. Because H is a function of large order $H(s)$ is nearly equal to $\langle H \rangle$ in all points of the region. To choose $\alpha\hat{A}(s)$ means to choose the $\alpha\hat{A}(s)$ value of each s. For S to be maximum, we must solve

$$\frac{\partial}{\partial \hat{A}(s)} [S - \beta\langle H \rangle] = 0 , \tag{7-39}$$

where β is the Lagrange multiplier. (See the discussion of Lagrange multipliers in Chapter 6.) The result of the differentiation is (after rearrangement):

$$\alpha(\hat{A}(s) - \langle \hat{A} \rangle) - \beta(H(s) - \langle H \rangle) = 0 ,$$

i.e., the best choice is $\alpha\hat{A}(s) = \beta H(s)$. This answer is just the rule of constant temperature. The other rules can be obtained by a similar procedure.

This rule of maximum value has its own utility, not for the derivation of the above rules but to derive approximate calculational schemes. For example, if $H(s)$ is very complicated, so that the calculation of Z by the method of constant temperature is too difficult, we can use a simpler $H'(s)$ for $H(s)$ where $H'(s)$ has several adjustable parameters $\alpha_1, \alpha_2, \ldots, \alpha_m$. Then we use

$$\frac{\partial}{\partial \alpha_i} [S - \beta\langle H' \rangle] = 0 , \qquad i = 1, 2, \ldots, m , \tag{7-40}$$

to determine the values of α_i so that S is a maximum and at the same time keeping $\langle H' \rangle$ constant. In the following, when we use the mean field method we employ this rule to derive it. Notice that if we let

$$\rho(s) \equiv \frac{e^{-\alpha\hat{A}(s)}}{\sum_s e^{-\alpha\hat{A}(s)}} , \tag{7-41}$$

then the S in (7-37) can be written as

$$S = - \sum_s \rho(s) \ln \rho(s) \ . \tag{7-42}$$

This is the common way of expressing the entropy.

We must remember that the $\hat{A}(s)$ above must be a variable of large number, i.e. of order N, so that the results will be meaningful. All the rules of calculation stem from this basic assumption. Besides this assumption, there are no additional hypotheses and we do not need any other new concepts. Some people prefer to start from (7-42) and then use the condition that S is a maximum to determine $\rho(s)$. This is of course just the rule of maximum value discussed here. Equation (7-42), although commonly employed, looks mysterious as a basic equation. The discussion here points out that it is the same as the rule for constant energy or that for restricted energy or all the above rules. It is just a special way of expressing the basic assumption.

All the above rules have the same consequences but with different details. They all calculate the number of states in the region of motion. The determination of this region of motion is independent of the way we use $\delta(E - H)$, or $e^{-(H-E)/T}$ or $\delta(V - \hat{V})$ or $e^{-(V-\hat{V})p/T}$, etc. The results are the same.

The derivation of these rules does not require the introduction of any new concept. We only wrote the basic assumption in different forms. The various thermodynamic potentials are the special combination of the various generalised coordinates. Of course, these combinations have special implications in thermodynamics. In Sec. 7.1 we discussed some of them. In the next chapter we shall use some examples to illustrate the application of these rules.

Problems

1. Review the method of Lagrange multipliers in Chapter 6, Sec. 6.1. Compare the discussion of Sec. 7.1 with that of the method of Lagrange multipliers.

2. The various rules of calculation in Sec. 7.2 are linked to the orders of magnitude of the various quantities (i.e. S, E, N, V are $O(N)$ and Γ is $O(e^N)$). Prove that if the order of magnitude is not right, then the various rules of calculation are contradictory and cannot be used.

3. The rules of calculation in this chapter have the same result as the method of ensembles in the other books. The rule of constant temperature is the same as the canonical ensemble. The rule of constant temperature for an open system is the grand canonical ensemble. The rules of this chapter are the result of the basic assumption that $S = \ln \Gamma$. If the reader prefers to use the method of ensembles, he may do so but he should be aware that the various concepts in

ensemble theory (such as infinitely many systems) are rather abstract and not realistic. He must also note that the time scale does matter.

4. Apply all the rules of calculation to the ideal gas.

5. The walls of a gas container have N_A adsorbent states and each state can at most adsorb one gas molecule, with binding energy B. If the pressure in the gas is p and the temperature is T, calculate the total number of adsorbed molecules.

6. An elastic metallic wire at a temperature T is bent by a force f. (Fig. 7-3). If f is suddenly released, what will the final equilibrium temperature of the wire be?

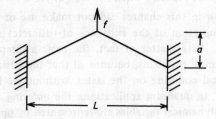

Fig. 7-3 A wire bent by a force f.

7. Use the rule of constant temperature to solve the one-dimensional Ising model.

$$H = -J \sum_{i=1}^{N-1} s_i s_{i+1} , \qquad s_i = \pm 1 .$$

8. A number of particles of radius a and mass m are dispersed in water. The temperature of water is T and gravity acts downwards. Calculate the distribution of molecules as a function of depth. Answer: $\rho(y) = \rho(0) e^{-mgy/T}$.

9. Calculate the entropy, the equation of state (i.e. the relation between pressure, volume and temperature), and the heat capacity of a system composed of N molecules is in one-dimensional space. The molecules are nonpenetrable (hence they are ordered); each molecule is rigid, with size a. The volume of the container is L. Which rule of calculation is the most convenient?

10. Suppose the interaction energy for the above problem is

$$U(x_i - x_{i+1}) = \epsilon , \qquad |x_i - x_{i+1}| < a ,$$
$$= 0 , \qquad |x_i - x_{i+1}| > a ,$$

where x_i is the position of the i-th particle, the interaction is limited to nearest neighbours. Calculate the equation of state, the entropy and the heat capacity.

11. N ions of positive charge q and N with negative charge $-q$ are constrained to move in a two-dimensional square of side L. The interaction energy is $\zeta(|\mathbf{r}_i - \mathbf{r}_j|) = -q_i q_j \ln |\mathbf{r}_i - \mathbf{r}_j|$ where $q_i, q_j = \pm q$. Write down the equation of state for this system and prove that when $T < q^2/2$ the partition function Z at constant temperature does not exist.

This model will reappear in Chapter 19.

12. Suppose the molecules in a three-dimensional space are flat discs, and that the moment of inertia of each disc is I.

(a) Calculate the entropy and energy as a function of temperature.

(b) Find the heat capacity when $T \ll 1/(2I)$ and $T \gg 1/(2I)$.

13. The discussion in this chapter did not make use of quantum mechanics except for the assumption of the existence of (discrete) states as a basis for statistical counting. As a matter of fact, the basic assumption can be applied to quantum-mechanical problems, because all that the basic assumption does is to perform statistical counting on the states, without the introduction of new concepts. However, in quantum applications, the ordering of physical variables is important, since dynamical variables are represented by operators.

The quantum nature is best exhibited in terms of the spin $\frac{1}{2}$ model, which has only two independent states. However, there are also other (not independent) states obtained by linear combination. These linear combinations may represent the spin pointing in different directions. When we count a state, we may choose any axis, and consider the two states in which the spin is pointing parallel or antiparallel to this axis. Readers should review the physics of spin $\frac{1}{2}$ systems.

14. Let $\mathbf{s}_1, \mathbf{s}_2, \ldots, \mathbf{s}_N$ be N spin $\frac{1}{2}$ operators with the Hamiltonian operator

$$H = -J \sum_{i=1}^{N-1} \mathbf{s}_i \cdot \mathbf{s}_{i+1}.$$

(a) Calculate the zero-point energy and the degeneracy of the ground state (i.e. the number of states with energy equal to the zero-point energy).

(b) Calculate the energy of the lowest excitation.

Chapter 8
ILLUSTRATIVE EXAMPLES

The aim of this chapter is to use simple examples to illustrate the rules of calculation derived in Chapter 7 and point out the importance of the observation period. First we use the rule for constant energy to analyse an ideal gas model. The molecules of this gas have two internal states. Then we use the rule for constant temperature to analyse the same problem. The rule for constant temperature is again applied to study the heat capacity of H_2 gas mainly to discuss the rotational degrees of freedom of the hydrogen molecules. The nuclei of the hydrogen molecule can have total spin $s = 1$ or 0. Under ordinary observation times and without the help of catalysts, the number of $s = 1$ and $s = 0$ states is constant, unaffected by temperature changes. If there is a catalyst, s changes rapidly, and the number of $s = 1$ and $s = 0$ states will change with temperature and the heat capacity will show a different behavior. The last example is a one-dimensional model which will illustrate the method for constant temperature and pressure, and show that a system under tension must eventually break up.

8.1. Time Scale and the Region of Motion

Suppose that a certain molecule has two internal states — the ground state of energy 0 and an excited state of energy ϵ. Let us now discuss a gas of such molecules.

The first step is to determine the region of motion in the phase space for this system. The molecules, besides moving and colliding, can change their internal state due to collision. Let n_i be the internal coordinate of the i-th molecule. If the molecule is in the ground state, $n_i = 0$; otherwise $n_i = 1$. Now there are two possibilities:

(1) The internal states are changing very rapidly, almost in every collision, or the observation period is very long, much longer than the period of change; (Here, "period" does not imply a regular periodic change. It means the approximate time required for every n_i to repeat itself once.)

(2) The internal states change very slowly, and are nearly unchanged during the observation period. Each n_i is unchanged, i.e. it is not easy for the collision to cause changes of state and the observation period is not long.

In situation (2), this gas can be regarded as the mixture of two kinds of molecules — one is the ground state molecules and the other the excited state molecules; the two are independent. Later we shall discuss this situation.

In situation (1) we must consider the change of the internal states, i.e. besides the positions and momenta we must also consider changes of the n_i.

The Hamiltonian is

$$H = \sum_{i=1}^{N} \left(\frac{\mathbf{P}_i^2}{2m} + \epsilon n_i \right) . \tag{8-1}$$

The region of motion can be determined by $E = H$, and the constraint that the positions of the molecules must be within the container. This is the same as (5-16) and (5-17) in Chapter 5, but H is slightly more complicated here. Now that the region of motion is determined the crucial step is accomplished.

The next step is to calculate the entropy and the other properties. We can use any one of the rules discussed in the last chapter. Let us use the rule for constant energy.

Let

$$\Gamma_0(E_0) = e^{S_0(E_0)} , \tag{8-2}$$

be the volume of the region of motion in the case $n_1 = n_2 = \ldots = n_N = 0$. Now besides integrating with respect to position and momentum, we have to sum over $n_1, \ldots, n_N$ also:

$$\Gamma(E) = \sum_{n_1=0}^{1} \sum_{n_2=0}^{1} \ldots \sum_{n_N=0}^{1} \frac{V^N}{N! \, h^{3N}} \int d^3 p_1 \ldots d^3 p_N \, \delta(E-H)$$

$$= \sum_{n_1=0}^{1} \ldots \sum_{n_N=0}^{1} \Gamma_0(E - \epsilon N_1) ,$$

$$N_1 = n_1 + n_2 + \ldots + n_N , \tag{8-3}$$

where N_1 is the number of excited molecules. The above equation can be written as

$$\Gamma(E) = \sum_{N_1=0}^{N} \frac{N!}{N_1!(N-N_1)!} \Gamma_0(E-\epsilon N_1)$$

$$\equiv \int dN_1 \, e^{S(E,N_1)}, \qquad (8\text{-}4)$$

$$S(E,N) \equiv S_0(E-\epsilon N_1) + N_1 \ln \frac{N}{N_1} + (N-N_1) \ln \frac{N}{N-N_1}. \qquad (8\text{-}5)$$

The combinatorial factor in (8-4) expresses the number of ways of dividing N molecules into two parts: N_1 and $N-N_1$. This is just

$$\sum_{n_1=0}^{1} \sum_{n_2=0}^{1} \cdots \sum_{n_N=0}^{1} \delta(n_1+n_2+\ldots+n_N-N_1)$$

$$= \frac{N!}{N_1!(N-N_1)!}. \qquad (8\text{-}6)$$

The number N_1 above is regarded as a large number and (8-5) is the result of simplifying (8-4) by the law of large numbers.

The integral of (8-4) can be done by the method of peak integration of Chapter 6. We first find the maximum of S:

$$\left(\frac{\partial S}{\partial N_1}\right)_E = 0 \,,$$

i.e.

$$-\epsilon \left(\frac{\partial S}{\partial E_0}\right) + \ln \frac{N-N_1}{N_1} = 0 \,, \qquad (8\text{-}7)$$

where $E_0 \equiv E - \epsilon N_1$ is the total energy of the molecules. The solution of (8-7) is

$$\bar{N}_1 = \frac{N}{e^{\epsilon/T}+1} \,,$$

$$\frac{1}{T} = \frac{\partial S_0}{\partial E_0} = \frac{\partial S}{\partial E} ,$$

$$\Gamma(E) \simeq e^{S(E,\bar{N}_1)} \int dN' e^{-N'^2/2K} ,$$

$$N' \equiv N_1 - \bar{N}_1 ,$$

$$\frac{1}{K} \equiv \frac{2\epsilon^2}{3NT^2} + \frac{N}{N_1(N-N_1)} . \tag{8-8}$$

The entropy of the system is

$$S = S(E, \bar{N}_1)$$

$$= S_0(E_0) + \frac{N\epsilon/T}{e^{\epsilon/T} + 1} + N \ln(1 + e^{-\epsilon/T}) , \tag{8-9}$$

where S_0 is the entropy due to the translational motion of the molecules and the rest is that due to internal motion. The various equilibrium quantities can be obtained by differentiation, e.g. the heat capacity is

$$C_V = T \frac{\partial S}{\partial T}$$

$$= C_{V0} + \frac{N(\epsilon/T)^2 e^{-\epsilon/T}}{(1 + e^{\epsilon/T})^2} \tag{8-10}$$

(see Fig. 8-1.) The curve has a hump near $T = \epsilon/2$. Note that the term $C_{V0} = \frac{3}{2}N$ is the heat capacity due to the translation degrees of freedom.

This is the result of situation (1), i.e. when the internal states of the molecules are changing very rapidly. In situation (2) each n_i is unchanged. Hence N_1 is unchanged. This gas is then a mixture composed of two different molecules with constant composition. Its heat capacity is then $C_{V0} = \frac{3}{2}N$. The last term of (8-10) no longer appears, and there is no hump like that in Fig. 8-1.

From the large scale viewpoint, the rapid change of the internal states of the molecules causes N_1 to become $N_1(T)$. When $N_1 = \bar{N}_1(T)$, the entropy is a maximum. This rapid change extends the region of motion in phase space.

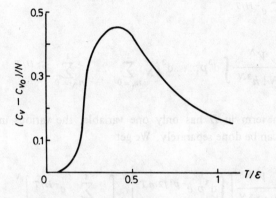

Fig. 8-1 Plot of Eq. (8-10).

The above result for the heat capacity can be analysed with the aid of the equation $C_V = \partial E/\partial T$. The total energy is

$$E = \frac{3}{2} NT + \epsilon N_1 \;. \tag{8-11}$$

If N_1 is a variable, i.e. in situation (1), then $N_1 = \bar{N}_1(T)$ and

$$C_V = \frac{3}{2} N + \epsilon \frac{\partial \bar{N}_1}{\partial T} \;. \tag{8-12}$$

We get (8-10) [see (8-8) for $\bar{N}_1$]. If N_1 is a constant, i.e. in situation (2), then

$$C_V = \frac{3}{2} N \;.$$

That is to say, if the internal states of the molecules are changing continually, then the basic assumption determines N_1. If N_1 is not changing, then the basic assumption can say nothing about the value of N_1. Thus the hump in the curve of Fig. 8-1 comes from $\partial \bar{N}_1/\partial T$.

We now use the rule of constant temperature to solve this problem. We would like to calculate

$$F = -T \ln Z \;,$$

138 STATISTICAL MECHANICS

$$Z = \sum_s e^{-H/T}$$

$$= \frac{V^N}{N! h^{3N}} \int d^3p_1 \ldots d^3p_N \sum_{n_1=0}^{1} \ldots \sum_{n_N=0}^{1} e^{-H/T} \quad . \quad (8\text{-}13)$$

Because each term in H has only one variable, the various integrations and summations can be done separately. We get

$$Z = \frac{V^N}{N! h^{3N}} \left[\int d^3p \, e^{-p^2/2mT} \right]^N \left[\sum_{n_N=0}^{1} e^{-H/T} \right]^N ,$$
(8-14)

$$F = N(f_0 + f_1) ,$$

$$f_0 = -T\left(\ln \frac{V}{N} + 1 + \frac{3}{2} \ln \frac{1}{\lambda} \right) ,$$

$$\lambda \equiv h/\sqrt{2\pi mT} ,$$

$$f_1 = -T \ln(1 + e^{-\epsilon/T}) . \quad (8\text{-}15)$$

The reader may obtain the entropy and the other quantities by differentiation. The above results apply for situation (1) when each n_i changes many times within the observation period. For situation (2) the n_i's are unchanged and the gas has N_1 excited molecules and $N-N_1$ ground state molecules. As n_1 is an invariant, the summation with respect to $n_1, \ldots n_N$ in (8-13) can be dropped, and $N!$ becomes $N_1! (N-N_1)!$. The thermodynamic potential obtained in this way is

$$F = Nf_0 - T\left[N_1 \ln \frac{N}{N_1} + (N-N_1) \ln \frac{N}{N-N_1} \right] + \epsilon N_1 .$$
(8-16)

See (8-15) for f_0. The second term on the right is caused by the entropy of mixing. (See (2-40).) The heat capacity calculated from (8-16) is just $(\frac{3}{2})N$.

Notice that at constant temperature if we change N_1 to minimize F, we get the results of (8-13); that is to say, the change of n_i minimizes F.

Obviously the rule of constant temperature is more convenient than the rule of constant energy. The reader should remember that in principle they give the same results. In the analysis, it uses temperature as the scale of energy and this is easier to handle than the energy E.

8.2. Gas of Hydrogen Molecules

Let us now look at the simplest molecule in nature, composed of two atoms. We first review some results from quantum mechanics.

The motion of the molecule as a whole and internally can be divided into several parts:

A. Motion of the electrons

As the electrons are light and the nuclei are heavy, as far as electrons are concerned the nuclei are essentially stationary. There is a large energy gap between the ground state of the electron and its first excited state, ($\sim 10\,\text{eV}$ $\sim 10^5$ K). If the temperature is not too high, the electrons will be mainly in the ground state.

B. Vibration of the nuclei

The two nuclei in the molecule repel each other because they are both positively charged. However they are enveloped and bound together by the electron cloud. The vibrational energy is

$$(n + \tfrac{1}{2})\omega \quad , \tag{8-17}$$

where $n = 0, 1, 2, 3, \ldots$, and ω is the vibrational frequency of about 6 000 K.

C. Rotation of the whole molecule

Because the amplitude of vibration is small, the molecule can be regarded as a rigid body with mass concentrated on the two nuclei. Let the angular momentum be J, then its energy is

$$\frac{1}{2I} J(J+1) \quad , \tag{8-18}$$

where $J = 0, 1, 2, 3, \ldots$ and $1/(2I)$ has a value of about 80 K.

D. The nuclear spin

The two nuclei of H_2 have a spin of value $\frac{1}{2}$ each, but the interaction is very small.

E. Spin of the electron

The two electrons of H_2 have spins that cancel each other in the ground state, and these need not be considered.

F. Translational motion of the whole molecule

The kinetic energy is $p^2/2m$.

Most diatomic molecules, not just H_2, have vibrational energies much higher than their rotational energies. (See Table 8-1.) Hence, over a large range, the

Table 8-1 Comparison of the vibrational and rotational energies (unit = K)

	H_2	N_2	O_2	NO	HCl
$\hbar\omega/k$	6 100	3 340	2 230	2 690	4 140
$\hbar^2/2Ik$	85.4	2.9	2.1	2.4	15.2

vibrational state is in the ground state ($n = 0$), while the rotational energy is usually excited. In the discussion following we ignore the vibrational energy as well as the translational energy which has been thoroughly discussed in the ideal gas case. We are left with the rotational energy and the nuclear spin energy. The energy is

$$H = \frac{1}{2I} J(J+1) \ . \tag{8-19}$$

The energy associated with nuclear spin is negligible. Now we calculate the thermodynamic potential according to the rule for constant temperature

$$F = N(f_1 + f_2) \ ,$$

$$f_1 = -T \ln \sum_J (2J+1) e^{-J(J+1)\alpha} \ ,$$

$$f_2 = -T \ln 4 \quad,$$

$$\alpha \equiv \frac{1}{2IT} \quad, \tag{8-20}$$

where f_1 is the thermodynamic potential for rotation, and f_2 is that for nuclear spin, which has four states, two for each nucleus.

This result above is applicable to the HD molecules (D means deuterium) but is not suitable for H_2 molecules. This is because the two nuclei of H_2 are identical fermions, and an exchange of position changes the sign of the wavefunction. Recall that the wavefunction is a product of the rotational part and the spin part:

$$\varphi \propto Y_{JM}(\hat{\mathbf{R}}) \chi_{sm} \quad, \tag{8-21}$$

where $\hat{\mathbf{R}}$ is the direction of the relative position of the nuclei, $M = -J$, $-J+1, \ldots, J-1, J$, and s is the total spin of the nuclei. The state $s = 1$ is a triplet:

$$m = 1 \quad, \qquad 0 \quad, \qquad -1 \quad,$$

$$\chi_{sm} = |\uparrow\uparrow\rangle \quad, \quad \frac{1}{\sqrt{2}}(|\uparrow\downarrow\rangle + |\downarrow\uparrow\rangle) \quad, \quad |\downarrow\downarrow\rangle \quad. \tag{8-22}$$

while $s = 0$ is a singlet:

$$\chi_{00} = \frac{1}{\sqrt{2}}(|\uparrow\downarrow\rangle - |\downarrow\uparrow\rangle) \quad. \tag{8-23}$$

where the arrows denote the direction of the nuclear spin. If we interchange the nuclei, $\hat{\mathbf{R}}$ is changed to $-\hat{\mathbf{R}}$, χ_{1m} is unchanged and χ_{00} changes sign, and we have

$$Y_{JM}(-\hat{\mathbf{R}}) = (-1)^J Y_{JM}(\hat{\mathbf{R}}) \quad. \tag{8-24}$$

Hence to make φ change sign, $s = 1$ can only have odd J while $s = 0$ can only have even J. This is a result from quantum mechanics. Therefore although the spin of the nucleus and the rotation the molecules have no mutual interaction, J and s are not independent and f_1 and f_2 should not be calculated separately but together:

$$f_1 + f_2 = \ln z$$

$$z = 3 \sum_{\text{odd } J} (2J+1) e^{-J(J+1)\alpha}$$

$$+ \sum_{\text{even } J} (2J+1) e^{-J(J+1)\alpha} \, . \tag{8-25}$$

From this result we can calculate the entropy and the heat capacity

$$C_V = T \frac{\partial S}{\partial T} \, ,$$

$$\frac{C_V}{N} = T \frac{\partial^2}{\partial T^2} (T \ln z) \, . \tag{8-26}$$

The curve in Fig. 8-2 is the result of (8-26) and the dots are experimental results. Obviously they do not agree. Why?

The reason is very simple. The case of H_2 is very similar to that of situation (2) discussed in the last section; but (8-25) is only true for situation (2), that is to say, the nuclear spin of H_2 during usual observation times is unchanged.

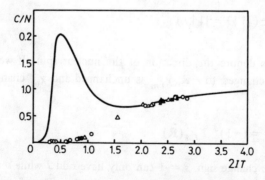

Fig. 8-2 The rotational capacity of H_2. The solid line is calculated from (8-25) and (8-26) and the dots are experimental results. Calculation and experiment do not agree at low temperatures.

However (8-25) assumes that it is changing continually. The interaction between the nuclear spins of H_2 molecules is very weak and the interaction with the electrons is also weak. At the same time the spins of the electrons cancel each other and it is not easy for them to interact with the nuclear spin. The nuclei are surrounded by the electrons and the collisions of the molecules do not affect the nuclear spins. Hence the experimentally measured quantity is a short-time result and (8-25) is not applicable (unless we use an effective catalyst to speed up the change of nuclear spins).

The correct calculation must consider N_1 the number of molecules at $s = 1$ and N_0 the number of molecules at $s = 0$ as invariant quantities. The region of motion is defined according to the invariant quantities. With this in mind, we find that the entropy is

$$C_V = N_0 \frac{\partial \epsilon_0}{\partial T} + N_1 \frac{\partial \epsilon_1}{\partial T} \quad , \tag{8-27}$$

$$\epsilon_0(T) = -\frac{1}{2I} \frac{\partial}{\partial \alpha} \sum_{\text{even } J} (2J+1) e^{-J(J+1)\alpha} \quad ,$$

$$\epsilon_1(T) = -\frac{1}{2I} \frac{\partial}{\partial \alpha} \sum_{\text{odd } J} (2J+1) e^{-J(J+1)\alpha} \quad , \tag{8-28}$$

where ϵ_0 and ϵ_1 are, respectively, the energy of each molecule for $s = 0$ and $s = 1$. The calculated results of $\partial \epsilon_0/\partial T$ and $\partial \epsilon_1/\partial T$ are shown in Fig. 8-3. The curve in Fig. 8-4 is the result of (8-27) assuming that $N_0/N = \frac{1}{4}$, $N_1/N = \frac{3}{4}$, and agrees with experiment.

Notice that the hump of the curve in Fig. 8-2 is similar to that in Fig. 8-1. If the nuclear spin is changing continually, then the values of N_0 and N_1 must be determined by the temperature, and

$$C_V = \frac{\partial}{\partial T} \left[\bar{N}_0(T) \epsilon_0(T) + \bar{N}_1(T) \epsilon_1(T) \right] \quad . \tag{8-29}$$

This is similar to (8-26). The numbers $\bar{N}_0$ and $\bar{N}_1 = \bar{N} - \bar{N}_0$ change with T, leading to the hump in Fig. 8-2. Notice that when T is very small ($T \ll \frac{1}{2}T$), $\bar{N}_1(T)$ will be very small, because the smallest value of $J(J+1)$ is 2:

$$\frac{\bar{N}_1(T)}{N} \approx e^{-2\alpha} = e^{-1/IT} \quad .$$

144 STATISTICAL MECHANICS

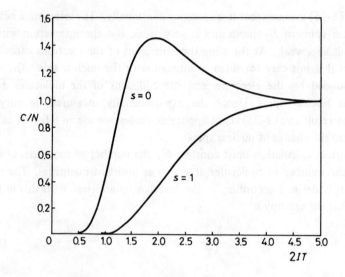

Fig. 8-3 The curves of the heat capacity of the H_2 nuclei with spin $s = 0$ and $s = 1$.

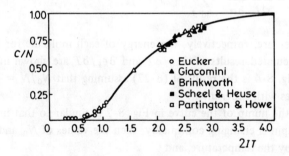

Fig. 8-4 The rotational heat capacity of H_2 calculated from (8-26) and (8-27), showing agreement with experiment. The three figures above are taken from Wannier (1966), p. 219, 220 and 221.

We can estimate the low temperature part of the curve in Fig. 8-2:

$$\frac{1}{I} \frac{\partial \bar{N}_1(T)}{\partial T} \propto \frac{e^{-1/IT}}{I^2 T^2} \ , \qquad (8\text{-}30)$$

$\partial N_1/\partial T$ is a curve with a hump, and the maximum is at $T = \frac{1}{2}I$. This is similar to (8-10). In fact the discussion of last section was meant as a preparation for this section.

8.3. One-Dimensional Model

The two examples considered above are models of various ideal gases. The effect of the interaction between the molecules is to make the variables of motion change continually; but its other effects are ignored. In the statistical calculation the interaction energy had not been included. A dilute gas is very similar to the ideal gas. But in the most substances, the interaction has many important effects. Once the interaction is included, the mathematics becomes extremely complex. Though rare, the simple and soluble models are most valuable. Though such a model may be very simple, it can nevertheless explain some effects of the interaction. One-dimensional models are the simplest and they will now be discussed.

We begin by considering a collection of molecules on a straight line. The positions of the molecules are x_i, $i = 0, 1, 2 \ldots N$, and $x_0 < x_1 < x_2 \ldots < x_N$. We fix x_0 at the origin. The volume of this system is x_N. Suppose the Hamiltonian is

$$H = \sum_{i=1}^{N} \left[\frac{p_i^2}{2m} + U(x_i - x_{i-1}) \right] . \tag{8-31}$$

The interactions of neighbouring molecules include a short distance repulsion and a long distance attraction, as in Fig. 8-5. We now use the rule for constant temperature and pressure to calculate the thermodynamic potential G:

$$G = -T \ln \bar{Z} ,$$

$$\bar{Z} = e^{-F_0/T} \int_0^\infty dx_N \int_0^{x_N} dx_{N-1} \ldots \int_0^{x_2} dx_1$$

$$\cdot \exp\left\{ -\frac{1}{T}\left[\sum_{i=1}^{N} U(x_i - x_{i-1}) + px_N \right] \right\} , \tag{8-32}$$

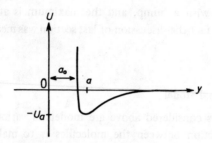

Fig. 8-5 Intermolecular potential showing short distance repulsion and long distance attraction.

$$e^{-F_0/T} \equiv \frac{1}{h^N} \int d^3p_1 \ldots d^3p_N \exp\left\{-\sum_{i=1}^{N} \frac{p_i^2}{2mT}\right\},$$

$$F_0 \equiv -\frac{1}{2} NT \ln\left(\frac{2\pi mT}{h}\right). \tag{8-33}$$

Here x_N is the $\hat{V}(s)$ of (7-30).

Now change the variables of integration to

$$y_1 = x_1,$$
$$y_2 = x_2 - x_1,$$
$$y_3 = x_3 - x_2,$$
$$\vdots$$
$$y_N = x_N - x_{N-1}. \tag{8-34}$$

These are the distances between neighbouring molecules and x_N is their sum: $x_N = y_1 + y_2 + \ldots + y_N$. The integral for $\bar{Z}$ becomes very simple:

$$\bar{Z} = \zeta^N e^{-F_0/T}, \tag{8-35}$$

$$\zeta = \int_0^\infty dy\, e^{-[U(y)+py]/T} \quad,$$

$$G = -\frac{1}{2} NT \ln\left(\frac{2\pi mT}{h}\right) - NT \ln \zeta \tag{8-36}$$

ζ is a function of p and T. If the temperature is high, i.e.

$$\frac{U_a}{T} \ll 1 \quad. \tag{8-37}$$

then

$$\zeta \simeq \int_{a_0}^\infty dy\, e^{-py/T} = \frac{T}{p} e^{-pa_0/T} \quad,$$

$$G \simeq -\frac{1}{2} NT \ln\left(\frac{2\pi mT}{h}\right) - NT \ln\left(\frac{T}{p}\right) + pa_0 N \quad. \tag{8-38}$$

The volume can be obtained by differentiation:

$$V = \left(\frac{\partial G}{\partial p}\right)_T = \frac{NT}{p} + Na_0 \quad,$$

i.e.

$$p(V - a_0 N) = NT \quad. \tag{8-39}$$

Hence at high temperatures this system is like an ideal gas with effective volume $V - a_0 N$ where a_0 is the shortest distance between two molecules. The entropy is also easily calculated:

$$S = -\left(\frac{\partial G}{\partial T}\right)_p$$

$$= N\left[\frac{1}{2} \ln\left(\frac{2\pi mT}{h}\right) + \frac{3}{2}\right] \quad. \tag{8-40}$$

148 STATISTICAL MECHANICS

Now consider the situation for low temperatures, i.e.,

$$\frac{U_a}{T} \gg 1 \ .$$

The main contribution of ζ comes from the neighbourhood of the minimum of $U(y) + py$. Suppose p is not too large, then the minimum occurs near a. Expanding $U(y)$ and retaining terms to third order term we obtain

$$U(y) \simeq U_a + \frac{1}{2}\alpha(y-a)^2 - \frac{1}{6}\alpha'(y-a)^3 \ . \qquad (8\text{-}41)$$

This is the approximation of $U(y)$ near a. The minimum of $U(y) + py$ occurs at $\partial U/\partial y = -p$, i.e.

$$\bar{y} \simeq a - \frac{p}{\alpha} + \frac{1}{2}\frac{\alpha'}{\alpha}\left(\frac{p}{\alpha}\right)^2 \ . \qquad (8\text{-}42)$$

Therefore,

$$\zeta \simeq \exp\left\{-[U(\bar{y}) + p\bar{y}]/T\right\}\left[\frac{2\pi T}{U''(\bar{y})}\right]^{\frac{1}{2}} \ ,$$

$$U''(\bar{y}) \simeq \alpha + \frac{\alpha'}{\alpha}p \ . \qquad (8\text{-}43)$$

Substituting this result into (8-36) and making some rearrangement, we get

$$G = \frac{1}{2}NT\ln\left(\frac{h}{2\pi mT}\right) - \frac{1}{2}NT\ln\left(\frac{2\pi T}{\alpha}\right) + NT\frac{\alpha' p}{2\alpha^2}$$

$$+ N\left[U_a + pa - \frac{p^2}{2\alpha} - \frac{1}{6}\alpha'\left(\frac{p}{\alpha}\right)^3\right] \ , \qquad (8\text{-}44)$$

$$V = \frac{\partial G}{\partial p} = N\left[a - \frac{p}{\alpha} - \frac{\alpha'}{2\alpha}\left(\frac{p}{\alpha}\right)^2 + \frac{T\alpha'}{2\alpha^2}\right] \ . \qquad (8\text{-}45)$$

These are the low temperature results. The statement $y \approx a$ means that distance between neighbouring molecules is a, hence these results describe a crystal. The high temperature results (8-38) and (8-39) describe a gas.

If we retain only the second order term of the expansion of $U(y)$ in (8-45), this becomes the potential energy of a simple harmonic motion. The lattice vibration discussed in Chapter 6 uses this approximate potential. From (8-45) it can be seen that if we use the simple harmonic potential and ignore α', then the relation of V with temperature is lost. The expansion of a solid with temperature is associated with the anharmonic terms.

Although this model exhibits some properties of the crystal and the gas, there is no phase transition here. When the temperature is decreasing, it does not suddenly change from the gas to the solid or liquid state at a certain temperature. Because (8-35) is a continuous, differentiable function of p, the isothermus in the p-V plane have continuous derivatives everywhere. Generally speaking, in one dimension, a system with short-range interaction (i.e. each molecule interacts with its few neighbouring molecules) does not exhibit phase transitions. This will be a topic for later discussion.

However there is the following important result. If

$$p \leqslant 0 \quad , \tag{8-46}$$

i.e. there is no pressure or a tension, then there is a problem in the integral of (8-35). This is because $U(y)$ must vanish at very large y (i.e. the interaction is short-ranged), so the convergence of the integral depends on the factor $e^{-yp/T}$. This is so no matter how strong the short-range interaction is. The situation for $p > 0$ is simple. Fig. 8-6(a) tells us that y should remain in the neighbouring of the minimum. The situation $p < 0$ is a valley, and once the barrier is overcome, it is an endless downward slope. (Of course, if the tension

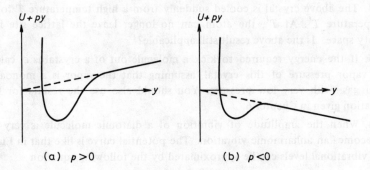

Fig. 8-6

is large enough, even the valley will disappear). This result can be explained as follows. If each y_i stays at the bottom of the valley, then these molecules are linked as a chain. If one or more y_i jump out of the valley, then the chain is broken and the tension vanishes. The conclusion of this example is appropriate for any system under tension: such a system will eventually break up.

Of course, the problem of tension is a question of the length of time. If the time is short, a solid under tension is an equilibrium state. Thus, in calculating the region of motion we should exclude the states of the broken system. How do we exclude them? This question still has no completely satisfactory answer. But in many cases this is not hard to manage, e.g. in (8-44) and (8-45) we did it by expansion. Even for $p < 0$, this result is still applicable, provided we ignore the problem in the integral of (8-35).

Problems

1. N atoms are arranged regularly to form a perfect crystal. If we move n atoms from the original lattice sites to the space in between the atoms, this forms an imperfect crystal with n defects. The number of such possible spaces N' is the same order as the number of atoms N. If ω is the energy to move an atom from a lattice site to the empty space, prove that

$$\frac{n^2}{(N-n)(N'-n)} = e^{-\omega/T} \qquad (8\text{-}47)$$

2. The above problem can be solved by the principle of detailed balance. An atom absorbing a phonon of energy ω will leave the lattice thereby creating a vacant site. From this basic process derive the result of the above problem.

3. Calculate the entropy and heat capacity due to the defects.

4. The above crystal is cooled suddenly from a high temperature T to a low temperature T'. At T', the atom can no longer leave the lattice site for an empty space. Is the above result still applicable?

5. If the energy required to kick a molecule out of a crystal is ϵ, calculate the vapor pressure of this crystal, assuming that the vapor is a monoatomic ideal gas, with very low pressure. You should also use the results for lattice vibration given in Chapter 6.

6. When the amplitude of vibration of a diatomic molecule is very large, it becomes an anharmonic vibration. The potential curve is like that in Fig. 8-7. The vibrational levels can be approximated by the following equation

$$\epsilon_n = \left(n + \frac{1}{2}\right)\omega - a\left(n + \frac{1}{2}\right)^2 \omega, \qquad n = 0, 1, 2, \ldots, \qquad (8\text{-}48)$$

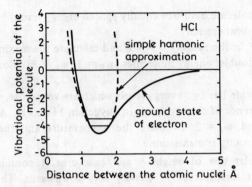

Fig. 8-7 Vibrational potential of a diatomic molecule.

where a represents the degree of anharmonicity. Calculate the heat capacity and indicate the effect due to the anharmonic term up to the first order in a.

7. Prove that if we neglect quantum effects, then the thermodynamic potential of charged particles will not be affected by a magnetic field. Hint: the Hamiltonian is

$$H = \sum_{i=1}^{N} \frac{1}{2m_i} \left[\mathbf{p}_i - \frac{e_i}{c} \mathbf{A}(\mathbf{r}_i) \right]^2 + U(\mathbf{r}_1, \mathbf{r}_2, \ldots, \mathbf{r}_N) \ . \qquad (8\text{-}49)$$

8. Some molecules have two nearby energy levels for the electronic motion (e.g. oxygen and nitric oxide), with the energy difference being far less than T. As a result, the higher level can have a considerable population and thus influence the thermodynamic properties of the substance.

(a) Prove that the heat capacity due to the motion of the electron is

$$C = \omega N e^{\epsilon/T} \left[\frac{\epsilon/T}{e^{\epsilon/T} + w} \right]^2 \ . \qquad (8\text{-}50)$$

The formula assumes that the low energy level is zero and that there are w_0 single particle states. The high energy level is ϵ and has w_1 single particle states. Here $w = w_0/w_1$.

(b) Obtain approximations for C in the limits of high and low temperatures and prove that the maximum of C occurs at

$$\frac{\epsilon}{RT} = \ln w + \ln\left(\frac{\epsilon/T + 2}{\epsilon/T - 2}\right) \ . \qquad (8\text{-}51)$$

(c) If each molecule has three equally spaced energy levels, and $w_i = 1$, what is the form of its heat capacity?

(d) Take ϵ/T in the range $1.0 \sim 4.0$ and calculate the maximum of the heat capacity of the doublet energy molecules ($w_0 = 1$, $w_1 = 3$). Compare the result with that of (c).

(e) Notice oxide has two energy levels which are very close, with $\epsilon = 174$ K. The energy difference of the two levels in oxygen in 11 300 K. At high temperatures $w_0 = 3$ and $w_1 = 2$. Estimate the temperature and the heat capacity when the heat capacity is maximum.

9. A rectangular box of height h and base area A contains N molecules forming a gas at temperature T. Gravity acts downwards. The mass of each molecule is m.

(a) Calculate the gas pressure at the top and at the bottom.

(b) Calculate the pressure at the walls.

10. A cylindrical container of radius R and length L is rotating round its axis with angular velocity ω. There is a gas in the container of temperature T and the mass of each molecule is m. Notice that in the frame rotating with the container this gas is in equilibrium with fixed boundary. Calculate the Hamiltonian in this frame.

11. Answer the questions of problem 9 for the cylindrical container of problem 10.

12. A one-dimensional system with N molecules is described by the Hamiltonian

$$H = \sum_{i=1}^{N} \frac{p_i^2}{2m} + \sum_{i=1}^{N} U(x_i - x_{i-1}),$$

$$x_0 < x_1 < x_2 < ,\ldots, < x_N,$$

$$U(y) = \frac{1}{2} m\omega^2 (y-a)^2, \quad y > 0, \quad (8\text{-}52)$$

where $x_0 = 0$ and a pressure p is exerted on the N-th molecule.

(a) Calculate the entropy and heat capacity as a functions of temperature for this system.

(b) Find $\langle x_N \rangle$.

In the above calculation, assume that $m\omega^2 a^2 \gg 1$, i.e. ignore

$$\alpha \equiv \exp\{-m\omega^2 a^2/2T\}. \quad (8\text{-}53)$$

(c) Calculate the coefficient of expansion of this system. Note that α cannot be completely ignored, but higher order terms of α may be neglected.

Chapter 9
YANG-LEE THEOREM

It can be seen from the previous analyses and examples that the application of the basic assumption to thermodynamics gives satisfactory results. The rules of calculation listed in Chapter 7 agree perfectly well with thermodynamics; that is to say, under given conditions the equilibrium property is fixed. Because a body is macroscopic, i.e. the number of molecules N is macroscopically large as pointed out in Chapter 7, we can use different statistical weights to perform the calculation and obtain the same result. However, our analysis is not rigorous. In the process of analysis we have not considered the mathematical meaning of the various equations. The examples are all over-simplified models, chosen for being readily soluble. Although these examples provide a measure of confidence and understanding for the rules of calculation and the basic assumption, they do not reveal the difficulties that would be encountered in more general cases. In this chapter we pay some attention to mathematical rigor. If the number of molecules N is finite, then the calculation of the various thermodynamic potentials does not pose any problem. Although N is not infinite, it is nevertheless a very large number like 10^{20}, hence the problem of the $N \to \infty$ limit becomes a very important problem for the application of the basic assumption in thermodynamics, i.e. the so-called problem of the macroscopic limit. The rigorous mathematical analysis of this limit is a branch of statistical mechanics. The pioneering work in this topic is the Yang-Lee theorem of thirty years ago.[a] The theorem was proposed for phase transitions. Following them, many have

---

[a] Yang-Lee (1952). One of the appendices of Huang (1963) gives quite a detailed introduction; see p. 458.

applied rigorous mathematical analyses to describe phase transitions, because the model problems of phase transitions are not easily solvable and less than rigorous analysis is not reliable. However the application of the Yang-Lee theorem is quite universal. In this chapter we shall discuss one of the Yang-Lee theorems, i.e. the existence of the thermodynamic potential Ω/V in the limit $V \to \infty$. The aim of this chapter is not to discuss phase transitions. (Phase transitions will be discussed in later chapters.) However, we want to point out the reason for the existence of this limit, and the relation of the macroscopic limit with the interaction of the molecules and the statistical procedure. From the proof of this theorem we can see that the most important feature of a macroscopic body in equilibrium is that it can be divided into many parts, with each part being in the equilibrium state. Here we do not emphasise a rigorous proof. Rather we want to have some further understanding of the basic assumption and to remind the reader that rigorous mathematics does not guarantee correct physical meaning. Care should be exercised in this respect.

9.1. Macroscopic Limit (i.e. Thermodynamic Limit)

We first set up a model. Let the volume of the body be V and the interaction potential between the molecules be as shown in Fig. 9-1. The necessary properties of the potential $u(r)$ are

(a) $u(r) = 0$,   if $r > b$,

(b) $u(r) = \infty$,   if $r < a$,

(c) $u(r) > -\epsilon$.   (9-1)

That is to say, the distance between the molecules cannot be smaller than a. Hence the number of molecules accommodated in this volume V is finite:

$$N < N_m = V \bigg/ \left(\frac{4\pi}{3} a^3 \right) \qquad (9\text{-}2)$$

The attraction is finite, i.e. ϵ is finite. The interaction is zero for molecules far apart, i.e. b is finite. If all the conditions in (9-1) are satisfied, the detailed form of $u(r)$ is unimportant.

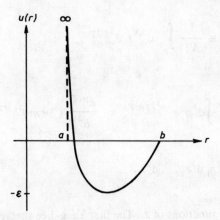

Fig. 9-1 A model interaction potential between molecules.

Using the rule for constant temperature we analyse the system. The definition of the thermodynamic potential Ω is

$$\xi \equiv e^{-\Omega/T} = \sum_s e^{-[H(s) - \mu N(s)]/T} , \qquad (9\text{-}3)$$

$$p = -\frac{\Omega}{V}, \qquad H = \sum_{i=1}^{N} \frac{p_i^2}{2m} + U ,$$

$$U = \frac{1}{2} \sum_{i,j=1}^{N} u(r_i - r_j) ,$$

$$\sum_s \equiv \sum_{N=0}^{N_m} \frac{1}{N! \, h^{3N}} \int_V d^3 r_1 \ldots d^3 r_N \, d^3 p_1 \ldots d^3 p_N . \qquad (9\text{-}4)$$

We first integrate out all the $\mathbf{p}_i$ and get

$$\xi(z, V) = \sum_{N=0}^{N_m} z^N \, Z_N(z, V) , \qquad (9\text{-}5)$$

$$Z_N(z,N) \equiv \frac{1}{N!} \int_V d^{3N}r \; e^{-U/T} \; , \tag{9-6}$$

$$z \equiv \frac{e^{\mu/T}}{h^3} \int d^3p \; e^{-p^2/2mT} = \frac{e^{\mu/T}}{h^3} (2\pi mT)^{3/2} \; ,$$

$$d^{3N}r \equiv d^3r_1 \, d^3r_2 \ldots d^3r_N \; . \tag{9-7}$$

ξ, Ω and Z all are functions of z. The first Yang-Lee theorem is:
If (9-1) is satisfied $z > 0$, then the limit

$$\lim_{V \to \infty} \left(-\frac{\Omega}{TV}\right) \equiv \lim_{V \to \infty} \frac{1}{V} \ln \xi(z,V) \; , \tag{9-8}$$

exists and is independent of the shape of the container. It is also a continuous, monotonically increasing function of z. The above assumes that the surface of the system is less than $V^{2/3}$ times a fixed constant, i.e. the surface does not curve in and out too much.

The existence of the limit (9-8) means that if V is sufficiently large, Ω/V will not change with V. If we divide V into n parts and each part is very large, then the result will be the same if we calculate Ω/V in each part separately. The proof starts from this viewpoint.

We divide the body into n parts, and set a "boundary region" of width b between each part, (Fig. 9-2). Let the volume of the boundary region be V'. If we eliminate the boundary region, then the remaining volume is nV_0. If we impose the restriction that there is no molecule in the boundary region, ξ becomes

$$[\xi(z,V_0)]^n \; , \tag{9-9}$$

because the effective distance of the interaction between the molecules is b, there is no interaction between the different parts. The integration of the different parts can be performed separately and we get (9-9).

From (9-6) we can see that if we diminish the region of integration, the integral becomes less because the integrand is positive definite. Hence if we

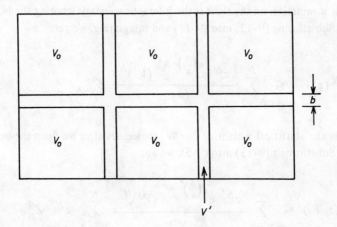

Fig. 9-2 Division of a body into parts with boundary regions between them.

eliminate the boundary region, then Z_N can only decrease and ξ decreases too, i.e.

$$[\xi(z, V_0)]^n \leq \xi(z, V) . \tag{9-10}$$

Thus we conclude that if we cut up a system into n parts and ignore the boundary regions, we make ξ smaller.

We put back the boundary region, and simplify the interaction so as to get an upper limit for ξ. We rewrite (9-6) as

$$Z_N(z, V) = \sum_{N'=0}^{N} \frac{1}{N'!(N-N')!} \int_{V'} d^{3N'}r \int_{V-V'} d^{3(N-N')}r \; e^{-U/T} , \tag{9-11}$$

i.e. we restrict $\mathbf{r}_1, \mathbf{r}_2, \ldots, \mathbf{r}_{N'}$ to be inside the boundary region and let the others be in the n parts of the volume V_0. Each molecule interacts at most with $(b/a)^3$ other molecules (conditions (9-1a) and (9-1b)). The interaction energy of each pair of molecules is not less than $-\epsilon$ (condition (9-1c)). If the interaction energy inside the boundary region is replaced by $-\epsilon(b/a)^3 N'$, the value of U becomes smaller, i.e.

$$U \geq \sum_{k,j=N'+1}^{N} u(\mathbf{r}_k - \mathbf{r}_j) - \epsilon N'(b/a)^3 . \tag{9-12}$$

The first summation on the right is the interaction energy ignoring the boundary region. Substituting (9-12) into (9-11) and integrating, we get

$$z^N(z, V) \leq \sum_{N'=0}^{N} \frac{z^N e^{\epsilon \left(\frac{b}{a}\right)^3 N'} (V')^{N'}}{N''!} Z_{N-N'}^0 . \qquad (9\text{-}13)$$

$Z_{N-N'}^0$ is the statistical weight of $N - N'$ molecules after we drop the boundary region. Substituting (9-13) into (9-5), we get

$$\xi(z, V) \leq \sum_{N'=0}^{\infty} \frac{z^{N'} e^{\epsilon \left(\frac{b}{a}\right)^3 N'} (V')^{N'}}{N'!} \sum_{N''=0}^{N} z^{N''} Z_{N''}^0 . \qquad (9\text{-}14)$$

In the above we have written $N - N'$ as N'' and taken away the constraint $N' < N$. Now the last summation is just $[\xi(z, V_0)]^n$ and the first summation can also be done. Equation (9-14) becomes

$$[\xi(z, V_0)]^n \leq \xi(z, V) \leq e^{\sigma V'} [\xi(z, V_0)]^n ,$$

$$\sigma \equiv z e^{\epsilon (b/a)^3} . \qquad (9\text{-}15)$$

We have included the lower limit in (9-10).

Now V' is the volume of the boundary region and is approximately equal to the boundary surface times the width b, i.e.

$$V' = c V_0^{2/3} , \qquad (9\text{-}16)$$

where c is a constant. Let

$$\phi(z, V_0) \equiv \frac{1}{(V/n)} \ln \xi(z, V_0) , \qquad (9\text{-}17)$$

then from (9-15), we get

$$\phi(z, V_0) \leq \frac{1}{V} \ln \xi(z, V) \leq \phi(z, V_0) + \sigma c V_0^{-1/3} ,$$

$$\left| \frac{1}{V} \ln \xi(z, V) - \phi(z, V_0) \right| \leq \sigma c V_0^{-1/3} . \qquad (9\text{-}18)$$

That is to say, if V_0 is sufficiently large, the difference between $\phi(z, V_0)$ and $V^{-1} \ln \xi(z, V)$ is very small. The thermodynamic potential per unit volume can be calculated from a part of the body; it does not matter which part. The error in the calculation is inversely proportional to the size of the portion. This has clearly proved the thermodynamic limit in this model. In the above analysis, V_0 need not be a cube; we only require that (9-16) be satisfied. Nor is it necessary that each part have the same volume.

From the point of view of physics, (9-18) is the result we seek. If we want complete mathematical rigor, we must supplement our discussion with some mathematical explanations. We shall not get into this here.

Because

$$\frac{\partial p}{\partial \mu} = \frac{N}{V}, \quad z \propto e^{\mu/T}, \tag{9-19}$$

we have

$$\frac{N}{V} = \frac{1}{T} z \frac{\partial p}{\partial z} = \frac{1}{V} \sum_{N=1}^{N_m} N z^N Z_N \Big/ \sum_{N=0}^{N_m} z^N Z_N . \tag{9-20}$$

Each term in (9-20) is positive, so p must increase with q whether V is large or small. As the limit $V \to \infty$ exists, $p(z)$ must increase with z in this limit. In addition

$$\frac{N}{V} < \frac{N_m}{V} = \frac{1}{a^3} . \tag{9-21}$$

So $\partial p/\partial z$ is finite, i.e. p is continuous. This completes the proof of the first Yang-Lee theorem. The second Yang-Lee theorem is very useful in the discussion of phase transitions. The reader is referred to the literature, as we will not discuss it in this book.

9.2. Generalisation of the Theorem

Now let us return to examine the three conditions in (9-1), i.e. (a) a finite effective interaction distance, (b) a short distance repulsion and (c) an overall attraction not too large in magnitude. If (a) is not satisfied then there is long-range interaction and there is no way to determine the boundary region. If (b) and (c) are not satisfied there will be instability, i.e., all the particles will be concentrated in a small region. In the problem of the macroscopic limit, we ask what we shall obtain if we relax the conditions above. We shall mention several results.

Condition (a) can be slightly relaxed. If $u(\mathbf{r})$ vanishes sufficiently rapidly, i.e. when $r > b$,

$$u(\mathbf{r}) \leq A r^{-\lambda}, \quad \lambda > d, \tag{9-22}$$

the theorem still holds. Here d is the spatial dimension of the system and is usually $d = 3$. Condition (b) can also be slightly relaxed. When $r < a$, if we have

$$u(\mathbf{r}) \geq B r^{-\lambda'}, \quad \lambda' > d, \tag{9-23}$$

then the theorem is still intact. Condition (c) cannot be relaxed.

Now the most important interaction does not fit into the above scheme. Matter is made up of the positively charged nuclei and the negatively charged electrons. There is also the dipole interaction. These electromagnetic interactions are long-ranged and the interaction energy of positive and negative charges when $r \to 0$ is $-\infty$. None of the three conditions are met. Hence the Yang-Lee theorem cannot be used directly to discuss the stability of matter and its macroscopic limit. In the following we shall mention the effects of the interaction between the charges, and the quantum nature of matter.

From our experience with electromagnetism we know that uncancelled charges always go to the surface. Here the energy and the other quantities will be dependent on the shape of the system. Hence, only those systems with no net charges can have a thermodynamic limit.

We must also consider the quantum effects, otherwise the separation between the positive and negative charges can be infinitesimally small. All the particles will be concentrated at a point. Quantum mechanics says that a particle restricted to a small region will have a large momentum. This theorem ensures that the particles do not collapse to a point. But, if the particles are bosons, we can well imagine that although the particles do not collapse to a point, they still prefer to cluster and the thermodynamic limit will encounter difficulties. It has been proven that if there were no exclusion principle, the zero-point energy of each particle will be lower than $-N^{2/5}$, i.e. as $N \to \infty$ the energy of each particle will tend to $-\infty$. If all the negatively-charged particles (or positively-charged particles, or both) are fermions (the types of particles are finite, otherwise the exclusion principle is still useless), then the zero-point energy of each particle is finite. These facts are quite difficult to prove. Much progress has been achieved in the proof of the macroscopic limit.[b] The validity of the basic assumption is quite well established, since the results of various rigorous analyses do not lead to contradiction.

Lastly we must emphasise the need for caution in such rigorous analysis, and we should not be too easily taken in by its conclusion. We must first understand the assumption in the analysis and the meaning of the various limits. For example, in the Yang-Lee theorem we can arrive at the conclusion that under a fixed temperature, and chemical potential there is one and only one equilibrium state. This excludes the existence of metastable states. In reality there are various types of metastable crystal structures and also supersaturated vapour configurations. This is because this theorem starts from the basic assumption without considering the time scale. This assumption is completely rigorous only under an infinitely long period of observation. A rigorous proof means that there is no approximation between the assumption and the conclusion. In an effort to simplify matters the assumption usually ignores some facts. The basic assumption of statistical mechanics is an oversimplified assumption. Its applications are extensive but it is not the absolute truth. Besides the assumption, some limits such as the volume $V \to \infty$ must be taken with care. A real body is finite. Hence the implication of the limit must not contradict reality. A common situation is

$$\lim_{V \to \infty} \frac{1}{\ln V} = 0 \ . \tag{9-24}$$

The meaning of this 0 is far from clear. Even for the largest system, the logarithm of its volume cannot be too large. Although (9-24) is mathematically impeccable, physically (9-24) is not rigorous enough and is questionable. Hence to understand a "rigorous conclusion" we must be careful in every detail of the proof. In general the true meaning of the conclusion usually becomes apparent from an examination of the steps of the proof.

Problems

1. The proof in Sec. 9.1 shows that once the Hamiltonian is fixed, all the thermodynamic properties are fixed. This is a rigorous proof. The proof pinpoints the obscurity in the basic assumption because it ignores the existence of metastable states; yet we regard a metastable state as being in equilibrium if the observation time is not too long.

---

[b] See Lebowitz and Lieb (1969), Dyson (1967). For a detailed introduction to the rigorous proof, see Griffiths (1972).

Consider the examples of graphite and diamond and show that in a collective model of carbon atoms, with a rigorous proof we can rule out the existence of diamond.

2. Derive all the results in Sec. 9.1.

3. For a d-dimensional model what are the changes in details of the proof?

The reader may encounter some $d \to \infty$ models in the literature. (Of course this is just a theoretical discussion. Nevertheless there are occasions when models with $d > 3$ can be very useful.) When $d \to \infty$, the proof in Sec. 9.1 is not correct, that is to say, the usual $d \to \infty$ statistical models are not dependable, since their thermodynamic limit is questionable.

PART III
PROBABILITY

Probability is a useful but often abused concept. This part repeatedly points out the inseparability of probability from statistics and discusses various applications. Probability is a number obtained from the statistics of the data, and does not stand for a conjecture or the degree of ignorance. Statistics is a procedure for compiling and interpreting the information at hand. It is a tool, not a theory. In statistical mechanics we have to compile the information about the various details of the molecular motion.

Chapter 10 discusses the relation of probability to statistics. Chapter 11 discusses the phenomenon of independence or chaos. This phenomenon is very important and not easy to understand. The discussion here is a description of the phenomenon and not a quest for its origin.

Chapter 12 discusses the central limit theorem, delving further into the meaning of independence and probability, and explains some results of the basic assumption. This assumption is inseparable from the phenomenon of independence. Chapter 13 presents some calculations of the correlation function and some examples of the concepts of fluctuation and response.

Chapter 10

PROBABILITY AND STATISTICS

We have introduced the basic assumption of statistical mechanics in the last five chapters. Although the examples are oversimplified models, it is not difficult to observe the following: (1) The basic assumption can be used to analyse and explain all the equilibrium phenomena and derive all the thermodynamic results. (2) This assumption and its applications do not require explicitly theorems or concepts in statistics other than those mentioned in the above chapters. (3) Not much information on the trajectory of the molecular motion is required. The role of the molecular motion is to make the basic assumption valid. Once the region of motion is fixed, the molecular motion can be ignored.

However, if we want to have a further understanding of the basic assumption, we have to look at the molecular motion more carefully, and for this we need more tools in statistics. Thus, although we have just started to discuss the applications of the basic assumption, we digress for a moment and discuss instead some topics in statistics. The main point in this chapter is the definition of probability, and we emphasise that statistical data are the fundamentals of the definition of probability.

10.1. The Use and Misuse of Probability

Statistics is the science of compiling information and probability is one of the basic concepts in statistics. In its mathematical formulation, probability is rigorously defined and its various properties are fixed by rules. In applications, the concept of probability is often misused, because its definition is not well understood. Now we discuss two different uses. Probability has different meanings, and if we are not careful we shall be led to incorrect conclusions and unnecessary trouble. This is especially true for students doing theoretical research. They are more liable to be confused by these two different uses of probability.

In the first use, probability is the ratio of two statistical numbers. For example, a die has been thrown 600 times, and the '1' turns up 98 times. Then we say that the probability of obtaining '1' is 98/600, or nearly 1/6. This is a fact or a natural phenomenon.

The second use of probability is to measure the degree of ignorance or uncertainty of a person about an event. For example, a person betting for money does not know which number on the die will appear. If you ask him what is the number on the die, he will say, "I do not know, because each different number has probability 1/6 to appear." In this use, the probability is not derived from statistical figures, but only indicates the complete ignorance of the result.

The above two uses cannot be mixed. Consider this example. Suppose a die is thrown 600 times and we ask how many times '1' will turn up. The reply might be, "As the die has six possibilities and I do not know anything else, hence '1' will appear 100 times." This result is a guess and it turns out to be correct, but the argument is not logical. Ignorance has been used to predict a natural phenomenon; this is surely impossible. Take another example. If I ask, "How many days in a year does it rain in Hsinchu?" One might reply, "As there are two possibilities, to rain or not to rain, and I am completely ignorant about Hsinchu, therefore it rains six months in a year." This reply relies on the same reason above, using ignorance to predict a natural phenomenon, and is equally invalid. Instead, the past record of the rainfall can be obtained from the meteorological office. The rainfall in the future can be predicted on the basis of the past records plus other information. We cannot use ignorance to make predictions.

The above emphasises the two different uses of probability. There is nothing wrong with the second use. Every experimental result should be accompanied by its error bars. This error does not say that the experiment has gone wrong, it only expresses the degree of accuracy of the experiment. This is an estimate by the experimenter. If he says that the error is 0.1%, this expresses that he is not sure of the figure to within 0.1%. Let us take an example to illustrate this.

A person uses a mercury barometer to measure the air pressure on the main street. His result reads, "At a certain year, month, day, hour and minute the air pressure is 1.02 ± 0.05 standard atmosphere." His estimation is that at that time the air pressure should be between 0.97 and 1.07, but where exactly in this range, he does not know. To estimate this error he must consider the accuracy of the barometer calibration and the accuracy of his eyesight in reading the scale.

Notice that this error is completely different from the fluctuation of the pressure and the two should not be confused. The pressure changes with time. This is fluctuation. The usual pressure measured by a mercury barometer is an average pressure, an average value over a second or longer. Long time pressure fluctuations can be measured by this barometer. Long time fluctuations are caused by the changes in weather while the short time rapid fluctuations are caused by the environmental disturbances such as traffic or road work. They cannot be measured by a mercury barometer, but we can use a recorder for the measurement. This pressure fluctuation caused by noise, etc. has a very small amplitude, about 10^{-10} atmosphere, and is totally unrelated to the 5% error in the barometer reading.

In most of this book, probability is used only with the first meaning. Unless otherwise stated, we shall not consider the second use. Later when we mention probability we shall have to discuss how it is obtained from statistical information.

Here we need to emphasise that probability is meaningful only in so far as its value is independent of the size of the data pool. (Of course we cannot have too few pieces of data.) For example, if we throw a die 1 200 times and count the number of appearances of '1', we can use the first 600 throws to get the probability, or the next 600 throws to get the probability, or the whole 1 200 throws to get the probability. The probability obtained should all be approximately 1/6. If we get 1/2 for the first 600 throws and 1/4 for the next 600 throws we can still say that the statistical result of 1 200 throws is 1/6 but the meaning of this result is questionable. That is to say, this result of 1/6 must be repeatable in order to be meaningful.

10.2. Distributions and the Statistics of the Average Value

In the last few chapters we have discussed distributions, like the age distribution of the population, or the energy distribution of the states of the particles. The distribution is a result of doing statistics by classification. For example, we classify the population according to age, each age group forming a class, and then we count the number in each class, obtaining an age distribution curve like Fig. 3.1. This is information about the population. Its statistical counting procedure can be expressed by the abstract equation

$$f(X') = \sum_s \delta(X' - X(s)) \quad , \tag{10-1}$$

where s is the name of each person, $X(s)$ is the age of person s and $\sum_s$ counts the number of the population. The δ function performs the classification, i.e. it collects all the names of the persons whose age is X'. The number of persons of age between X_1 and X_2 is

$$N(X_1 < X < X_2) = \int_{X_1}^{X_2} dX' f(X') \quad . \tag{10-2}$$

The reader might review Chapter 3 at this point.

Equation (10-1) is the standard procedure of statistics for classification. The analysis is usually made on the characteristic property $X(s)$ of a collection of "samples", with each of the samples identified by a assigned number (or name) s. We collect all those with the same $X(s)$ into a class. The number of samples in each class is called the statistical weight of the class. If we plot the statistical weight in a curve, we get a distribution of this special property X. This provides one group of information about the collection of samples. Information so obtained may sometimes be uncorrelated. We then use average values to describe the essential properties of distribution. For example the average value of X is

$$\langle X \rangle \equiv \int dX' \rho(X') X'$$

$$= \sum_s X(s)/N \quad , \tag{10-3}$$

$$\rho(X) \equiv f(X) \Big/ \int dX' f(X') \quad . \tag{10-4}$$

Here N is the total number of samples and $\rho(X) dX$ is the normalised percentage corresponding to $f(X) dX$.

The average value of $(X - \langle X \rangle)^2$ is

$$\langle (\Delta X)^2 \rangle = \langle (X - \langle X \rangle)^2 \rangle$$

$$= \int dX' (X' - \langle X \rangle)^2 \rho(X')$$

$$= \sum_s (X(s) - \langle X \rangle)^2 / N \quad . \tag{10-5}$$

The average value $\langle X \rangle$ is the 'centre of mass' of the distribution, and $\langle (\Delta X)^2 \rangle$ expresses the dispersion of the distribution. The function $\rho(X)$ is called the probability of the X distribution. It is the statistical weight of the class divided by the total number. This is the definition. We do not enquire into the relation between $\rho(X)$ and the concept of "chance" or "possibility". This is often possible, depending on the application of $\rho(X)$.

Any function $\vartheta(X)$ of $X(s)$ has its own average value, e.g. $\rho(X')$ is the average value of $\delta(X' - X(s))$, and the average value of 1 is 1:

$$\langle \vartheta(X) \rangle = \int dX' \, \vartheta(X') \rho(X')$$

$$= \frac{1}{N} \sum_s \vartheta(X(s)) \quad , \tag{10-6}$$

$$\rho(X') = \langle \delta(X' - X(s)) \rangle$$

$$= \frac{1}{N} \sum_s \delta(X' - X(s)) \quad , \tag{10-7}$$

$$\langle 1 \rangle = \int \rho(X) dX = 1 = 100\% \quad . \tag{10-7'}$$

Equation (10-7') is the normalisation condition, expressing the fact that the total percentage must be 100%.

We can combine some classes for our purpose. For example, the class of age below 3.0 can be called the "youth" class and those between 70 and 80 as the class of "the beginning of life", etc. The number in each class is its statistical weight. Generally speaking, the statistical weight of a group R of the samples (i.e. a subset in the sample space) can be defined as follows. Let the "counting function" $\theta_R(s)$ of R be

$$\theta_R(s) = 1 , \quad \text{if} \quad s \in R \quad ,$$

$$\theta_R(s) = 0, \quad \text{if} \quad s \notin R \quad . \tag{10-8}$$

Therefore,

$$N_R \equiv \sum_s \theta_R(s) \quad , \tag{10-9}$$

is the statistical weight of the R class. Let

$$P(R) \equiv \frac{N_R}{N} = \langle \theta_R \rangle \quad , \tag{10-10}$$

be the probability of the R class. This is a definition. Notice that $\delta(X' - X(s))dX'$ is an example of R, i.e. the set of the samples with $X(s)$ between X' and $X' + dX'$ is R.

If R is the set consisting of samples with $X(s) < X'$, then

$$P(R) = P(X < X') = \langle \theta(X' - X) \rangle = \int_{-\infty}^{X'} dX \rho(X) \quad , \qquad (10\text{-}11)$$

is the probability that $X < X'$.

Those samples to be counted may be a group of persons or objects, or the record of events, experimental data, states in the phase sapce, etc. Let us now look at some examples.

The following are the last digits of the telephone numbers listed on p. 205, 206, 214 of the directory of Hsinchu, for subscribers with the surnames Ma and Liang:

5 3 6 3 1 8 3 6 1 1 7 7 1 1 8 3 6 0 8 0 6 6 6 8 7 1 3 7 6 8 9 8

5 7 8 1 5 2 9 1 5 7 3 1 5 5 5 7 7 8 9 8 8 0 9 7 0 3 5 8 3 5 5 5

6 5 6 5 2 7 9 3 0 6 9 2 2 5 3 0 9 0 3 9 6 7 7 8 0 0 5 9 6 7 5 5

$$(10\text{-}12)$$

There are $N = 96$ households, and the above members are one characteristic $X(s)$ of each household s, $s = 1, 2, \ldots, 96$. Define the probability distribution as

$$\rho(X') = \frac{\sum_{s=1}^{N} \delta(X' - X(s))}{N} = \frac{N_{X'}}{N} \quad , \qquad (10\text{-}13)$$

where N_X is the number of times X appears, and $N = 96$.

Here the argument of the δ function takes discrete values, hence the δ function is defined as

$$\delta(M) = 1, \qquad \text{if} \quad M = 0 \quad ,$$
$$= 0, \qquad \text{if} \quad M \neq 0 \quad . \qquad (10\text{-}14)$$

The normalized probability $\rho(X)$ can be calculated from the above information:

X	0	1	2	3	4	5	6	7	8	9
$N(X)$	9	9	4	11	0	17	12	13	12	9
$\rho(X)\%$	9.37	9.37	4.16	11.46	0	17.71	12.50	13.54	12.50	9.37

(10-15)

From $\rho(X)$, the various average values can be calculated, e.g.

$$\langle X \rangle = 4.95 \quad ,$$

$$C_0 \equiv \langle (\Delta X)^2 \rangle = \langle X^2 \rangle - \langle X \rangle^2 = 7.94 \quad . \tag{10-16}$$

Equations (10-15) and (10-16) come from the statistics of (10-12). The meaning of these results has yet to be understood. For example, why does the digit '4' not appear in any of the 96 households? Another equally interesting and important question is why do the other digits have more or less similar $\rho(X)$ and what accounts for their differences? The answer to the first question is easily discovered but the second question is not easy to answer.

Consider another example. From the data (10-12) we define

$$Y(s) = X(s)X(s+1) \quad , \qquad s = 1, 2, \ldots, N' \quad , \tag{10-17}$$

where $N' = N - 1 = 95$. These are the product of the neighbouring numbers in the sequence of (10-12). The average value of Y is

$$\langle Y \rangle = \frac{1}{N'} \sum_{s=1}^{N'} X(s) X(s+1)$$

$$= 24.05 \quad . \tag{10-18}$$

Now define

$$C_1 \equiv \langle Y \rangle - \langle X \rangle^2 = -0.45 \quad ,$$

$$\frac{C_1}{C_0} = \frac{C_1}{\langle (\Delta X)^2 \rangle} = -0.057 \quad . \tag{10-19}$$

C_1 is the so-called correlation value of the neighbouring numbers. In the same way we can define

$$C_k \equiv \frac{1}{N''} \sum_{s=1}^{N''} X(s) X(s+k) - \langle X \rangle^2 ,$$

$$k = 1, 2, 3, \ldots,$$

$$N'' = N - k . \qquad (10\text{-}20)$$

Of course $C_0 = \langle (\Delta X)^2 \rangle$. Moreover N'' is not N, and hence k must be much smaller than N so that the definitions of the average values are essentially the same. C_k is the correlation function of the distance k. The statistical results show that C_1 is much less than $\langle (\Delta X)^2 \rangle$. (See Eqs. (10-18) and (10-19).) This result means that neighbouring numbers are essentially uncorrelated or independent. According to our 'intuition', if these numbers are 'random', i.e. some $X(s)$ will be larger than $\langle X \rangle$ and some smaller than $\langle X \rangle$, but there is no regularity, the correlations $C_1, C_2 \ldots$ should all be zero. In the next chapter we shall introduce these concepts systematically.

Notice that this correlation function is of course dependent on the order of arrangement of the numbers. The label takes the order of the directory and has no statistical meaning. This ordering is a newly introduced concept and the distance between different s's is determined by this ordering.

Our aim is to use statistics to deal with the information on the molecular motion, and many experiments are performed to measure the average value of the various variables of motion. Our statistical procedure is based on the molecular motion and the average values defined by various probabilities must be defined in terms of the molecular motion. The ordering of time and spatial position is a basic concept. Various correlation functions are most important tools in discussing the regularity of the motion. Let us now look at a slightly different example.

Let us consider a vertical xy plane, with gravity acting downwards, the y axis pointing upwards and the x axis horizontal. A particle moves in this plane above the x-axis, which is a rigid floor from which the particle will rebound without loss of energy. The horizontal motion is restricted to $0 < x < L$. For simplicity we identify the two points $x = 0$ and $x = L$. Hence the motion of this particle is on a cylindrical surface with period L. This is the periodic or cyclic boundary condition (Fig. 10-1).

The motion of this particle obeys the equation of a motion of mechanics. The trajectory is very simple. The horizontal motion is a circle around the

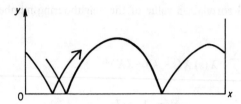

Fig. 10-1 Trajectory of a particle in 2 dimensions under a constant gravitational field.

cylinder and the vertical motion is up and down incessantly. Any property can be calculated from this trajectory and the observation time is $0 < t < \mathfrak{J}$.

We first ask: What is the distribution of height of the particle? This problem is not clear because a distribution must have a precise statistical meaning and we have not made this definition yet. In the usual statistics about the variables of motion, we use equal time intervals to be the statistical weight, because the results of experiments are usually the time average. We now calculate the time the particle spends in a certain height and define it to be the statistical weight of that height. Let

$$\rho(y') = \frac{1}{\mathfrak{J}} \int_0^{\mathfrak{J}} dt\, \delta(y' - y(t)) \quad, \tag{10-21}$$

where $y(t)$ is the height at time t and $\rho(y')$ is the probability distribution of the height. Equation (10-21) is an example of the general definition in (10-7). Here the sample space is the trajectory. At each instant t there is one sample, i.e. the state of the particle at t. The probability $\rho(y')$ can be easily calculated ($\rho \propto |dt/dy|$, i.e. ρ is inversely proportional to the vertical velocity).

$$\rho(y) = \tfrac{1}{2}\left[(h-y)h\right]^{-\frac{1}{2}}\left[1 + O(\tau/\mathfrak{J})\right] \quad, \qquad \text{if } y < h \quad,$$

$$= 0 \quad, \qquad \text{if } y > h \quad, \tag{10-22}$$

where h is the maximum height attained by the particle. The reader can derive (10-22) for himself. Here τ is the period of the vertical motion:

$$\tau = \frac{2 p_{y_0}}{mg} \quad, \tag{10-23}$$

and p_{y_0} is the perpendicular momentum at $y = 0$.

If $\mathfrak{J} \gg \tau$, then $O(\tau/\mathfrak{J})$ can be neglected. The integration of $\rho(y)$ is

$$\int_{y_1}^{y_2} \rho(y) \, dy = \mathfrak{J}(y_1 < y < y_2)/\mathfrak{J}$$

$$= P(y_1 < y < y_2) \quad , \qquad (10\text{-}24)$$

can be defined as the probability that the height is between y_1 and y_2. $\mathfrak{J}(y_1 < y < y_2)$ is the time spent by the particle between these heights.

The correlation function of the height (similar to C_k in (10-20)) is

$$C(t') = \frac{1}{\mathfrak{J}} \int_0^{\mathfrak{J}} dt \, y'(t) \, y'(t+t') \quad,$$

$$y'(t) \equiv y(t) - \langle y \rangle \quad, \qquad t' \ll \mathfrak{J} \quad, \qquad (10\text{-}25)$$

which is not difficult to calculate (see Fig. 10-2):

$$C(t') = \frac{g\tau^2}{8} \left[\frac{4}{25} - \frac{2}{3} t''^2 + \frac{2}{3} t''^3 - \frac{1}{6} t''^4 \right] \quad,$$

$$t'' = t'/2\tau \quad, \qquad 0 < t' < \tau \quad. \qquad (10\text{-}25')$$

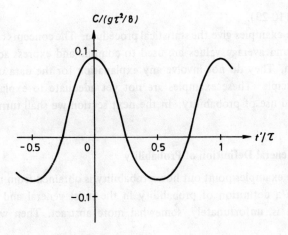

Fig. 10-2 Graph of $C(t')$. $C(t')$ is a periodic function, because $y(t)$ is a periodic function.

Next consider the average force acting on the floor between the time $0 < t < \mathcal{J}$. When the particle is in the air, the floor is not subject to any force. When the particle hits the floor, the momentum change is $2p_{y_0}$. If the particle hits the floor once in every time interval, then the n-th time occurs at $n\tau$. The force acting on the floor is

$$W(t) = \sum_{n=-\infty}^{\infty} 2p_{y_0}\, \delta(t - n\tau) \quad . \tag{10-26}$$

The average force is

$$\langle W \rangle = \frac{1}{\mathcal{J}} \int_0^{\mathcal{J}} dt\, W(t)$$

$$= 2p_{y_0} \frac{N}{\mathcal{J}}$$

$$= \frac{2p_{y_0}}{\tau}\left(1 + O\!\left(\frac{\tau}{\mathcal{J}}\right)\right) \quad , \tag{10-27}$$

where N is the number of times the particle hits the floor. If $\mathcal{J} \gg \tau$, $O(\tau/\mathcal{J})$ can be neglected:

$$\langle W \rangle = 2p_{y_0}/\tau = mg \quad , \tag{10-28}$$

τ is given by (10-23).

The above examples give the statistical procedure. The concepts of probability distribution and average values are used to compile and express accurately the available data. They do not involve any explanation for the data or matters of physical principle. These examples are not yet adequate to explain the usual definition and use of probability. In the next section we shall turn to a general discussion.

10.3. The General Definition of Probability

The above examples point out how probability is obtained from the statistics. Now we give a definition of probability in the most general and widely used terms, which is, unfortunately, somewhat more abstract. Then we give some examples.

To define probability, we require three things: (a) sample set; (b) classes and (c) the statistical weight of each class. These should satisfy some conditions:

(a) The sample set E must be nonempty. (We have at least one sample.)

(b) Each class is a subset of E and each subset is a class. Classes should satisfy the following conditions: if R_1 is a class and R_2 is another then their union $R_1 \cup R_2$ is also a class. Their intersection $R_1 \cap R_2$ is also a class. In addition, if R is a class then its complement is also a class.

(c) Each class is assigned a statistical weight $P(R)$, satisfying

$$P(R) \geqslant 0 \quad ,$$

$$P(R_1 \cup R_2) = P(R_1) + P(R_2) - P(R_1 \cap R_2) \quad ,$$

$$P(E) = 1 \quad .$$

The last condition $P(E) = 1$ is the normalisation of the statistical weights. If it is not satisfied, we only need to divide each $P(R)$ by $P(E)$. The above definition is very general. Any function $P(R)$ satisfying the above conditions is a probability. This is the mathematical definition of probability and is the starting point of modern probability theory.[a] Notice that this general definition does not mention statistics, although the probability defined by statistics always satisfies this general definition. In many applications of probability, the statistical meaning is always ignored and regarded as obvious. But sometimes problems may arise. Our main point here concerns the statistics. Hence in applying the theory of probability we should always bear in mind the statistical meaning of various definitions and conclusion.

Example 1: Throwing a die

Let the sample set be $E = \{1, 2, 3, 4, 5, 6\}$. The classes are all the subsets of E. The probability $P(R)$ is the number of samples in R divided by 6. The definition of this probability does not involve statistics but we know its statistical meaning. Notice that the sample here is not the experimental recordings of the example in the last section; nor the trajectory of motion but an abstract set of six possibilities. We can think of throwing a die many times and classify the events into six classes. The six samples of E are these six classes.

Example 2: Throwing a die

Using the same E, we use another classification (R_1, R', E):

$$R = \{1, 2, 3\} \quad , \qquad R' = \{4, 5, 6\} \quad ,$$
$$P(R) = 0.55 \quad , \qquad P(R') = 0.45 \quad . \tag{10-29}$$

---

[a] The mathematical analysis of probability is discussed in Yang (1979).

The probability thus defined also satisfies the conditions (a), (b) and (c). This probability can be used on a biased die. This example indicates that classification need not be too refined, and it only needs to satisfy (b). The subsets in this example are fewer than in the above, subset like $\{1, 3\}$, $\{2, 5, 6\}$, $\{2\}$ being excluded.

Example 3:

Take the set of all integers: what is the probability of obtaining an even integer? The answer obviously is $\frac{1}{2}$. What is the probability of getting a multiple of 10? Answer: $\frac{1}{10}$. The answers seem obvious. The statistical meaning is that from 0 to N (N is a large number) about $N/2$ are even integers and $N/10$ are multiples of 10.

Example 4:

We draw any chord on a circle of radius 1, what is the probability of the length of the chord being less than 1?

This question seems simple but is meaningless, because there is no obvious statistical meaning in drawing any chord. If we draw ten thousand chords, how do we draw them? Are they parallel and uniformly distributed in the circle? Or are they uniformly distributed on the circle? Each situation will give a different probability. Generally speaking, a continuous-valued sample space does not have obvious statistical weights.

Now let us look at two important examples.

Example 5:

The distribution of time in a trajectory. Let E be the region of the trajectory in the phase space of a system, i.e. the region of motion.

Now we define the states in the region as the samples and classify the states by a certain variable of motion $X(s)$, $s \in E$. If R is a class, then define the probability of R as

$$P(R) = \mathcal{T}_R / \mathcal{T} \quad , \tag{10-30}$$

where R is a subset of E i.e. a region inside the region of motion and $\mathcal{T}_R$ is the time the trajectory spends in R, where $\mathcal{T}$ is the total observation time.

Example 6:

The situation is as above but we now define the probability as

$$P'(R) = \frac{\Gamma(R)}{\Gamma(E)} \quad , \tag{10-30'}$$

where $\Gamma(R)$ is the volume of R, i.e. the total number of states in R and $\Gamma(E)$ is the total number of states in the region of motion.

Notice that (10-30) is based directly on the trajectory of motion, while (10-30′) does not mention the trajectory (of course E itself is determined by the trajectory). Remember that in Chapter 5 on the basic assumption we used quantum mechanics to make the region of motion into a discrete rather than a continuous sample space. Hence $\Gamma(E)$ and $\Gamma(R)$ are not ambiguous. Of course, whether the states are continuous or discrete, Γ_R and Γ have definite meanings. From the general definition of probability we can define the average values. Let

$$X_0 < X_1 < X_2 < X_3 \ldots < X_n \quad ,$$

$$R_k = \{s; X_{k-1} < X(s) < X_k\} \quad , \qquad k = 1, 2, 3, \ldots, n \quad .$$

(10-31)

That is to say all the samples with $X(s)$ between X_{k-1} and X_k are collected in R_k. X_0 should be smaller than the minimum of $X(s)$ and X_n should be larger than its maximum. Let

$$X_k - X_{k-1} = \mathrm{d}X \quad , \qquad k = 1, 2, \ldots, n$$

(10-32)

be a small number. Then the average value $\langle X \rangle$ and distribution function $\rho(X)$ of X are

$$\langle X \rangle = \sum_{k=1}^{n} X_k P(R_k) = \int \mathrm{d}X \, \rho(X) X \quad ,$$

$$\rho(X_k) \mathrm{d}X \equiv P(R_k) \quad .$$

(10-33)

Example 7:

Use (10-30) as the definition of probability. Let $X(s)$ be a macroscopic variable, (whose value is of order N). Then

$$\rho(X) = \frac{1}{\sqrt{2\pi K}} e^{-(X-\bar{X})^2/2K} \quad .$$

(10-34)

The definitions of X and K are like those of A and K_A in the last section of Chapter 6. Equation (10-34) is just (6-86) with $A = X$ and $K_A = K$.

Here we must emphasise that a classification is a collection of subsets of the sample space but does not necessarily include all the subsets; it only needs to satisfy (b). (The second example of throwing a die demonstrates this point.)

If we classify the region of motion according to the values of macroscopic variables, then each class R is a large region within the whole region, i.e. $\Gamma(R)$ and $\Gamma(E)$ are of order e^N so that $\ln \Gamma(R)$ and $\ln \Gamma(E)$ are of order N. Hence the definition of (10-30) for $P(R)$ can have a clear equilibrium meaning. The reason is as follows.

The trajectory is the set of states actually occuring in the observation period. The total number of states in a trajectory is determined by the motion. If in each interval τ, each molecule changes its state once, then the total number of states in the trajectory is about

$$N\mathcal{J}/\tau \quad . \tag{10-35}$$

This is negligible when compared with $\Gamma(E) \sim e^N$. Therefore, if R is a very small subset, the trajectory may not pass through R and $\mathcal{J}_R$ is zero. If it accidentally passes through R in a certain instant, $\mathcal{J}_R \neq 0$. This kind of probability is of no use for the analysis of equilibrium properties. But if the volume of R is comparable with E, then the time the trajectory stays in R is comparable with $\mathcal{J}$, and $P(R)$ will not depend on changes of $\mathcal{J}$, thus becoming an equilibrium property. Hence the classification indicated in (10-30) only includes this kind of large regions.

Problems

1. A street lamp is at a height h above the ground. Find the distribution of the intensity of illumination on the ground.

2. There is a small hole of radius a on the wall of a gas container. The wall is of thickness b. Calculate the distribution of the velocity of the leaking molecules. Discuss the influence of a, b and the mean free distance on the answer.

3. Change of variables.

If we know the distribution $\rho(x)$, and $x' = f(x)$, what is the distribution of $\rho(x')$? Answer:

$$\rho(x') = \langle \delta(x' - f(x)) \rangle$$

$$= \int dx\, \rho(x)\, \delta(x' - f(x)) \quad . \tag{10-36}$$

Notice that each x' may have several x's satisfying $x' = f(x)$. This must be taken into account.

Let $r = (x, y)$ be distributed uniformly in the square $(0 < x < 1, 0 < y < 1)$, i.e. $\rho(x, y) = 1$. Calculate the distribution $\rho(x, y)$ in the polar coordinates as well as $\rho(r)$ and $\rho(\theta)$.

4. Take the last 4 digits from 100 numbers in a directory and call them $x_1, x_2, \ldots, x_N$, $N = 100$.
   (a) Calculate $\langle x \rangle$, $\langle x^2 \rangle$.
   (b) Calculate C_k, $k = 0, 1, 2$.
   (c) Discuss the randomness of this sequence.

5. Use a calculator to calculate the coordinates of the particle bouncing on the floor in the last example of Sec. 10.2. Calculate the distribution of these points. This distribution naturally is dependent on the relative magnitude of L and l where $l \equiv v_x \tau$, v_x is the horizontal velocity, and τ is the period of the vertical motion. (See Eq. (10-23).)

Chapter 11

INDEPENDENCE AND CHAOS

Independence means an absence of interference between phenomena, which therefore show no correlation of causal relation, a situation often described by the words chaos or randomness. Statistics provides a quantitative method of analysis for such phenomena. Independence is a very important phenomenon and it can simplify many problems; however its origin is not clear. This chapter discusses the definition of independence and gives examples to illustrate its effects. Many variables of motion have essentially mutually independent fluctuations but the test of independence is by no means simple. The correlation value is a very important concept and it indicates the degree of independence of the different variables. The correlation value of many variables of motion can be measured by scattering experiments, as discussed in the last section in this chapter.

11.1. The Definition and Consequence of Independent Phenomena

Let us first review two definitions:

The joint probability of R_1 and R_2 is denoted by

$$P(R_1 \cap R_2) \equiv P(R_1, R_2) \ . \tag{11-1}$$

Given condition R_2, the probability of R_1 is

$$P(R_1 | R_2) \equiv \frac{P(R_1 \cap R_2)}{P(R_2)} \ . \tag{11-2}$$

The condition of given R_2 means that we take R_2 as the sample set and then classify within R_2. From these definitions we can determine the joint distribution $\rho(X, Y)$ and the conditional distribution $\rho(X|Y)$ of two variables X and Y. The definitions are similar to (10-31) and (10-33). We classify the sample set with values of (X, Y) [see (10-31) and (10-33)]. Let

$$R_k = \{s; X_{k-1} < X(s) < X_k\} ,$$

$$Q_l = \{s; Y_{l-1} < Y(s) < Y_l\} . \tag{11-3}$$

Define

$$\rho(X_k, Y_l) dX dY \equiv P(R_k \cap Q_l) ,$$

$$\rho(X|Y) \equiv \rho(X, Y)/\rho(Y) . \tag{11-4}$$

The reader can prove that

$$\rho(X) = \int dY \, \rho(X, Y) . \tag{11-5}$$

In a similar manner we can define the joint distribution of more variables and their conditional distribution.

The definition of independence is as follows: If

$$P(R_1, R_2) = P(R_1) P(R_2)$$

or

$$P(R_1|R_2) = P(R_1) , \tag{11-6}$$

then R_1 and R_2 are mutually independent, i.e. R_2 does not affect R_1, or vice versa.

The definition of independence for the variables X and Y is

$$\rho(X, Y) = \rho(X) \rho(Y) \tag{11-7}$$

or

$$\rho(X|Y) = \rho(X) .$$

That is, the statistical data on X does not depend on Y. If X and Y are mutually independent, then any functions f and g will satisfy

$$\langle f(X) g(Y) \rangle = \langle f(X) \rangle \langle g(Y) \rangle . \tag{11-8}$$

Let the correlation function of X and Y be

$$C \equiv \langle XY \rangle - \langle X \rangle \langle Y \rangle$$
$$= \langle (X - \langle X \rangle)(Y - \langle Y \rangle) \rangle \quad . \tag{11-9}$$

Hence if X and Y are mutually independent, then $C = 0$.

Independence is a characteristic phenomenon. It seems simple but its origin is not obvious. If there is no such phenomenon, the application of probability will be severely limited. Many applications in probability merely manipulate the combinations or permutations of some simple, independent variables to obtain the distribution or average values for other more complicated variables. Independent phenomenon is chaotic or random. The correlation function is a measure of the degree of regularity between the variables. We must emphasise that independence must be tested by statistical data. It is not *a priori* obvious. To test whether several variables are mutually independent is not easy. We shall discuss it in the next section. Now consider several applications of some independent phenomena.

Example 1:

N gas molecules are in a container of volume V. Let v be the volume of a small region, and n the number of molecules in this region. Calculate $\rho(n)$, the distribution of n. Define n_i as follows: $(i = 1, 2, \ldots, N)$. If the i-th particle is in the small region, then

$$n_i = 1 \quad . \tag{11-10}$$

Otherwise $n_i = 0$. Therefore $n = \sum_{i=1}^{N} n_i$.

Now let us assume that (1) the n_i's are mutually independent and (2) the probability of $n_i = 1$ is

$$p = \langle n_i \rangle = \frac{v}{V} \quad . \tag{11-11}$$

From these two assumptions we get

$$\rho(n) = \binom{N}{n} p^n (1-p)^{N-n} \quad . \tag{11-12}$$

(This is the binomial distribution.) If N, V are both large and the density is

$N/V = \alpha$, then (11-12) can be simplified as

$$\rho(n) = \frac{(\alpha v)^n}{n!} e^{-\alpha v} \quad . \tag{11-13}$$

(This is the Poisson distribution.)

Example 2:

In a gas of density α, what is the distribution of distance r between two molecules?
Answer:

$$\rho(r) = 4\pi r^2 \alpha e^{-4\pi r^3 \alpha/3} \quad , \tag{11-13'}$$

where $\alpha 4\pi r^2 dr$ is the probability of having one molecule between r and $r + dr$ while $e^{-4\pi r^3 \alpha/3}$ is the probability of having no other molecules in r.

Example 3:

A long-chain molecule (free to move in water) is composed of N links, each of length l. Suppose the direction of each link is an independent variable, distributed uniformly. Calculate $\langle R^2 \rangle$, where $\mathbf{R}$ is the relative position vector of the first and the last molecules.

$$\mathbf{R} = \mathbf{r}_1 + \mathbf{r}_2 + \ldots + \mathbf{r}_N \quad ,$$

$$|\mathbf{r}_i| = l \quad ,$$

$$\langle R^2 \rangle = \sum_{i,j} \langle \mathbf{r}_i \cdot \mathbf{r}_j \rangle = \sum_i \langle \mathbf{r}_i^2 \rangle = Nl^2 \quad . \tag{11-14}$$

The value $\sqrt{\langle R^2 \rangle} = \sqrt{Nl^2}$ may be regarded as the size of this molecule.

Let us now look at a more complicated example.

Example 4:

Let $\mathbf{u} + \mathbf{R}$ be the position of an atom in a crystal where $\mathbf{R}$ is its equilibrium position, i.e. $\langle \mathbf{u} \rangle = 0$. Calculate $\langle \mathbf{u}^2 \rangle$. The displacement $\mathbf{u}$ is the linear combination of the fundamental modes of vibrations $\boldsymbol{\phi}_j$:

$$\mathbf{u} = \sum_j \eta_j \, \boldsymbol{\phi}_j(\mathbf{R}) \quad , \tag{11-15}$$

$$\boldsymbol{\phi}_j(\mathbf{R}) = \frac{1}{\sqrt{N}} \mathbf{e}_j e^{i\mathbf{k}_j \cdot \mathbf{R}} \quad . \tag{11-16}$$

The fundamental vibration is a plane wave, $\mathbf{e}_j$ is the polarisation, $\mathbf{k}_j$ is its wave vector, and N is the total number of atoms. The variable of motion is η_j. Let us assume that the η_j's are mutually independent. Then

$$\langle u^2 \rangle = \sum_j \langle \eta_j^2 \rangle |\phi_j|^2 = \frac{1}{N} \langle A \rangle \quad,$$

$$A \equiv \sum_i \eta_i^2 \quad. \tag{11-17}$$

The value A is a large number. We can use the methods of Chapters 5 and 6 to calculate it. The total energy of vibration is

$$H = \sum_i \left(\frac{p_i^2}{2m} + \frac{m \omega_i^2 \eta_i^2}{2} \right) \quad, \tag{11-18}$$

where m is the mass of the atom and ω_i is the frequency of the i-th element. Because each frequency is shared by many modes, we can directly use the result of the equipartition of energy (if we assume the temperature is high so that we can ignore quantum effects), and we get

$$\frac{1}{2} N_\alpha T = \sum_{i=1}^{N_\alpha} \frac{1}{2} m \omega_\alpha^2 \eta_i^2 \quad,$$

$$A_\alpha = \sum_{i=1}^{N_\alpha} \eta_i^2 = \frac{N_\alpha T}{m \omega_\alpha^2} \quad. \tag{11-19}$$

Now assume that these N_α modes have essentially the same frequency ω_α. Then we can add the different α groups and get

$$A = \sum_\alpha A_\alpha = \sum_\alpha \frac{N_\alpha T}{m \omega_\alpha^2}$$

$$= \int d\omega \, g(\omega) \frac{T}{m \omega^2} V \quad, \tag{11-20}$$

where $g(\omega)$ is the distribution of the vibrational frequencies; see Chapter 6, (6-36). Substituting into (11-17), we get

$$\langle u^2 \rangle = \frac{V}{N}\int d\omega \; g(\omega) \frac{T}{m\omega^2} . \qquad (11\text{-}21)$$

The reader may ask: Why do we bother to go through (11-19) and introduce N_α? Why do we not directly say that each mode satisfies

$$\langle \tfrac{1}{2} m \omega_i^2 \eta_i^2 \rangle = \frac{T}{2} , \qquad \text{(equipartition of energy)}$$

$$\therefore \; \langle \eta_i^2 \rangle = \frac{T}{m\omega_i^2} . \qquad (11\text{-}22)$$

Would that not be simpler? There is nothing wrong with this procedure, but we must be careful here about the meaning of the average value $\langle \ldots \rangle$. The meaning of $\langle \mathbf{u}^2 \rangle$ is a time average of $\mathbf{u}^2$. In the discussion of the equipartition of energy in Chapter 5, we only used the basic assumption to define entropy and then calculated the total energy of N modes and the result was NT. There N must be a larger number and this is a necessary condition for calculation. Hence the meaning of the $\langle \ldots \rangle$ symbol in (11-22) is very unclear.

Equation (11-20) comes from the basic assumption, since it is a large number. Hence there is no point in taking the average $\langle \ldots \rangle$ on the right-hand side of (11-17). If the reader is not clear about this point, he should review the last section of Chapter 6.

Equation (11-21) is a very important and interesting result. It shows that the low frequency vibrations have a significant influence on the displacement of the atoms (because of the factor of $1/\omega^2$ in the integrand). If ω is small, then $g(\omega) \propto \omega^2$ [see (6-42)] and this just cancels $1/\omega^2$. But if the crystal is in a two dimensional space (e.g. a layer of atoms attached to a plane), then $g(\omega) \propto \omega$ and we run into difficulty with this integral. Generally speaking, d dimensional crystals have $g(\omega) \propto \omega^{d-1}$ and (11-21) is essentially

$$\langle \mathbf{u}^2 \rangle \propto \frac{T}{\theta_D} \lambda^2 \int_0^1 \frac{dx}{x} x^{d-2} ,$$

$$\lambda \equiv \frac{h}{\sqrt{2\pi m k \theta_D}} , \qquad (11\text{-}23)$$

where θ_D is the Debye temperature, and λ is the wavelength of the atom at this temperature. The $g(\omega)$ in (11-23) is a generalisation of (6-42) ($x = \omega/\omega_D$). When $d < 2$, $\langle \mathbf{u}^2 \rangle \to \infty$ i.e. the low frequency vibrations cause the displacement of the atoms to increase without bound. This result will be discussed later (in Chapter 29). Besides this, the above result shows that the displacement of the atoms is not very large because θ_D is about 300 K (see Table 6-1), and λ is about 0.5 Å.

11.2. Test of Independence

The examples in the above section obviously demonstrate the importance of the independence property. Mutually independent variables can be considered separately, simplifying many complicated problems. Now we discuss some fundamental problems of independent phenomena.

The terms "independent" and "uncorrelated" do not mean spatial separation or disconnection. For example, we have two unconnected simple pendula oscillating separately. Let their displacements be $X_1(t) = \cos \omega t$, $X_2(t) = \cos(\omega t + \phi)$. These two pendula, according to usual terminology, are independent and uncorrelated. But if they have the same frequency, then (suppose $\langle X_1 \rangle = \langle X_2 \rangle = 0$),

$$C_{12} = \langle X_1 X_2 \rangle = \frac{1}{\mathcal{T}} \int_0^{\mathcal{T}} dt\, X_1(t)\, X_2(t)$$

$$= \frac{1}{2} \cos \phi + O\left(\frac{1}{\omega \mathcal{T}}\right) \quad , \tag{11-24}$$

i.e. X_1 and X_2 are not independent variables (assume $\phi \neq \pi/2$). Hence, separation, lack of connectedness or the absence of interaction do not guarantee independence. In modern terms, if the frequencies of two laser beams are exactly the same, they can interfere.

Notice that (1) we have customarily taken the observation time $\mathcal{T}$ as finite. The magnitude of $\mathcal{T}$ depends on the problem under study. (2) Here we repeat that statistics is the basis for calculating the probability and average values. Different statistical procedures determine different probabilities and average values. Each problem is different and should be treated by its own statistical procedure appropriate to the situation. Many experiments require repeated observation. In addition to the time average, we have also to take a grand average. Perhaps the whole process is repeated on several different bodies, then we take an average again. However, it will pose no problem if we clarify the statistical procedure.

To satisfy the definition of independence in the above section is not a simple task. To test whether a group of variables are independent is very difficult. In fact, the usual practice does not require complete independence. That is to say, even if a group of variables do not completely satisfy the definition of independence, many useful results are not affected too much.

We first look at an example to illustrate that independence is not easy to satisfy.

Example 5:

A sample has 4 balls labelled by $S = 1, 2, 3, 4$. The classes are all the subsets. The probability for each subset is the number of balls in the subset divided by 4. Let X, Y, Z be three variables (three qualities of the sample). The variables have two values $+1$ or -1. We define X to be the colour, $+1$ for yellow, -1 for red; Y to be the weight, $+1$ for heavy, -1 for light and Z to be the hardness, $+1$ for hard, -1 for soft.

The data are listed as follows:

Property \ Ball	1	2	3	4
colour (X)	yellow (1)	yellow (1)	red (-1)	red (-1)
weight (Y)	heavy (1)	light (-1)	heavy (1)	light (-1)
hardness (Z)	hard (1)	soft (-1)	soft (-1)	hard (1)

Obviously, each quality is shared by two of the four balls, therefore

$$\rho(X) = \rho(Y) = \rho(Z) = \frac{1}{2} \quad . \tag{11-25}$$

In addition, of the yellow balls, half are heavy and half are light, half are hard and half are soft. The variables X, Y, Z are pairwise independent, i.e.

$$\rho(X|Y) = \rho(X) \quad ,$$
$$\rho(X|Z) = \rho(X) \quad ,$$
$$\rho(Y|Z) = \rho(Y) \quad , \quad \text{etc.} \tag{11-26}$$

But X, Y, Z are not mutually independent, because

$$\rho(X|Y, Z) \neq \rho(X) \quad . \tag{11-27}$$

A heavy and hard ball can only be yellow and a light and soft one must be yellow too.

In addition a light and yellow ball must be soft and a heavy and red ball must be hard, etc. Therefore, although any two variables among the three are mutually independent, the three variables together will be no longer so.

From this example it is not difficult to see that if there are many variables, the test of independence is a major problem. If we have proved that any two variables are mutually independent, this does not mean that any three variables are mutually independent. For the same reason if we can prove the case for three variables, we have no guarantee for the case of four.

The variables that we want to examine naturally are the variables related to the motion of the molecules. From the above example we can see that if we have too many variables, it is a very difficult job to test their independence. But this test is not always necessary. For example, in the case of the long chain molecule discussed above, we want to calculate $\langle R^2 \rangle$ and we only need

$$\sum_{i \neq j} \langle \mathbf{r}_i \cdot \mathbf{r}_j \rangle \ll N l^2 \quad . \tag{11-28}$$

Then the result $\langle R^2 \rangle = N l^2$ can be established. That is to say, we only require the $\mathbf{r}_i$ to be pairwise independent. There is no need to require the independence of three variables, etc. In many applications, the pairwise independence is sufficient. Of course, there are also many cases requiring more.

Independent phenomena possess a very important property, i.e. the central limit theorem. We shall discuss it in the next chapter. This theorem can be used to estimate the mutual independence in the case of many variables.

11.3. Random Sequences

In Sec. 10.2 (see Eq. (10-12)), we discussed a process of copying out a sequence of numbers from the directory. This sequence of numbers is an example of a random sequence.

The numbers obtained by throwing a die are a sequence of random numbers. We know roughly what is meant by randomness: each case of throwing the die is an independent event. Each throw is uncorrelated with any other throw. The sequence thus obtained is random.

We always mix the term "random" with "at will", or "indefinite". We throw a die at will and do not know a definite result before hand. This way of thinking is not correct. For example if we simply write down a sequence of numbers "at will", then it would probably not be completely random. To obtain a truly random sequence, we must consult a random number table or obtain it from a computer programme. But note the following: A sequence generated by a table or by a programme is definite enough, yet it is "random". However, any sequence written "at will" may not pass the test.

The sequence obtained from a die is random enough. But we must notice that we need a die as our tool. The sequence given by the die cannot be altered at will.

Of course we need to have a precise definition of randomness. We use this definition to check a sequence, determining its degree of randomness. Statistics is the basis of our definition. The sequence must first of all be very long. The definition is very simple: Let $X_1, X_2, X_3, \ldots$ be a certain sequence. If each number is independent of every other number, then this sequence is random. The definition of mutual independence must be determined by statistics: Let

$$\langle f(X) \rangle \equiv \frac{1}{N} \sum_{i=k}^{N+k} f(X_i) \quad . \tag{11-29}$$

This average value must be essentially independent of N and k. Here N is a large number. Assume that the sequence is much longer than N. Let

$$X_i(r) = X_{i+r} \quad , \qquad i = 0, 1, 2, \ldots \quad . \tag{11-30}$$

If

$$\frac{1}{N} \sum_{i=k}^{N+k} f(X_i) \, g(X_{i+r}) \equiv \langle f(X(0)) \, g(X(r)) \rangle$$

$$= \langle f(X) \rangle \langle g(X) \rangle \quad , \qquad r = 1, 2, 3, \ldots, \tag{11-31}$$

then each $X(r)$ is independent and the sequence is random. Here f, g are any two functions. In many applications, we only need to consider the correlation function

$$C(r) \equiv \langle X(r) \, X(0) \rangle - \langle X \rangle^2 \quad . \tag{11-32}$$

If
$$r \neq 0, \quad C(r) = 0, \qquad (11\text{-}33)$$
then the sequence can be regarded as random.

We write a sequence of numbers at will. This will certainly pass the test (11-31) or (11-33). The sequence in (10-12) can more or less pass the test. (See Eq. (10-19).) Sequences associated with statistical mechanics are often functions of time, i.e. the various variables of motion of the system. Although it is not a discrete sequence, its randomness can still be defined. For example, if $n(t)$ is the number of molecules in a certain small region of the gas, we can define its time correlation functions:

$$\begin{aligned}
C(t) &\equiv \langle n(t)\, n(0) \rangle - \langle n \rangle^2 \\
&\equiv \frac{1}{\mathcal{J}} \int_0^{\mathcal{J}} dt'\, [n(t'+t)\, n(t') - \langle n \rangle^2], \\
\langle n \rangle &\equiv \frac{1}{\mathcal{J}} \int_0^{\mathcal{J}} dt'\, n(t').
\end{aligned} \qquad (11\text{-}34)$$

The observation time $\mathcal{J}$ must be suitably long. If $C(t)$ is only nonzero for $t \leq \tau$, i.e.

$$C(t) = 0, \quad t \gg \tau, \qquad (11\text{-}35)$$

then we call τ the correlation time.

On a time scale larger than τ, $n(t)$ is random, i.e. the value of $n(t)$ at different times are independent.

According to the results of many experiments the variables for an equilibrium system all have this property of independence between different times. (The origin of this property is still not clear.) Because of the development of the computer, random sequences of enormous length (up to millions) can be obtained. We can make use of these sequences to simulate the various variables in the molecular motion in order to calculate the equilibrium properties. This kind of calculation shall be discussed in Chapter 22.

11.4. Scattering Experiments and the Measurement of the Correlation Functions

If we want to observe some details of the motion of a system, we can shoot in some particles (e.g. photons, neutrons or electrons) and measure the momentum, energy and the angular distribution of the spins of the particles scattered by the system. From the result of the measurements, we can deduce certain details of the structure of the system. The scattering experiments directly measure some correlation values. Let us now review some basic ideas.

Suppose we use neutrons to irradiate a certain body. The influence of the molecules of the body on the neutron is represented by a potential energy $U(\mathbf{x}, t)$. Assume this interaction is

$$U(\mathbf{x}, t) = \sum_i v(\mathbf{x} - \mathbf{r}_i(t)) \qquad (11\text{-}36)$$

where $\mathbf{x}$ is the position of the neutron, $\mathbf{r}_i(t)$ is the position of the i-th molecule at time t, and the interaction of the neutron with one molecule is $v(\mathbf{x} - \mathbf{r})$, where $\mathbf{x} - \mathbf{r}$ is the relative distance between the neutron and the molecule. Then $U(\mathbf{x}, t)$ is the interaction of the neutron with all the molecules. Assume that the positions of the molecules $\mathbf{r}_i(t)$ are known and we neglect the spin of the neutron. The wave equation for the neutron is

$$i \frac{\partial}{\partial t} \psi(\mathbf{x}, t) = -\frac{1}{2m} \nabla^2 \psi(\mathbf{x}, t) + U(\mathbf{x}, t) \psi(\mathbf{x}, t) \quad, \qquad (11\text{-}37)$$

where m is the mass of the neutron. Let $a_\mathbf{q}(t)$ be the coefficient of expansion of the plane wave ψ:

$$\psi(\mathbf{x}, t) = \sum_{\mathbf{q}'} a_{\mathbf{q}'}(t) \, e^{-i\epsilon' t} \, e^{i\mathbf{q}' \cdot \mathbf{x}} \quad,$$

$$\epsilon' \equiv q'^2 / 2m \quad. \qquad (11\text{-}38)$$

Let the momentum of the incoming neutron be $\mathbf{q}$. When $t = 0$,

$$\psi(\mathbf{x}, 0) = e^{i\mathbf{q} \cdot \mathbf{x}} \quad. \qquad (11\text{-}39)$$

Assume that U is very weak, then (11-37) can be written as

$$i \frac{\partial}{\partial t} a_{\mathbf{q}'}(t) = \langle \mathbf{q}' | U | \mathbf{q} \rangle \, e^{i\omega t} + O(U^2) \quad, \qquad (11\text{-}40)$$

$$\langle \mathbf{q}' | U | \mathbf{q} \rangle = \int d^3 x \, e^{-i\mathbf{k} \cdot \mathbf{x}} \, U(\mathbf{x}, t)$$

$$\equiv U_k(t) \quad, \qquad (11\text{-}41)$$

$$\mathbf{k} \equiv \mathbf{q}' - \mathbf{q} ,$$

$$\omega \equiv \epsilon' - \epsilon = (q'^2 - q^2)/2m$$

$$= (\mathbf{q} + \mathbf{q}') \cdot \mathbf{k}/2m . \tag{11-42}$$

where $\mathbf{k}$ is the momentum transferred from the neutron and ω is the energy. Hence, in time $\mathcal{J}$, the amplitude that the neutron changes its momentum to $\mathbf{q}'$ is the solution of (11-40)

$$a_{\mathbf{q}'}(\mathcal{J}) = -i \int_0^{\mathcal{J}} dt \, e^{i\omega t} \, U_{\mathbf{k}}(t) . \tag{11-43}$$

Its square divided by $\mathcal{J}$ is the scattering probability per unit time:

$$P(\mathbf{q} \to \mathbf{q}') = |a_{\mathbf{q}'}(\mathcal{J})|^2/\mathcal{J}$$

$$= \frac{1}{\mathcal{J}} \int_0^{\mathcal{J}} dt_1 \int_0^{\mathcal{J}} dt_2 \, e^{i\omega(t_1 - t_2)} \, U_{\mathbf{k}}^*(t_2) \, U_{\mathbf{k}}(t_1)$$

$$\equiv P(\mathbf{k}, \omega) . \tag{11-44}$$

The probability $P(\mathbf{q} \to \mathbf{q}')$ can be obtained by experiment. Different values of $\mathbf{q}, \mathbf{q}'$ can lead to different $\mathbf{k}, \omega$. From $P(\mathbf{k}, \omega)$ we can calculate some correlation functions. We first look at the result of integration with respect to ω.

A. *Equal time correlation value*

Because

$$\int_{-\infty}^{\infty} d\omega \, e^{i\omega t} = 2\pi \delta(t) ,$$

the integral of (11-44) with respect to ω is

$$\int_{-\infty}^{\infty} d\omega \, P(\mathbf{k}, \omega) = \frac{2\pi}{\mathcal{J}} \int_0^{\mathcal{J}} dt \, U_{\mathbf{k}}^*(t) \, U_{\mathbf{k}}(t)$$

$$\equiv 2\pi \langle U_{\mathbf{k}}^* \, U_{\mathbf{k}} \rangle . \tag{11-45}$$

Unless otherwise specified, the bracket $\langle \ldots \rangle$ means time average. The number $U_\mathbf{k}$ is directly related to the density of the molecules. From (11-36) we get

$$U_\mathbf{k}(t) = v_\mathbf{k}\, \rho_\mathbf{k}(t) \quad,$$

$$v_\mathbf{k} := \int d^3\mathbf{x}\, e^{-i\mathbf{k}\cdot\mathbf{x}}\, v(\mathbf{x}) \quad,$$

$$\rho_\mathbf{k}(t) = \int d^3\mathbf{x}\, e^{-i\mathbf{k}\cdot\mathbf{x}}\, \rho(\mathbf{x},t) \quad,$$

$$\rho(\mathbf{x},t) = \sum_i \delta(\mathbf{x} - \mathbf{r}_i(t)) \quad. \tag{11-46}$$

Therefore from (11-45) we obtain

$$A \equiv \frac{1}{2\pi |v_\mathbf{k}|^2} \int d\omega\, P(\mathbf{k},\omega) = \langle \rho_\mathbf{k}^* \rho_\mathbf{k} \rangle \quad. \tag{11-47}$$

Let

$$\langle \rho_\mathbf{k}^* \rho_\mathbf{k} \rangle \equiv |\langle \rho_\mathbf{k} \rangle|^2 + \langle \rho_\mathbf{k}^* \rho_\mathbf{k} \rangle_c \quad,$$

$$\langle \rho(\mathbf{x}) \rho(\mathbf{x}') \rangle \equiv \langle \rho(\mathbf{x}) \rangle \langle \rho(\mathbf{x}') \rangle + \langle \rho(\mathbf{x}) \rho(\mathbf{x}') \rangle_c \quad, \tag{11-48}$$

then

$$A = |\langle \rho_\mathbf{k} \rangle|^2 + \int d^3\mathbf{x}\, d^3\mathbf{x}'\, e^{-i\mathbf{k}\cdot(\mathbf{x}-\mathbf{x}')} \langle \rho(\mathbf{x})\rho(\mathbf{x}') \rangle_c \quad. \tag{11-49}$$

Hence, the correlation function of the fluctuation of the density of the molecules can be obtained from scattering experiments. Assume that $\rho(\mathbf{x})$ and $\rho(\mathbf{x}')$ are uncorrelated if $|\mathbf{x} - \mathbf{x}'| \gg \xi$ (the correlation length). Next assume that ξ is a microscopic scale and the body is homogeneous. Then (11-49) can be simplified as

$$A = V \int d^3\mathbf{r}\, e^{-i\mathbf{k}\cdot\mathbf{r}} \langle \rho(\mathbf{r})\rho(0) \rangle_c \quad, \tag{11-50}$$

assuming that $\mathbf{k} \neq 0$. V is the volume of the system.

B. Different time correlation value

If we do not integrate over ω, we can obtain the different time correlation value of the density of the molecules from (11-44). The analysis is as follows:

let $t = (t_1 + t_2)/2$, $t' = t_1 - t_2$, then

$$P(\mathbf{q} - \mathbf{q}') = \frac{1}{\mathfrak{I}} \int dt' \, e^{i\omega t'} \int dt \, U_\mathbf{k}^*(t - \frac{t'}{2}) U_\mathbf{k}(t + \frac{t'}{2}) \, . \qquad (11\text{-}51)$$

See Fig. 11-1 for the integration region of t, t'.

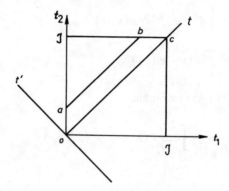

Fig. 11-1

If t' is fixed, the integration of t along the 45° line such as ab in the figure. Define

$$\frac{1}{t_b - t_a} \int_{t_a}^{t_b} dt \, U_\mathbf{k}^*(t - \frac{t'}{2}) U_\mathbf{k}(t + \frac{t'}{2})$$

$$\equiv |\langle U_\mathbf{k} \rangle|^2 + \langle U_\mathbf{k}^* U_\mathbf{k} \rangle_c \, . \qquad (11\text{-}52)$$

The average value is the time average between t_a and t_b. The last term of (11-52) is the correlation value of $U_\mathbf{k}$ with its value after a time t'. Now let us suppose that t' is much longer than the correlation time τ, so that this correlation value tends to zero. In addition assume that $\langle U_\mathbf{k} \rangle$ is approximately a constant. If these assumptions are valid, then $t_b \approx \mathfrak{I}$, $t_a \approx 0$. Substitute (11-52) into (11-51) and we get

$$P(\mathbf{k}, \omega) = 2\pi \delta(\omega) |\langle U_\mathbf{k} \rangle|^2 + \int dt' \, e^{i\omega t'} \langle U_\mathbf{k}^* U_\mathbf{k}(t') \rangle_c \, .$$

$$(11\text{-}53)$$

Using (11-46), we can simplify these equations as

$$\frac{1}{|v_k|^2} P(\mathbf{k}, \omega) = 2\pi \delta(\omega) |\langle \rho_k \rangle|^2$$
$$+ V \int dt \, d^3\mathbf{r} \, e^{i(\omega t - \mathbf{k} \cdot \mathbf{r})} \langle \rho(\mathbf{r}, t) \rho(0, 0) \rangle_c \, .$$

(11-54)

This is a generalisation of (11-49). Equation (11-49) is an integral of this equation with respect to ω. These results connect the scattering amplitude with the density fluctuation. The correlation value is an average with respect to time. The definition is very clear. These results involve neither thermodynamic concepts nor the basic assumption of statistical mechanics.

Problems

1. Let $x_1, x_2, \ldots, x_n$ be independent variables. Calculate the distribution of the product

$$y = x_1 x_2 \ldots x_n \, , \qquad (11\text{-}55)$$

assuming that the distribution of each x_i is normal and $\langle x_i^2 \rangle = 1$, $\langle x_i \rangle = 0$.

2. Calculate the distribution of

$$y = \left(\frac{1}{n} \sum_{i=1}^{n} x_i^2 \right)^{1/2} \, , \qquad (11\text{-}56)$$

where x_i are as in the above problem. (Yang 1979.)

3. Let the sequence $x_1, x_2, x_3, \ldots$ be generated by the formula

$$x_{n+1} = 4\lambda x_n (1 - x_n) \, , \qquad n = 1, 2, 3, \ldots \, . \qquad (11\text{-}57)$$

The values of x_1 and λ are fixed in advance. Use a calculator to find this sequence for $0 < x_1 < 1$ and $0 < \lambda < 1$. If λ is close to 1, this sequence seems to be random. (For further analysis, see Feigenbaum (1978).)

4. Let

$$x_n = \sin(2^n) \, , \qquad n = 1, 2, \ldots \, . \qquad (11\text{-}58)$$

Discuss the randomness of this sequence.

5. Let a random variable x be uniformly distributed in $0 < x < 1$. Calculate the distribution of

$$t \equiv \tau \ln \frac{1}{x} \, , \qquad (11\text{-}59)$$

where τ is a constant. (Answer: $(1/\tau)e^{-t/\tau}$, i.e. an exponential distribution.) Therefore, we can obtain an exponentially distributed random sequence from a uniformly distributed random sequence.

6. The normal distribution will be discussed in detail in Chapter 12. We first make a preparation here. This distribution should be familiar. The velocity distribution of gas molecules is normal:

$$\rho(v_x) = (2\pi\sigma^2)^{-1/2} e^{-v_x^2/2\sigma^2} \quad ,$$

$$\sigma^2 \equiv \langle v_x^2 \rangle = T/m \quad . \tag{11-60}$$

Calculate (a) $\langle y^{2n} \rangle$

(b) $\langle e^{-iky} \rangle$

of a normal distribution $\rho(y)$.

7. From a uniformly distributed sequence to a normally distributed sequence. This is not so simple as the exponential distribution (see problem 5) but it is not too difficult either. The steps are as follows:

 (a) Take the uniformly distributed random sequences $0 < r < 1, 0 < \theta < \pi$.
 (b) Let

$$Z = \left[2 \ln \frac{1}{r} \right]^{1/2} \quad ,$$

$$t = \sigma Z \cos\theta \quad , \tag{11-61}$$

then t is a normally distributed random sequence, with $\langle t^2 \rangle = \sigma^2$, $\langle t \rangle = 0$. Derive the above. Notice that $r = e^{-Z/2}$ and hence Z has a distribution like that of the velocity of the gas molecules in two dimensions, t is the projected velocity in a specified direction. The reader can think of other methods.

8. If (11-36) is changed to

$$U(\mathbf{x}, t) = \sum_i \delta(\mathbf{x} - \mathbf{r}_i(t)) \, \mathbf{s}_i(t) \cdot \boldsymbol{\sigma}(t) \quad , \tag{11-62}$$

i.e. the spin σ of the neutron interacts with the spin $\mathbf{s}_i$ of the molecules, then what is the relation of the scattering amplitude with the spin correlation function of the molecules? Notice that here U, σ are all 2×2 matrices and the wavefunction of the neutron has two components. In the process of scattering some spins change while others do not. The spins of the molecules can be regarded as classical variables.

9. Analyse the content of Sec. 11.4 using quantum mechanics.

10. Prove that

(a) $P(A_1, A_2, A_3, \ldots, A_n)$
$= P(A_1|A_2, A_3, \ldots, A_n) P(A_2|A_3, A_4, \ldots, A_n)$
$\ldots P(A_{n-2}|A_{n-1}, A_n) P(A_{n-1}|A_n) P(A_n)$. (11-63)

(b) Let $A_1 \cup A_2 \ldots \cup A_n = E$ and all classes are disjoint, i.e. $A_i \cap A_j = A_i \delta_{ij}$. Prove that

$$P(A_i|B) = \frac{P(A_i) P(B|A_i)}{\sum_j P(A_j) P(B|A_j)} .$$ (11-64)

11. In an examination, there is a probability p that the student knows the answers. There is a probability $(1-p)$ that he does not know the answer, but simply makes a guess, in which case there is a probability r that the guess is correct. It is found that the answer he gives is indeed correct. What is the probability that the student actually knows the answer?

12. The elements y_{ij} of an $n \times n$ matrix are either $+1$ or -1, with equal probabilities. The different y_{ij}'s are mutually independent. Find the average value of the square of the determinant

$$\xi_n = [\text{Det}(y)]^2 .$$ (11-65)

(The last three problems are taken from Yang (1979), and the solutions may be found therein.)

Chapter 12

SUM OF MANY INDEPENDENT VARIABLES

Independence leads to a very important result, i.e. the central limit theorem. This theorem points out that if N is a large number, then the sum of N essentially independent variables will have an "essentially" normal distribution, with the width proportional to $\sqrt{N}$. We shall carefully discuss the meaning of the word "essentially" later. We use two methods to derive this theorem, the first being a direct proof and the second containing some subtleties. The first gives a general idea of the theorem while the second probes into the essence of the theorem. Then we give examples to illustrate the applications of this theorem. We also present some material about the cumulant, which is very useful in calculations by expansion. In this chapter we use it to analyse the central limit theorem. Then we use this theorem to discuss the basic assumption of statistical mechanics, and extend this assumption so that it is more effective both in principle and application.

12.1. Normal Distributions

The definition of a normal distribution is

$$\rho(Y) = \frac{1}{\sqrt{2\pi\sigma^2}} e^{-Y^2/2\sigma^2} \quad . \tag{12-1}$$

Some of its properties can be seen from Fig. 12-1.

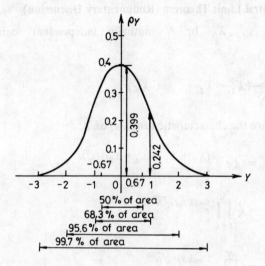

Fig. 12-1 Normal distribution with $\sigma = 1$.

Its properties can be expressed by various average values:

$$\langle Y^2 \rangle = \sigma^2 \quad,$$

$$\langle Y^{2n} \rangle = \frac{(2n)!}{2^n n!} \sigma^{2n} \quad,$$

$$\langle Y^{2n+1} \rangle = 0 \quad, \qquad n = 1, 2, \ldots \quad. \tag{12-2}$$

These results can be directly integrated from (12-1). They can be expressed by a very simple formula

$$f(k) = \langle e^{-ikY} \rangle = e^{-k^2 \sigma^2 / 2} \quad,$$

$$\rho(Y) = \frac{1}{2\pi} \int dk \, e^{ikY} f(k) \quad, \tag{12-3}$$

where $f(k)$ is called the characteristic function of $\rho(Y)$. Expanding this formula in powers of k, we get (12-2).

12.2. The Central Limit Theorem (Rudimentary Discussion)

Let $X_1, X_2, \ldots, X_N$ be N mutually independent variables. Suppose $\langle X_i \rangle = 0$. Let

$$Y \equiv \frac{1}{\sqrt{N}}(X_1 + X_2 + \ldots + X_N) \quad . \tag{12-4}$$

Let us now derive the characteristic function of Y:

$$\langle e^{-ikY} \rangle = \langle e^{-ik(X_1 + X_2 + \ldots + X_N)/\sqrt{N}} \rangle$$

$$= \langle \prod_{j=1}^{N} e^{-ikX_j/\sqrt{N}} \rangle$$

$$= \prod_{j=1}^{N} \langle e^{-ikX_j/\sqrt{N}} \rangle$$

$$= \exp \sum_{j=1}^{N} A_j(k/\sqrt{N}) \quad , \tag{12-5}$$

$$A_j(k/\sqrt{N}) \equiv \ln \langle e^{-ikX_j/\sqrt{N}} \rangle \quad . \tag{12-6}$$

Suppose N is a large number and we expand $A_j(k/\sqrt{N})$, getting

$$A_j(k/\sqrt{N}) \simeq A_j(0) + \frac{k^2}{2N} A_j''(0) + \frac{k^3}{N} O(N^{-\frac{1}{2}}) \quad . \tag{12-7}$$

Notice that the first order term in k is 0, because it is proportional to $\langle X_j \rangle = 0$. The quadratic term is

$$A_j''(0) = - \langle X_j^2 \rangle \quad . \tag{12-8}$$

These can be directly obtained from (12-6). Substituting (12-7) into (12-5), we immediately obtain

$$\langle e^{-ikY} \rangle = \exp\left\{ -\frac{1}{2} k^2 \sigma^2 + O\left(\frac{1}{\sqrt{N}}\right) \right\} \quad ,$$

$$\sigma^2 \equiv \frac{1}{N} \sum_{j=1}^{N} \langle X_j^2 \rangle \quad . \tag{12-9}$$

This is the central limit theorem, i.e. Y is normally distributed, since (12-9) is the same as (12-3) if we neglect terms of $O(1/\sqrt{N})$. If $\langle X_j \rangle \neq 0$, then

$$\langle Y \rangle = \sum_{j=1}^{N} \langle X_j \rangle / \sqrt{N} \quad , \tag{12-10}$$

$Y - \langle Y \rangle$ is a normal distribution and we only need to change Y to $Y - \langle Y \rangle$ in (12-9).

As Y is a normal distribution with width σ, $Y/\sqrt{N}$ is also normal with width $\sigma/\sqrt{N}$ in the variable $Y/\sqrt{N}$:

$$\frac{Y}{\sqrt{N}} = (X_1 + X_2 + \ldots + X_N)/N \quad . \tag{12-11}$$

This can be regarded as the mean of the N different X_i's and if N is large, $\sigma/\sqrt{N}$ is small. That is, the mean differs very little from the average value of this mean

$$\langle (X_1 + X_2 + \ldots + X_N)/N \rangle \quad , \tag{12-12}$$

i.e. the fluctuation is very small. In other words, if we add N independent variables and divide by N, then the fluctuation of the X_i's is largely cancelled. Of course $\sqrt{N} Y = X_1 + X_2 + \ldots + X_N$ has a distribution width of $\sqrt{N} \sigma$. If we add up N independent variables, the increase of fluctuation is only $\sqrt{N}$ not N.

Notice that k in (12-9) cannot be too large. If $k > \sqrt{N}$, then (12-9) cannot be established. Hence the so-called normal distribution of Y is limited to the calculation of the average values of lower powers of Y. Let us now consider some examples.

Example 1: Binomial distribution

In Chapter 11 (see discussion around Eq. (11-10)) we discussed the distribution

$$n = \sum_{i=1}^{N} n_i \quad ,$$

$$n_i = 0 \quad \text{or} \quad 1 \quad . \tag{12-13}$$

The probability of $n_i = 1$ is p and the n_i are mutually independent. The resulting distribution of n is the binomial distribution

$$\rho(n) = \binom{N}{n} p^n q^{n-1} \quad ,$$

$$q = 1 - p \quad . \tag{12-14}$$

Using the central limit theorem, we get

$$\rho(n) \simeq \frac{1}{\sqrt{2\pi N\sigma^2}} e^{-(n-\langle n \rangle)^2/2N\sigma^2} \quad , \tag{12-15}$$

$$\sigma^2 = \frac{1}{N} \sum_{i=1}^{N} (\langle n_i^2 \rangle - \langle n_i \rangle^2)$$

$$= p(1-p) = pq \quad . \tag{12-16}$$

Table 12-1 shows $\rho(n)$ with $N = 2500$ and $p = 0.02$. It tells us how close are (12-14) and (12-15). Notice that

$$\langle n \rangle = 50 \quad ,$$

$$N\sigma^2 = 49 \quad . \tag{12-17}$$

Table 12-1 Values of $\rho(n)$ calculated using (12-14) and (12-15).

n	Eq. (12-14)	Eq. (12-15)
25	0.0000	0.0001
30	0.0006	0.0010
35	0.0052	0.0057
40	0.0212	0.0205
45	0.0460	0.0442
50	0.0569	0.0570
55	0.0424	0.0442
60	0.0199	0.0205
65	0.0061	0.0057
70	0.0013	0.0010
75	0.0002	0.0001

Example 2:

Let the distribution of the variables X_i, $i = 1, 2, \ldots, N$ be

$$\rho(X_i) = \frac{a_i}{\pi(a_i^2 + X_i^2)} , \qquad (12\text{-}18)$$

where each a_i is a positive constant. Find the distribution of the sum of the X_i.

Let the sum be

$$X = X_1 + X_2 + \ldots + X_N . \qquad (12\text{-}19)$$

Because $\langle X_i^2 \rangle$ is infinite, the central limit theorem is not applicable. The reader can prove that the distribution of X is

$$\frac{Na}{\pi(N^2 a^2 + X^2)} ,$$

$$a \equiv \frac{1}{N} \sum_{i=1}^{N} a_i . \qquad (12\text{-}20)$$

12.3. Higher Order Averages and the Cumulant

We first generalise the definition of the correlation value. We have defined before

$$\langle X^2 \rangle = \langle X \rangle^2 + \langle X^2 \rangle_c ,$$

i.e.

$$\langle X^2 \rangle_c = C \equiv \langle X^2 \rangle - \langle X \rangle^2 . \qquad (12\text{-}21)$$

$\langle X^2 \rangle_c$ can be called the second correlation value. We give it another name — the second cumulant. The second correlation value or cumulant $\langle X_i X_j \rangle_c$ of different variables is defined as

$$\langle X_i X_j \rangle = \langle X_i \rangle \langle X_j \rangle + \langle X_i X_j \rangle_c . \qquad (12\text{-}22)$$

If X_i and X_j are mutually independent, then its correlation value is 0 or the cumulant is 0. Equation (12-21) is a special case of (12-22), i.e. the case $i = j$. To generalise, let

$$\langle X_i X_j X_k \rangle = \langle X_i \rangle \langle X_j \rangle \langle X_k \rangle$$
$$+ \langle X_i X_j \rangle_c \langle X_k \rangle + \langle X_i X_k \rangle_c \langle X_j \rangle$$
$$+ \langle X_j X_k \rangle_c \langle X_i \rangle + \langle X_i X_j X_k \rangle_c . \qquad (12\text{-}23)$$

The last term is the third cumulant.

204   STATISTICAL MECHANICS

The case of three variables is much more complicated than that of two. If among the three variables, two are not mutually independent, then the three are not independent. The correlation of X_i, X_j and X_k comes from pairwise correlation, and besides these is the "pure" correlation among the three, i.e. the third cumulant $\langle X_i X_j X_k \rangle_c$. If X_i, X_j, X_k are mutually independent, then any two among the three must be mutually independent i.e. $\langle X_i X_j \rangle_c = \langle X_j X_k \rangle_c = \langle X_k X_i \rangle_c = 0$. But if any two variables are independent, this does not guarantee that the three variables are mutually independent (see Sec. 11-2 of the last chapter). Therefore in discussing the average values of the product of more than two variables, we can regard the correlation as the combination of the products of the cumulants.

The fourth cumulant $\langle X_i X_j X_k X_l \rangle_c$ can be similarly defined as:

$$\begin{aligned}
\langle X_i X_j X_k X_l \rangle &= \langle X_i \rangle \langle X_j \rangle \langle X_k \rangle \langle X_l \rangle \\
&\quad + \langle X_i X_j \rangle_c \langle X_k \rangle \langle X_l \rangle + \ldots \quad \text{(6 terms)} \\
&\quad + \langle X_i X_j \rangle_c \langle X_k X_l \rangle_c + \ldots \quad \text{(3 terms)} \\
&\quad + \langle X_i X_j X_k \rangle_c \langle X_l \rangle + \ldots \quad \text{(4 terms)} \\
&\quad + \langle X_i X_j X_k X_l \rangle_c \quad .
\end{aligned} \qquad (12\text{-}24)$$

The special case when the variables in (12-23) and (12-24) are all the same gives

$$\langle X^3 \rangle = \langle X \rangle^3 + 3 \langle X^2 \rangle_c \langle X \rangle + \langle X^3 \rangle_c \quad ,$$

$$\begin{aligned}
\langle X^4 \rangle &= \langle X \rangle^4 + 6 \langle X^2 \rangle_c \langle X \rangle^2 + 3 \langle X^2 \rangle_c^2 \\
&\quad + 4 \langle X^3 \rangle_c \langle X \rangle + \langle X^4 \rangle_c \quad .
\end{aligned} \qquad (12\text{-}25)$$

Equations (12-22) to (12-25) are the cumulant expansion of the average value of products. These expansion can be represented by simple diagrams. (See Fig. 12-2.)

These diagrams are helpful for memorising and analysis. Each cumulant is represented by a group of connected points and isolated points represent separate average values.

The expansion of higher products is more lengthy. In the next section we shall discuss the expansion theorem for any product. Let us first look at the role of $\langle X^2 \rangle_c$ in the high order average value. Let $\langle X_i \rangle = 0$. (If $\langle X_i \rangle \neq 0$; we

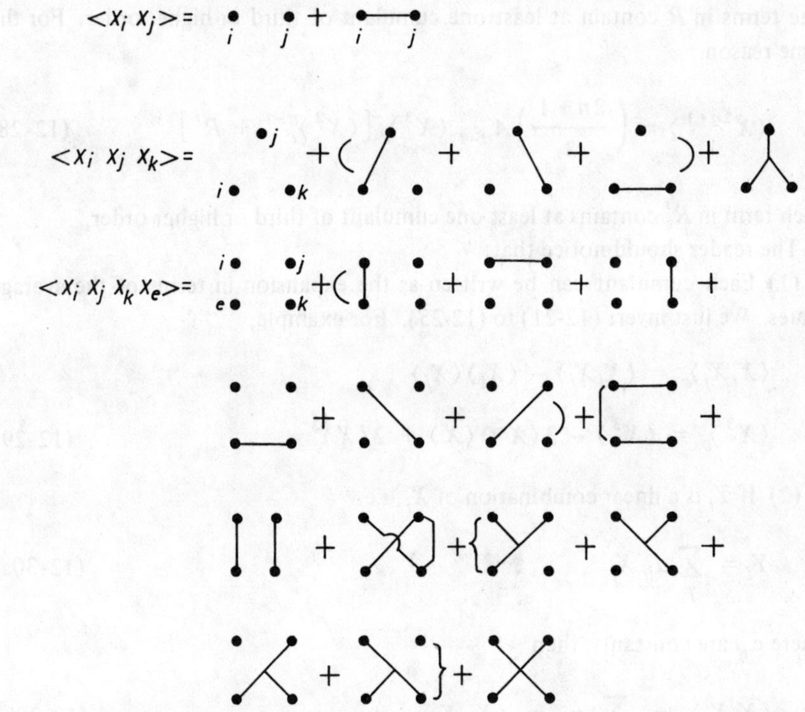

$\langle X_i X_j \rangle =$ [diagram] $+$ [diagram]

$\langle X_i X_j X_k \rangle =$ [diagram] $+ ($ [diagram] $+$ [diagram] $+$ [diagram] $) +$ [diagram]

$\langle X_i X_j X_k X_e \rangle =$ [diagram] $+ ($ [diagram] $+$ [diagram] $+$ [diagram] $+$

[diagram] $+$ [diagram] $+$ [diagram] $) + ($ [diagram] $+$

[diagram] $+$ [diagram] $+$ [diagram] $+$ [diagram] $+$

[diagram] $+$ [diagram] $\} +$ [diagram]

Fig. 12-2 The expansion of the cumulants.

can subtract $\langle X_i \rangle$ from X_i, then the average value of the new variable is 0. Therefore it is always possible to put $\langle X_i \rangle = 0$.) In the expansion of the $2n$ order average values, there are A_n terms containing the second cumulant, where

$$A_n = \frac{(2n)!}{2^n n!} \quad . \tag{12-26}$$

The number A_n is just the number of ways of arranging $2n$ variables in pairs. Therefore, if there is only one variable, then the expansion of $\langle X^{2n} \rangle$ is

$$\langle X^{2n} \rangle = \frac{(2n)!}{2^n n!} \langle X^2 \rangle_c^n + R \quad . \tag{12-27}$$

The terms in R contain at least one cumulant of third or higher order. For the same reason

$$\langle X^{2n+1} \rangle = \left(\frac{2n+1}{3} \right) A_{n-1} \langle X^3 \rangle_c \left[\langle X^2 \rangle_c^{n-1} + R' \right] \quad . \quad (12\text{-}28)$$

Each term in R' contains at least one cumulant of third or higher order.

The reader should notice that:

(1) Each cumulant can be written as the expansion in terms of the average values. We just invert (12-21) to (12-25). For example,

$$\langle X_i X_j \rangle_c = \langle X_i X_j \rangle - \langle X_i \rangle \langle X_j \rangle \quad ,$$

$$\langle X^3 \rangle_c = \langle X^3 \rangle - 3 \langle X^2 \rangle \langle X \rangle + 2 \langle X \rangle^3 \quad (12\text{-}29)$$

(2) If Y_i is a linear combination of X_i, i.e.

$$Y_i = \sum_j \alpha_{ij} X_j \quad , \quad (12\text{-}30\text{a})$$

where α_{ij} are constants, then

$$\langle Y_i Y_j \rangle_c = \sum_{k,l} \alpha_{ik} \alpha_{jl} \langle X_k X_l \rangle_c \quad . \quad (12\text{-}30\text{b})$$

The same is true for higher products. The reader can prove this.

(3) If $Y = f(X)$ is not a linear function of X, then the result like (12-30b) cannot be established. For example,

let $\quad \langle X \rangle = 0 \quad ; \quad Y = X^2$

then $\quad \langle Y^2 \rangle_c = \langle Y^2 \rangle - \langle Y \rangle^2$

$\quad\quad\quad\quad = \langle X^4 \rangle - \langle X^2 \rangle^2$

but $\quad \langle X^4 \rangle_c = \langle X^4 \rangle - 3 \langle X^2 \rangle^2 \quad .$

Hence $\quad \langle Y^2 \rangle_c \neq \langle X^4 \rangle_c \quad . \quad\quad\quad (12\text{-}31)$

That is to say the c in the subscript of $\langle \ldots \rangle_c$ is defined by the variable. If the variable is changed, the definition is also changed. Only in the case of the linear combination (12-30) is it unchanged.

12.4. The Cumulant Expansion Theorem

This theorem states that

$$f_c(k) \equiv \ln \langle e^{-ikX} \rangle$$

$$= \sum_{n=1}^{\infty} \frac{(-ik)^n}{n!} \langle X^n \rangle_c$$

$$= \langle (e^{-ikX} - 1) \rangle_c \quad , \tag{12-32}$$

where $f_c(k)$ can be called the cumulant characteristic function of $\rho(X)$.

The proof of this theorem is as follows: We first expand the characteristic function:

$$f(k) \equiv \langle e^{-ikX} \rangle$$

$$= \sum_{m=0}^{\infty} \frac{(-ik)^m}{m!} \langle X^m \rangle \quad . \tag{12-33}$$

Then we write $\langle X^m \rangle$ as the linear combination of the cumulants as in Fig. 12-1. We draw m points and divide these into groups, and connect all the points within a graph, with each group being disconnected. The linked points of a group is a cumulant. Let m_n be the number of groups with n points, then m_1 groups have 1 point, m_2 groups have 2 points, etc. The sum of the total groups is

$$m = \sum_n n \, m_n \quad . \tag{12-34}$$

The m_n can be zero or a positive integer. If the m_n's are fixed there are still many ways of grouping, namely

$$\frac{m!}{m_1!(1!)^{m_1} \, m_2!(2!)^{m_2} \, \ldots \, m_n!(n!)^{m_n} \, \ldots} \tag{12-35}$$

ways. (Notice that if $m_i = 0$, then 0! is defined as 1.) This result is easily understood. The m points have $m!$ permutations. The interchange of points inside a group does not produce any change. So we have to divide by the $n!$.

208 STATISTICAL MECHANICS

Interchange of the groups with the same number of points also produces no change. So we divide by $m_n!$. Hence

$$\langle X^m \rangle = m! \sum_{m_1, m_2, \ldots} \frac{1}{m_1!} \left(\frac{\langle X \rangle}{1!} \right)^{m_1}$$

$$\cdot \frac{1}{m_2!} \left(\frac{\langle X^2 \rangle_c}{2!} \right)^{m_2} \cdots \frac{1}{m_n!} \left(\frac{\langle X^n \rangle_c}{n!} \right)^{m_n}$$

$$\cdots \delta(m - m_1 - 2m_2 - 3m_3 - \ldots - nm_n - \ldots) \,.$$

(12-36)

Sum over various $m_1, m_2, \ldots$ subject to the condition (12-34). Substitute (12-36) into (12-33) and we get

$$f(k) = \langle e^{-ikX} \rangle$$

$$= \sum_m (-ik)^m \sum_{m_1, m_2, \ldots} \delta(m - m_1 - \ldots - nm_n - \ldots)$$

$$\cdot \frac{1}{m_1! m_2! \ldots m_n! \ldots} \left(\frac{\langle X \rangle}{1!} \right)^{m_1} \cdots \left(\frac{\langle X^n \rangle_c}{n!} \right)^{m_n} \cdots$$

$$= \prod_{n=0}^{\infty} \sum_{m_n=0}^{\infty} \frac{1}{m_n!} \left(\frac{\langle X^n \rangle_c}{n!} \right)^{m_n} (-ik)^{nm_n}$$

$$= \exp\left\{ \sum_{n=1}^{\infty} \frac{(-ik)^n}{n!} \langle X^n \rangle_c \right\}$$

$$= \exp\{ f_c(k) \} \,.$$

(12-37)

Hence (12-32) is proved. This is an extremely important result.

The above theorem can be generalised to the expansion of many variables:

$$\ln \left\langle \exp\left\{-i \sum_\alpha k_\alpha X_\alpha\right\}\right\rangle$$

$$= \left\langle \exp\left\{-i \sum_\alpha k_\alpha X_\alpha\right\} - 1\right\rangle_c \quad . \tag{12-38}$$

This is left as an exercise for the reader.

12.5. Central Limit Theorem (General Exposition)

Now let us examine the cumulants of the normal distribution. The various average values (see Eq. (12-2)) can be compared with (12-27) and (12-28); we can use (12-2) and (12-32) to get

$$f_c(k) = -\frac{1}{2} k^2 \sigma^2 \quad , \tag{12-39}$$

i.e.

$$\langle Y^2 \rangle_c = \sigma^2 \quad ,$$

$$\langle Y^n \rangle_c = 0 \quad , \qquad n = 3, 4, 5, \ldots \quad . \tag{12-40}$$

That is to say, except for the second cumulant, all the other cumulants are zero. This is the essence of the normal distribution.

Now let us return to the central limit theorem. The following derivation is to clarify the conditions of the theorem and its flexibility. Let $X_1, X_2, \ldots, X_N$ be N variables and N is sufficiently large. In addition, suppose

(a) $\langle X_i \rangle = 0$,

(b) $\dfrac{1}{N} \sum_{i,j} \langle X_i X_j \rangle_c \equiv \sigma^2 = O(1)$,

(c) $\dfrac{1}{N} \sum_{i,j,k} \langle X_i X_j X_k \rangle_c \equiv \mu_3 \leqslant O(1)$,

$$\frac{1}{N} \sum_{j_1,\ldots,j_n} \langle X_{j_1} X_{j_2} \ldots X_{j_n} \rangle_c \equiv \mu_n \leqslant O(N^{(n-3)/2}) \quad ,$$

$$n \geqslant 4 \quad . \tag{12-41}$$

Let

$$Y = \frac{1}{\sqrt{N}}(X_1 + \ldots + X_n) \qquad (12\text{-}42)$$

Theorem: If (a), (b) and (c) are valid, then

(i) $\langle Y \rangle = 0$,

(ii) $\langle Y^2 \rangle_c = \sigma^2$,

(iii) $\langle Y^n \rangle_c \leqslant O\left(\dfrac{1}{\sqrt{N}}\right)$, $\qquad n = 3, 4, \ldots$. $\qquad (12\text{-}43)$

The first two results (i) and (ii) are obvious; (iii) is the main part of this theorem. We first consider $\langle Y^3 \rangle$: From (12-42) we get

$$\langle Y^3 \rangle_c = \frac{1}{N^{3/2}} \sum_{i,j,k} \langle X_i X_j X_k \rangle_c = \frac{\mu_3}{\sqrt{N}} \qquad (12\text{-}44)$$

The cases with $n > 3$ are just generalisation of (12-44), e.g.

$$\langle Y^n \rangle_c = \frac{1}{N^{n/2}} \sum_{j_1, \ldots, j_n} \langle X_{j_1} X_{j_2} \ldots X_{j_n} \rangle_c$$

$$= \frac{\mu_n}{N^{n/2-1}} . \qquad (12\text{-}45)$$

Therefore, provided (12-41) is valid, (12-43) is established too. The theorem is thus proved. Now let us carefully examine its implications.

(1) The implication of the conclusion (12-43) is as follows. Except for terms of $O(1/\sqrt{N})$, (12-40) is the same as (12-43); hence the average values of Y can be calculated from the normal distribution, to within $(1/\sqrt{N})$. There is an obvious restriction, i.e. the high order average values cannot use this result because the cumulant expansion of $\langle Y^n \rangle$ has many terms. When the number of terms is close to $\sqrt{N}$, then $O(1/\sqrt{N})$ terms cannot be neglected. Therefore the distribution of Y can be regarded as normal if we only limit our calculation to average values of low powers of Y.

(2) If $X_1, X_2, \ldots, X_N$ are mutually independent variables, then (12-41) is immediately established. $\sigma^2, \mu_n, n = 3, 4, \ldots$ are all $O(1)$.

(3) If they are not independent, but each variable is only correlated to at most m other variables and $m \ll N$, then the sum in (b) involves at most Nm^2 terms, and in (c) at most Nm^n terms. Equation (12-41) is established and the theorem can be used.

(4) From a practical point of view, we usually need to calculate only the low order average values of Y. Hence if (a), (b) and (c) of Eqs. (12-41) are established for a few n, the theorem can be used. We do not need to consider whether $X_1, \ldots, X_N$ are independent. This is the reason why the theorem is so useful. If we do not demand too much, then the conditions for the validity are likewise few.

In fact, to examine whether a group of variables are independent is a difficult task. If N is large, then it is indeed impossible. Hence, we use the central limit theorem always with the practical viewpoint. This is why in the beginning of the chapter we said that the variables are "essentially" independent of the distribution is "essentially" normal.

This theorem not only has practical value but it also has important impact on some fundamental concepts. Let us look at some examples.

12.6. Repeated Trials and the Determination of Probability

Repeated experiments are very important in many applications of probability, for example the probability of getting '1' in throwing a die. The value $1/6$ means that if we have a repeated trial of N throws, '1' will turn up approximately $N/6$ times. What is the meaning of the term "approximately"? According to the central limit theorem, "approximately" means a percentage error of $(1/\sqrt{N})$. To be more explicit, suppose we throw a die N times and let the number of times '1' turns up to be Y:

$$Y = \sum_{i=1}^{N} n_i \quad ,$$

$n_i = 1$, if the i-th throw gives 1,

$ = 0$, otherwise. \hfill (12-46)

If N is very large, then Y/N approaches $1/6$. How does it approach the limit? We repeat the experiment on Y, say M times (each with N throws), then we get M values of Y: $Y_1, Y_2, \ldots, Y_M$. According to the central limit theorem, the distribution of these values is normal, and the width of the distribution of Y/N is

$$\sigma = \frac{1}{\sqrt{N}} \left(\frac{5}{36}\right)^{1/2} . \qquad (12\text{-}47)$$

This value can be determined by M tries of the experiment:

$$\sigma^2 \equiv \left\langle \left(\frac{Y}{N} - \frac{1}{6}\right)^2 \right\rangle$$

$$= \frac{1}{M} \sum_{\alpha=1}^{M} \left(\frac{Y_\alpha}{N} - \frac{1}{6}\right)^2 . \qquad (12\text{-}48)$$

That is to say, if N is larger, Y/N is closer to $1/6$. If it is not, then the meaning of the value $1/6$ is questionable. Each throw of the die is regarded as an independent event. The general definition of probability usually assumes many repeated independent experiments. The probability of an event is defined up to an accuracy of $(1/\sqrt{N})$, where N is the number of repetitions of the trial. In the absence of a large number of repetitions, this probability is just a mathematical terminology with little physical content.

The above discussion uses the central limit theorem. Seeing this, we can deduce that the determination of probability requires this theorem, and the derivation of this theorem must have probability defined first. Therefore, this theorem and the concept of probability are intimately related.

The central limit theorem is inapplicable in some cases. In such cases, is the definition of probability questionable? We can use an example to illustrate this problem. Let the distribution of a variable Y be

$$\rho(Y) = \frac{1}{\pi(Y^2 + 1)} . \qquad (12\text{-}49)$$

The mean square value $\langle Y^2 \rangle$ of this distribution is infinite. Suppose we repeat N experiments, getting $Y_1, Y_2, \ldots, Y_N$. Let

$$Z = \frac{1}{N} \sum_{i=1}^{N} Y_i . \qquad (12\text{-}50)$$

According to the example of Sec. 12.2 (see Eq. (12-18) to (12-20)) the distribution of Z is the same as (12-49), i.e.

$$\rho(Z) = \frac{1}{\pi(Z^2 + 1)} \quad , \tag{12-51}$$

and is surprisingly independent of N. That is to say no matter how many trials we have performed the average value of Y cannot be determined more accurately. This is completely different from the error of order $(1/\sqrt{N})$ discussed above. If we cannot determine the average value of Y, then what is the meaning of probability here?

The above result (12-51) tells us that we cannot use the result of repeated trials to calculate the average value of Y. However, repeated trials can still be used to determine $\rho(Y)$. We do N experiments and count the number of times $n(a, b)$ that Y occurs between a and b:

$$n(a, b) = \sum_{i=1}^{N} n_i \quad ,$$

$$n_i = 1 \quad , \quad \text{if } a < Y_i < b \quad ,$$

$$= 0 \quad , \quad \text{otherwise} \quad . \tag{12-52}$$

Now n is the sum of N independent variables, and $\langle n_i^2 \rangle < 1$ and is finite. Hence the central limit theorem can be used:

$$\frac{n(a, b)}{N} = \int_a^b dY \, \rho(Y) + O\left(\frac{1}{\sqrt{N}}\right) \quad . \tag{12-53}$$

From different values of a, b and (12-53) we can determine $\rho(Y)$. Although $\rho(Y)$ is determined, it is not accurate enough to yield the average value

$$\int_{-\infty}^{\infty} dY \, \rho(Y) Y \quad , \tag{12-54}$$

because the probability of $Y > N$ is $O(1/N)$ but its influence on (12-54) is $NO(1/N) = O(1)$. This example tells us that if a variable X has $\langle X^2 \rangle_c \to \infty$ or $\langle X^3 \rangle_c \to \infty$, then the meaning of $\langle X \rangle$ is questionable.

12.7. Random Motion and Diffusion

Suppose a collection of particles, such as pollens or big molecules, are moving in water and continually colliding with the water molecules. Let **p** be the momentum of a certain particle, then

$$\frac{d\mathbf{p}}{dt} = -\alpha \mathbf{p} + \mathbf{f}(t) \quad , \tag{12-55}$$

is the force experienced by the particle. This force naturally is rather complicated and (12-55) is only an assumption, where $\alpha \mathbf{p}$ is the frictional force, and $\mathbf{f}(t)$ is assumed to be random. The solution of (12-55) is

$$\mathbf{p}(t) = \int_0^\infty dt' \, e^{-\alpha t'} \mathbf{f}(t - t') \quad . \tag{12-56}$$

Now assume that **f** fluctuates rapidly so that $\langle \mathbf{f} \rangle = 0$; in addition, assume it has a correlation time τ satisfying

$$\frac{1}{\alpha} \gg \tau \quad . \tag{12-57}$$

That is to say, in a time interval of $1/\alpha$, **f** fluctuates many times. The correlation time has been explained in the last section of the preceding chapter. Let

$$C(t) = \frac{1}{3} \langle \mathbf{f}(t) \cdot \mathbf{f}(0) \rangle$$

$$\equiv \frac{1}{3\mathcal{J}} \int_0^{\mathcal{J}} dt' \, \mathbf{f}(t' + t) \cdot \mathbf{f}(t') \quad , \tag{12-58}$$

where $\mathcal{J}$ is a very long observation time, i.e. $\mathcal{J} \gg 1/\alpha$. If $|t| > \tau$, then $C(t) \approx 0$, because **f** evaluated at different times is independent outside a time interval τ. Because $1/\alpha \gg \tau$, (12-58) shows that $\mathbf{p}(t)$ can be regarded as the sum of many independent variables. The values of **f** at different times (separated by at least τ) are independent. Hence we use the central limit theorem to define the distribution of p_x, p_y and p_z.

$$\rho(\mathbf{p}) = \left(\frac{1}{\sqrt{2\pi\sigma^2}}\right)^3 e^{-p_x^2/2\sigma^2} \, e^{-p_y^2/2\sigma^2} \, e^{-p_z^2/2\sigma^2}$$

$$= (2\pi\sigma^2)^{-3/2} \, e^{-p^2/2\sigma^2} \quad , \tag{12-59}$$

$$\sigma^2 = \langle p_x^2 \rangle = \langle p_y^2 \rangle = \langle p_z^2 \rangle$$

$$= \int_0^\infty dt_1'\, dt_2'\, e^{-\alpha(t_1' + t_2')} C(t_1' - t_2') \quad . \tag{12-60}$$

The above equations are obtained from (12-58) and the square of (12-56). Because $1/\alpha \gg \tau$, the above integral is

$$\langle p_x^2 \rangle = \int_0^\infty dt\, e^{-2\alpha t} \int_{-2t}^{2t} dt''\, C(t'')$$

$$\simeq \frac{1}{2\alpha} \int_{-\infty}^\infty dt''\, C(t'') \quad . \tag{12-61}$$

According to the theorem of equipartition of energy, $\langle p_x^2 \rangle = mT$. Hence from (12-61) we get

$$\alpha = \frac{1}{2Tm} \int_{-\infty}^\infty dt''\, C(t'') \quad . \tag{12-62}$$

This is a very important result. It relates the coefficient of friction with the fluctuation caused by the molecular motion of the liquid. This is called the fluctuation — dissipation theorem. We shall talk about it in detail later.

Using the central limit theorem, we can also obtain the distribution of the displacement of the particle. Let $\mathbf{v} = \mathbf{p}/m$ be the speed of a certain particle, then its displacement from t to $t + t'$ is

$$\Delta \mathbf{r} \equiv \mathbf{r}(t + t') - \mathbf{r}(t)$$

$$= \int_0^{t'} dt''\, \mathbf{v}(t'' + t) \quad . \tag{12-62'}$$

Now $\mathbf{v}(t'')$ is a random velocity and its time correlation function is

$$\langle \mathbf{v}(t')\, \mathbf{v}(0) \rangle = \langle v^2 \rangle e^{-\alpha t'} \quad . \tag{12-63}$$

This result is from the solution of (12-55):

$$\mathbf{v}(t' + t) = \mathbf{v}(t) e^{-\alpha t'} + \frac{1}{m} \int_0^{t'} dt''\, e^{-\alpha t''} \mathbf{f}(t' + t - t'') \quad .$$

Assume that $\mathbf{v}(t)$ is uncorrelated with $\mathbf{f}$ at later times. If t' in (12-62') is much larger than $1/\alpha$, then $\Delta\mathbf{r}$ is the sum of many independent variables. According to the central limit theorem, the distribution of $\Delta\mathbf{r}$ is normal:

$$\rho(\Delta\mathbf{r}) = (2\pi\sigma^2)^{-3/2} e^{-(\Delta\mathbf{r})^2/2\sigma^2} \quad,$$

$$\sigma^2 = \frac{1}{3}\langle(\Delta\mathbf{r})^2\rangle \quad, \tag{12-64}$$

and σ^2 can be obtained from (12-62) and (12-63):

$$\frac{1}{3}\langle(\Delta\mathbf{r})^2\rangle = \frac{1}{3}\int_0^{t'} dt_1 \int_0^{t''} dt_2 \, \langle \mathbf{v}(t_1)\cdot\mathbf{v}(t_2)\rangle$$

$$= \frac{1}{3}\int_0^{t'} dt_1 \int_{-2t}^{2t} dt'' \, \langle v^2\rangle e^{-\alpha|t''|}$$

$$\simeq \frac{2}{3}\langle v^2\rangle \frac{t'}{\alpha} \quad. \tag{12-65}$$

This result can be written in the more familiar form of the diffusion equation:

$$\rho(\Delta\mathbf{r}, t') = (4\pi D t')^{-3/2} e^{-(\Delta\mathbf{r})^2/4Dt'} \quad,$$

$$D \equiv \frac{\langle v^2\rangle}{3\alpha} = \frac{T}{m\alpha} \quad, \tag{12-66}$$

where D is the coefficient of diffusion.

The above averages are all time averages. If we observe many particles simultaneously, then the average values may also be regarded as averages over these particles. If in a small region $d^3 r'$ the number of particles is $n(\mathbf{r}')d^3 r'$, then after a time t, the distribution of the positions of these particles is

$$n(\mathbf{r}, t) = \int d^3 r' \, \rho(\mathbf{r}-\mathbf{r}', t)\, n(\mathbf{r}') \quad. \tag{12-67}$$

This $n(\mathbf{r}, t)$ is a solution of the diffusion equation:

$$\frac{\partial}{\partial t} n(\mathbf{r}, t) = D \nabla^2 n(\mathbf{r}, t) \quad.$$

12.8. Fluctuation of Macroscopic Variables

In the last section of Chapter 6, we mentioned that the distribution of large value variables is always normal (see Eq. (6-86)). This distribution is the statistical result of the states in the region of motion. This conclusion is only applicable to large value variables (of order N). Is this kind of normal distribution related to the central limit theorem?

Let us give an example to discuss this problem. Divide the system into two halves 1 and 2. Let

$$\hat{A}(s) = H_1(s_1) - H_2(s_2) \quad , \tag{12-69}$$

where H_1 and H_2 are the energies of the two parts. This variable has been analysed in Chapter 6 (see Eqs. (6-84) and (6-85) and the discussion). Its distribution is

$$\frac{\Gamma(E, A)}{\Gamma(E)} = \frac{1}{\sqrt{2\pi K_A}} e^{-(A-\bar{A})^2/2K_A} \quad , \tag{12-70}$$

where $\bar{A}$ is the solution of (6-85). Here $\bar{A} = 0$ because the two parts are of the same size. The definition of K_A is

$$-\frac{1}{K_A} = \frac{\partial^2 S}{\partial A^2} = -\frac{1}{4} \left(\frac{\partial^2 S_1}{\partial E_1^2} + \frac{\partial^2 S_2}{\partial E_2^2} \right) \quad . \tag{12-71}$$

Rearranging, we get

$$K_A = T^2 C_v \quad , \tag{12-72}$$

where $C_v = \partial E / \partial T$ is the heat capacity of the system.

Now we analyse the distribution of $\bar{A}$ directly from the point of view of the molecular motion. We first write $\hat{A}$ as an integral of the energy density.

$$\hat{A}(s(t)) = \int d^3 r \, f(\mathbf{r}) \, \epsilon(\mathbf{r}, t) \quad , \tag{12-73}$$

$$f(\mathbf{r}) = 1 \quad , \quad \text{if } \mathbf{r} \text{ lies in part 1} \quad ,$$
$$= -1 \quad , \quad \text{if } \mathbf{r} \text{ lies in part 2} \quad , \tag{12-74}$$

and $\epsilon(\mathbf{r}, t)$ is the energy density. For example, if this system is a collection of particles, then

$$s \equiv (\mathbf{r}_1, \mathbf{r}_2, \ldots, \mathbf{r}_N, \mathbf{p}_1, \mathbf{p}_2, \ldots, \mathbf{p}_N) \quad,$$

$$\epsilon(\mathbf{r}, t) = \sum_i \delta(\mathbf{r} - \mathbf{r}_i(t)) \frac{p_i^2(t)}{2m}$$

$$+ \frac{1}{2} \sum_{i,j} U(\mathbf{r}_i(t) - \mathbf{r}_j(t)) \, \delta\left(\mathbf{r} - \frac{\mathbf{r}_i(t) + \mathbf{r}_j(t)}{2}\right) \quad,$$

(12-75)

where $p_i^2/2m$ is the kinetic energy of the particle, $U(\mathbf{r}_i - \mathbf{r}_j)$ is the interaction energy between i and j, $\mathbf{r}_i(t)$ and $\mathbf{p}_i(t)$ are the position and momentum of the particle at time t. Hence (12-73) is in fact the sum of a large number of variables. Each term in the sum is a function of the momenta and positions of the particles. Each variable changes with t. Now let us calculate $\langle \hat{A}^2 \rangle_c$:

$$\langle \hat{A}^2 \rangle_c = \int d^3r \, d^3r' \, f(\mathbf{r}) \, f(\mathbf{r}') \langle \epsilon(\mathbf{r}) \epsilon(\mathbf{r}') \rangle_c \quad. \tag{12-75'}$$

The average here is that over the observation time:

$$\langle \hat{A} \rangle = \frac{1}{\mathfrak{I}} \int_0^{\mathfrak{I}} dt \, \hat{A}(s(t)) \quad,$$

$$\langle \hat{A}^2 \rangle_c = \frac{1}{\mathfrak{I}} \int_0^{\mathfrak{I}} dt \, \hat{A}^2(s(t)) - \langle \hat{A} \rangle^2 \quad. \tag{12-76}$$

Now assume that the energy density is uncorrelated for points very far apart, i.e.

$$\langle \epsilon(\mathbf{r}) \epsilon(\mathbf{r}') \rangle = 0 \quad, \quad \text{if } |\mathbf{r} - \mathbf{r}'| \gg \xi \quad, \tag{12-77}$$

where ξ is the correlation length and is on a microscopic scale. If the system is homogeneous, then

$$\langle \epsilon(\mathbf{r}) \epsilon(\mathbf{r}') \rangle_c = \langle \epsilon(\mathbf{r} - \mathbf{r}') \epsilon(0) \rangle_c \quad. \tag{12-78}$$

The origin 0 can be taken to be in the centre of the system. From these assumptions, we obtain

$$\langle \hat{A}^2 \rangle_c = V \left[\int d^3r'' \, \langle \epsilon(\mathbf{r}'') \epsilon(0) \rangle_c + O\left(\frac{\xi}{L}\right) \right] \quad, \tag{12-79}$$

where $V=L^3$ is the volume of the system. The last term is related to the surface of the system, and the error is determined by the boundary of the two parts. Because $\xi \ll L$, (12-73) is the sum of a large number of essentially independent variables.

Now use the central limit theorem to obtain the distribution of A:

$$\rho(\hat{A}) = \frac{1}{\sqrt{2\pi N\sigma^2}} \exp\left\{-\frac{(A-\langle A\rangle)^2}{2N\sigma^2}\right\},$$

$$\langle \hat{A} \rangle = 0,$$

$$\sigma^2 = \frac{V}{N}\int d^3 r \,\langle \epsilon(\mathbf{r})\epsilon(0)\rangle_c . \qquad (12\text{-}80)$$

The distribution is obtained from the viewpoint of the molecular motion and the independent assumption of $\xi \ll L$. The energy density correlation function can be measured from experiment. Such methods of measurement are discussed in the last section of the preceding chapter.

12.9. Fluctuations and an Extension of the Basic Assumption

Now we can use the result of the last section to perform a further analysis of the basic assumption. This assumption is introduced for calculating the entropy. With the entropy known, the other thermodynamic quantities can be calculated. In Chapters 6, 7 and 8 we gave many examples and introduced some rules of calculation to illustrate the application of this basic assumption.

Simplicity and convenience are the outstanding features of this assumption. With this assumption we do not need detailed information on the molecular motion in order to calculate thermodynamic properties.

We have repeatedly emphasised that the region of motion must be determined by the motion. It represents the amplitude of change of the various variables. But the basic assumption does not specify how to calculate the magnitude of the fluctuation. From the analysis of Chapter 5 to Chapter 7, it indicates that the fluctuation of macroscopic variables is very small, but there is no definite quantitative conclusion. Indeed the distribution (12-70) is calculated statistically from the number of states in the region of motion. It indicates that almost all the $\hat{A}(s)$ in the region of motion have a value equal to $\overline{A}$ with an error of $O(1/\sqrt{N})$. But the relation of the normal distribution function (12-70) and the molecular motion is not clear. (See the last section of Chapter 16.)

In the last section of the last chapter we have pointed out that the correlation function can be measured by scattering experiments. In the preceding section we have discussed the relation of the fluctuation of A with the correlation function. This leads us to the conclusion that fluctuation can be determined by scattering experiments. The experiments measure an average value with respect to time. The distribution of the variable means a time distribution. The meaning of the distribution (12-80) is as follows. Let

$$R = \{s; \; A_1 < \hat{A}(s) < A_2\} \; , \qquad (12\text{-}81)$$

i.e. all the states s with value of $\hat{A}(s)$ between A_1 and A_2 are included in R. Let $\mathfrak{I}_R$ be the time the trajectory stays in R, then the meaning of (12-80) is

$$\int_{A_1}^{A_2} dA \, \rho(A) = \frac{\mathfrak{I}_R}{\mathfrak{I}} \; , \qquad (12\text{-}82)$$

where $\mathfrak{I}$ is the total observation time. This can be regarded as a quantity determined purely from the time distribution of the trajectory. It does not involve the entropy or the basic assumption.

The distribution (12-70) comes from the sizes of the regions in the region of motion. Let $\Gamma(R)$ be the volume of R, then

$$\int_{A_1}^{A_2} dA \, \frac{\Gamma(E, A)}{\Gamma(E)} = \frac{\Gamma(R)}{\Gamma(E)} \; . \qquad (12\text{-}83)$$

This is purely a calculation of the volume of the region and does not involve the concept of the trajectory or the time.

The $\rho(E)$ in (12-80) and the ratio $\Gamma(E, A)/\Gamma(E)$ in (12-70) obviously have different meanings. They are both normally distributed. But are they equal? The condition of equality is $K_A = N\sigma^2$ (see Eqs. (12-72) and (12-80)), i.e.

$$T^2 C_v = V \int d^3 r \, \langle \epsilon(\mathbf{r}) \, \epsilon(0) \rangle_c \; . \qquad (12\text{-}84)$$

The left-hand side can be measured by thermodynamic experiments while the right-hand side can be measured by scattering experiments. We can also use the basic assumption to calculate the left-hand side and other methods to calculate the right-hand side, and check whether the two results are the same. We do not discuss in full the details of these experiments or calculations. The result is that they are equal. These results indicate that

$$\frac{\mathfrak{I}_R}{\mathfrak{I}} = \frac{\Gamma(R)}{\Gamma(E)} \; , \qquad (12\text{-}85)$$

i.e. the time that the trajectory stays in R is proportional to the volume of R. Notice that R is determined by classification according to A, where A is a large variable. Here R is a large region. We have mentioned that R must be a large region in order for $\mathcal{J}_R$ to have meaning. (See the discussion from Eqs. (10-34) to (10-35).)

Although the above discussion concerns a special example, the general conclusion is similar. If R is any large region, Eq. (12-85) is valid. This conclusion can be stated as, "The trajectory is uniformly distributed in the region of motion" or "Any state in the region of motion has equal probability of occurring". In spite of all these words, we must remember that the R in (12-85) must be a large region. If R is too small, $\mathcal{J}_R$ has no meaning.

Equation (12-85) can be regarded as an extension of the basic assumption. It essentially tells us that our trajectory is dispersed throughout the region of motion. Not only can we use the basic assumption to calculate entropy and the average values of the thermodynamic coordinates $\hat{A}$, but we can also use it to calculate the fluctuation of $\hat{A}$. The average value of $\hat{A}$ is $O(N)$ and its fluctuation is $O(\sqrt{N})$. From $O(N)$ to $O(\sqrt{N})$ is a very important step.

We emphasise again that the fluctuation of $\hat{A}$ discussed here represents the change of $\hat{A}$ with time and has nothing to do with the inaccuracy in measuring $\hat{A}$. Generally speaking, this inaccuracy in measurement comes from the surface of the system. The boundary between the interior and exterior of the system is not definite, making a measuring inaccuracy of $O(N^{2/3})$. The number of molecules around the surface of the body is $O(N^{2/3})$. This error is much larger than the fluctuation $O(N^{1/2})$. The two are completely different. The measurement of the average value of $\hat{A}$ and the measurement of its fluctuation are by completely different methods. The former has an inaccuracy of $O(N^{2/3})$ but does not indicate that the latter $O(N^{1/2})$ is not measurable. The former is an average in the observation time and is an unchanged "D.C." part while the latter is a fluctuation, the changing part, i.e. the part with non-zero frequency.

The reader may already notice that the above example concerns $A = H_1 - H_2$ (see Eq. (6-9)). Why do we not use $H_1 + H_2 = H$ as the example? The reason is that H is a conserved quantity. If the interface is adiabatic, then the fluctuation of H is very small. If it is open, then the fluctuation is larger. When we use $H_1 - H_2$, we can avoid this problem, because $H_1 - H_2$ is not a conserved quantity. The fluctuation of the energy density is better analysed by $H_1 - H_2$. The scattering experiments analyse the wavevector:

$$\epsilon_k = \int d^3r \, e^{-i\mathbf{k}\cdot\mathbf{r}} \, \epsilon(\mathbf{r}) \quad , \tag{12-86}$$

where $H_1 - H_2$ may be regarded as the nonzero limit as $k \to 0$, i.e. $k \sim 1/L$ where L is the size of the system. If $k = 0$, then $\epsilon_0 = H$.

Notice that the fluctuation experiments tell us of the existence of molecules or atoms. That is to say, matter is not continuous but is composed of discrete elements. If it is continuous, the N must be infinite, and the fluctuation is zero because $O(N^{1/2})/N = O(N^{-1/2})$. There is fluctuation only because N is finite. Because we have used units with $k = 1$, it is easy to overlook the role of the Boltzmann constant in the various equations. The energy kT is a microscopic quantity. It is about the energy of a molecule. If matter is continuous a "molecule" would be infinitesimal, and k must be zero. The fact that k is nonzero indicates the existence of discrete elements. The equations in thermodynamics do not involve k. But the equations concerning fluctuation and involving T must also involve k. For example in (12-62) and (12-84)

$$m\alpha = \frac{1}{2kT} \int dt' \, C(t') \quad , \tag{12-87}$$

$$\frac{TC_v}{V} = \frac{1}{kT} \int d^3r \, \langle \epsilon(\mathbf{r}) \, \epsilon(0) \rangle_c \quad , \tag{12-88}$$

where α, m and the correlation function are concepts not involving temperature. The combination TC_v/V is a macroscopic quantity. The remaining T must be accompanied by k, otherwise the units are not consistent. The correlation function is of course a special feature of fluctuation.

Problems

1. Calculate the cumulant characteristic function

$$\ln \langle e^{-ikn} \rangle$$

of the binomial distribution

$$\rho(n) = \binom{N}{n} p^n (1-p)^{N-n} \quad . \tag{12-89}$$

Ans.: $N \ln [1 - p(1 - e^{-ik})]$

2. In the above problem, when N and n are very large, what is the difference between $\rho(n)$ and the normal distribution?

3. Calculate the characteristic function of (12-18) and then derive (12-20).

4. The cumulant expansion theorem is an extremely important discovery for statistical mechanics. Let

$$H = H_0 + U$$

(a) Prove that

$$F = F_0 - T \ln \langle e^{-U/T} \rangle_0 \quad ,$$

where $\langle \ \rangle_0$ is defined as an average over $e^{-H_0/T}$:

$$\langle A \rangle_0 = \sum A \, e^{-H_0/T} / Z_0 \quad , \tag{12-90}$$

and F_0 is the thermodynamic potential for H_0.

(b) From the cumulant expansion theorem prove that

$$F = F_0 - T \langle e^{-U/T} - 1 \rangle_{0c} \quad , \tag{12-91}$$

i.e. the correction U on F_0 is the various cumulant values of U/T.

(c) Let H_0 be the kinetic energy of a certain gas and U the interaction energy, which is assumed to be short range. Prove that the expansion of F with respect to U has terms all proportional to the volume. This result would be hard to prove without the cumulant expansion theorem. (Hint: Use the model in Sec. 9.1.) Write U as

$$U = \frac{1}{2} \int d^3r \, d^3r' \, \rho(r) \, \rho(r') \, u(r - r') \quad . \tag{12-92}$$

Notice that:

$$\langle \rho(r) \rho(0) \rangle_0 = Z \delta(r - r') \quad . \tag{12-93}$$

The density of an ideal gas is uncorrelated in different places. First find $\langle U \rangle_{0c}$, $\langle U^2 \rangle_{0c}$, then the cases of high order terms can be seen.

5. Effective energy.

Let the state of a system be $s = (s_1, s_2)$ where s_1 and s_2 are respectively the states of the two parts of the system. Let the total energy be

$$H(s) = H_1(s_1) + H_2(s_2) + H_{12}(s_1, s_2) \quad . \tag{12-94}$$

If a variable $A(s_1)$ is only dependent on s_1, prove that its average value can be calculated by an effective energy $H_1'(s_1)$, defined as

$$H_1'(s_1) = H_1(s_1) - T \ln \langle e^{-H_{12}/T} \rangle_2 \quad ,$$

$$\langle e^{-H_{12}/T} \rangle_2 \equiv \sum_{s_2} e^{-H_{12}/T} e^{-H_2/T} \Big/ \sum_{s_2} e^{-H_2/T} \quad . \tag{12-95}$$

That is to say, H_1' is H_1 plus a correction term. The correction is due to the interaction with the second part. If H_{12} is comparable to H_1, then this correction cannot be neglected.

The above steps for calculating the effective energy is again an example of classification and also an example of projection. (Review the last section of Chapter 5 and the first section of Chapter 6.)

6. Notice that in the above problem s_1 in the phase space (set of $s = (s_1, s_2)$) represents a region $R(s_1)$, i.e. $\{s_1$ fixed, all the set of $s_2\}$. If the second part has N_2 particles, where N_2 is a large number, then this region is a large region with a volume of order e^{N_2} :

$$\Gamma(R(s_1)) = O(e^{N_2}) \quad . \tag{12-96}$$

Let

$$P(s_1) = \Gamma(R(s_1))/\Gamma \quad , \tag{12-97}$$

where Γ is the volume of the region of motion of the whole system. Discuss the meaning of $P(s_1)$. Can it represent the probability of occurrence of s_1? Let N_1 be the number of particles in part 1. If N_1 is not a large number, what is the order of magnitude of $P(s_1)$? If N_1 is a large number, what is the order of magnitude of $P(s_1)$ then? Notice that $\Gamma = O(e^{N_1 + N_2})$.

Chapter 13
CORRELATION FUNCTIONS

The previous chapters discussed the importance of the concept of independence and correlation in statistical mechanics. The main point of this chapter is on some examples of the correlation function. Hence there is a bit more mathematical manipulation. From these examples we can have a more concrete understanding of the concept of correlation. First, we briefly summarise the relation between response and fluctuation discussed in the last chapter, and define the response function. Then we give examples of calculating the correlation function and response function in some gas models, including the dilute gas, the dense gas (one-dimensional model) and the quantum fermion gas. Some special properties arising from quantum mechanics are also discussed.

13.1. Response and Fluctuation

Using the central limit theorem, the analysis of the last chapter relates the distribution of the variable $\hat{A}$ in the region of motion with the correlation of its fluctuation. The relation of heat capacity with the fluctuation of the energy density is an important example. Heat capacity is a kind of susceptibility. It measures the response of the body with respect to the temperature change. Now we use a somewhat abstract procedure to derive the relation between the general response with the fluctuation. The following result further shows that the correlation value can in fact be regarded as a large value variable and its value is about the same throughout the region of motion.

Suppose that the Hamiltonian of a system is H. Now we add an external force λ so that the Hamiltonian becomes

$$H - \lambda \hat{A} \quad . \tag{13-1}$$

For example λ is the magnetic field and $\hat{A}$ is the total magnetic moment, or λ is the tension and $\hat{A}$ is the length of the system. The molecules inside the system respond to this external force and changes occur. This response can be determined by the change of the various variables. For example, the value of a variable B can be calculated from the average

$$B = \langle \hat{B} \rangle = \frac{\sum_s \hat{B} e^{-(H-\lambda\hat{A})/T}}{\sum_s e^{-(H-\lambda\hat{A})/T}} . \tag{13-2}$$

B is of course a function of λ. Differentiating the above equation, we get

$$\chi \equiv \frac{\partial B}{\partial \lambda} = \frac{1}{T}[\langle \hat{B}\hat{A} \rangle - \langle \hat{B} \rangle \langle \hat{A} \rangle]$$

$$\equiv \frac{1}{T} \langle \hat{B}\hat{A} \rangle_c . \tag{13-3}$$

This is the so-called differential susceptibility.

Example 1:

Let $\hat{A} = \hat{B} = H$, i.e. the total energy itself. Then

$$(H - \lambda A)/T \approx H/((1 + \lambda)T)) \tag{13-4}$$

and λ represents the fractional increase of temperature (assume that $\lambda \ll 1$). From (13-3) we get

$$T \frac{\partial E}{\partial T} = \frac{1}{T} \langle H^2 \rangle_c ,$$

i.e.

$$T^2 C_v = \int d^3r \, d^3r' \, \langle \epsilon(\mathbf{r}) \, \epsilon(\mathbf{r'}) \rangle_c$$

$$= V \int d^3r'' \, \langle \epsilon(\mathbf{r''}) \, \epsilon(0) \rangle_c . \tag{13-5}$$

This is just (12-84), and $\epsilon(\mathbf{r})$ is the energy density.

Example 2:

Let λ be the magnetic field h, $\hat{B} = \hat{A}$ be the total magnetic moment $\hat{M}$. Then from (13-3) we get

$$\chi = \frac{\partial M}{\partial h} = \frac{1}{T}\langle \hat{M}^2 \rangle_c$$

$$= \frac{V}{T}\int d^3r'' \langle m(\mathbf{r}'')\, m(0) \rangle_c \quad , \tag{13-6}$$

where $m(\mathbf{r})$ is the magnetic moment density and χ is the magnetic susceptibility.

From these results it can be seen that the correlation value can be regarded as a large value variable because it can be obtained by differentiating B, which is the value of the large value variable $\hat{B}$.

Let us now return and ask: What is the influence of the external force λ on the thermodynamic potential? The thermodynamic potential is

$$F = -T \ln Z \quad ,$$

$$Z = \sum_s e^{-(H - \lambda \hat{A})/T} \quad . \tag{13-7}$$

Let Z_0 be the value of Z when $\lambda = 0$. Then

$$Z = Z_0 \langle e^{\lambda \hat{A}/T} \rangle_0 \quad . \tag{13-8}$$

The average value $\langle \ldots \rangle_0$ means the average when $\lambda = 0$:

$$\langle \hat{B} \rangle_0 \equiv \sum_s \hat{B}\, e^{-H/T} \Big/ \sum_s e^{-H/T} \quad . \tag{13-9}$$

Now use the cumulant expansion theorem of the last chapter (see Eqs. (12-32) to (12-38)) to get

$$\ln \langle e^{\lambda \hat{A}/T} \rangle_0 = \frac{\lambda}{T}\langle \hat{A} \rangle_0 + \frac{\lambda^2}{2T^2}\langle \hat{A}^2 \rangle_{0c} + \ldots \quad . \tag{13-10}$$

Substitute (13-10) into $\ln Z$, and we get

$$F = -T \ln Z_0 - T \ln \langle e^{\lambda \hat{A}/T} \rangle_0$$

$$= F_0 - \lambda \langle \hat{A} \rangle_0 - \frac{\lambda^2}{2T}\langle \hat{A}^2 \rangle_{0c} + \ldots \quad . \tag{13-11}$$

228 STATISTICAL MECHANICS

Therefore,

$$-\frac{\partial F}{\partial \lambda} = \langle \hat{A} \rangle \quad,$$

$$-\frac{\partial^2 F}{\partial \lambda^2} = \frac{\partial}{\partial \lambda} \langle \hat{A} \rangle = \frac{1}{T} \langle \hat{A}^2 \rangle_c \quad, \tag{13-12}$$

etc. There is no subscript 0 in $\langle \ldots \rangle$ of (13-12) because this is unnecessary as the above expansion can take any value of λ as the origin.

Hence, all the average values and correlation values can be related to the differential of the thermodynamic potential. From this point of view, as the basic assumption specifies the rules of calculating entropy and the thermodynamic potential, it also specifies the rules of calculating its differentials. Hence the basic assumption also specifies the rules of calculating all the average values and correlation values.

The above discussion has not taken quantum mechanics into account. Hence, some results may need corrections. If $[\hat{A}, H] \neq 0$, then (13-8) must be amended, because

$$e^{-(H - \lambda \hat{A})/T} \neq e^{-H/T} e^{\lambda \hat{A}/T} \quad.$$

This will be rediscussed in the last part of this chapter. At the moment we do not consider any quantum mechanical effects.

Example 3: The response function and the correlation function.

These are only examples of (13-3). Let

$$\hat{B} = \int d^3 r \, a(\mathbf{r}) \, e^{-i\mathbf{k} \cdot \mathbf{r}} \equiv a_\mathbf{k} \quad,$$

$$\hat{A} = \hat{B}^* \quad. \tag{13-13}$$

Substitute in (13-3) and we get

$$VG_\mathbf{k} \equiv \frac{1}{T} \int d^3 r \, d^3 r' \, \langle a(\mathbf{r}) \, a(\mathbf{r}') \rangle_c \, e^{-i\mathbf{k} \cdot (\mathbf{r} - \mathbf{r}')}$$

$$= \frac{V}{T} \int d^3 r'' \, \langle a(r'') \, a(0) \rangle_c \, e^{-i\mathbf{k} \cdot \mathbf{r}''}$$

$$\equiv \frac{V}{T} C_\mathbf{k} \quad, \tag{13-14}$$

where C_k is the Fourier transform of the correlation function of the energy fluctuation and G_k is called the Fourier transform of the response function. Its Fourier integral is

$$G(\mathbf{r}-\mathbf{r}') \equiv (2\pi)^{-3} \int d^3k \ G(\mathbf{k}) \ e^{i\mathbf{k}\cdot(\mathbf{r}-\mathbf{r}')} \quad . \tag{13-15}$$

Its meaning is as follows. If $\lambda(\mathbf{r})$ represents a distribution of external force, then

$$\langle a(\mathbf{r}) \rangle = \int d^3r' \ G(\mathbf{r}-\mathbf{r}') \ \lambda(\mathbf{r}') \quad . \tag{13-16}$$

The above assumes λ to be very small, and $\langle a \rangle$ is zero for $\lambda = 0$. (This result is obtained from $\langle a_\mathbf{k} \rangle = \lambda_\mathbf{k} G_\mathbf{k}$.) The response function shows the relation of causality: $\lambda(\mathbf{r}')$ is the cause and $\langle a(\mathbf{r}) \rangle$ is the effect and $G(\mathbf{r}-\mathbf{r}')$ shows how the cause at $\mathbf{r}'$ affects the response at $\mathbf{r}$. It can also be called the propagation function.

Example 4:

The above result naturally can be applied to the distribution in an open system. Let $\hat{A} = \hat{B} = N$, i.e. the number of particles, and λ is the chemical potential μ. The result of (13-3) becomes

$$\frac{\partial N}{\partial \mu} = \frac{1}{T} \langle \hat{N}^2 \rangle_c$$

$$= \frac{V}{T} \int d^3r \ \langle \rho(\mathbf{r}) \rho(0) \rangle_c \quad . \tag{13-17}$$

Here $\rho(\mathbf{r})$ is the density of the molecules. Notice that

$$\frac{\partial N}{\partial \mu} = V \frac{\partial n}{\partial \mu} = V \frac{\partial n/\partial p}{\partial \mu/\partial p} = V n \frac{\partial n}{\partial p} \quad . \tag{13-18}$$

In the above we have made use of $n = \partial p/\partial \mu$, $n \equiv N/V = \langle \rho(\mathbf{r}) \rangle$. Therefore

$$n \frac{\partial n}{\partial p} = \frac{1}{T} \int d^3r \ \langle \rho(\mathbf{r}) \rho(0) \rangle_c \quad . \tag{13-19}$$

Hence the response of the density to pressure is related to the density fluctuation.

13.2. Density Correlation Function

Now let us consider the evaluation of the density correlation function. Let

$$n(\mathbf{r}|0) \equiv \langle \rho(\mathbf{r}) \rho(0) \rangle / n \quad . \tag{13-20}$$

This is the density distribution given that a particle is at $\mathbf{r} = 0$. Notice that

$\langle \rho(\mathbf{r}) \rho(0) \rangle$

= (probability of a molecule at $\mathbf{r} = 0$) × (the conditional probability that there is a particle at $\mathbf{r}$ given that there is a molecule at 0).

$$\equiv n \times n(\mathbf{r}|0) \quad . \tag{13-21}$$

Therefore, (13-20) can be regarded as a conditional probability.

If the system is an ideal gas, then

$$n(\mathbf{r}|0) = \delta(\mathbf{r}) + n \quad . \tag{13-22}$$

The first term on the right is the density of the molecules at $\mathbf{r} = 0$ and the second term n is the average distribution of the other molecules.

We can apply (13-22) to (13-19) to obtain the relation between pressure and density. From (13-22) and (13-20) we get

$$\langle \rho(\mathbf{r}) \rho(0) \rangle_c = n \, \delta(\mathbf{r}) \quad . \tag{13-23}$$

Subsituting in (13-19), we get

$$\frac{dn}{dp} = \frac{1}{T} \quad . \tag{13-24}$$

This of course is the ideal gas law $p = nT$.

If there is an interaction $u(\mathbf{r}_i - \mathbf{r}_j)$ between the molecules ($\mathbf{r}_i$, $\mathbf{r}_j$ are the positions of the i-th and j-th molecules), then on putting a molecule at $\mathbf{r} = 0$ the distribution of the molecules will no longer be (13-22), but becomes

$$n(\mathbf{r}|0) = n \, e^{-u(\mathbf{r})/T} + \delta(\mathbf{r}) \quad . \tag{13-25}$$

This equation tends to n when $\mathbf{r}$ is very large, and (13-23) now becomes

$$\langle \rho(\mathbf{r}) \rho(0) \rangle_c = n \left[\delta(\mathbf{r}) + n(e^{-u(\mathbf{r})/T} - 1) \right] \quad . \tag{13-26}$$

Equation (13-26) is valid only when the density of the gas is low, because (13-25) is the distribution of an ideal gas in a potential $u(\mathbf{r})$. If the density is high, then we must take into account of the interaction with other molecules. Hence the last term of (13-26) is the first correction term for low density case.

Substitute (13-26) into (13-19) and we get

$$T \frac{\partial n}{\partial p} = 1 + n \int d^3 r \, [e^{-u(\mathbf{r})/T} - 1] \quad . \tag{13-27}$$

After integrating we get

$$p = nT + n^2 T \int d^3 r \, [1 - e^{-u(\mathbf{r})/T}] \quad . \tag{13-28}$$

This is the ideal gas law plus a correction. This correction term will be discussed in detail in the next chapter. This just offers an application of (13-19).

The correlation length ξ is defined as follows. The correlation function tends to zero at distances larger than ξ. From (13-23) it can be seen that the correlation length of an ideal gas is zero. From (13-26) we see that for a dilute gas ξ is just the effective interaction distance of the molecular interaction $u(\mathbf{r})$. Strictly speaking, different correlation functions can have different correlation lengths. For example, the density correlation function can have a correlation length different from the spin correlation function. But generally speaking, the various correlation lengths are more or less the same.

The above example is a dilute gas. Let us consider an example in one dimension where the density can be high or low. This is the example in Sec. 8.3. Now let us calculate $\langle \rho(x) \, \rho(0) \rangle$. Fix a particle at the origin and let the position of the first particle on the right be y_1, the position of the second be $y_1 + y_2$, and the position of the n-th particle be $y_1 + y_2 + \ldots + y_n$. Therefore,

$$n(x|0) = \delta(x) + \langle \delta(x-y_1) \rangle + \langle \delta(x-y_1-y_2) \rangle$$
$$+ \ldots + \langle \delta(x-y_1-y_2-\ldots-y_n) \rangle + \ldots \quad . \tag{13-29}$$

The definition of $n(x|0)$ is (13-20). The y_i are the distances between neighbouring molecules. Equation (13-29) is correct for $x > 0$. If $x < 0$, a similar equation can be written. From (13-29) we get

$$C_k = \int dx\, e^{-ikx}\, \langle \rho(x)\rho(0)\rangle$$

$$= n\int dx\, e^{-ikx}\, n(x|0)$$

$$= n\left[1 + \sum_{m=1}^{N/2}(\phi_k^m + \phi_{-k}^m)\right]. \tag{13-30}$$

$$\phi_k \equiv \langle e^{-iky}\rangle = \frac{1}{\zeta}\int_0^\infty dy\, e^{-(py+u(y))/T}\, e^{-iky},$$

$$\zeta \equiv \int_0^\infty dy\, e^{-(py+u(y))/T}, \tag{13-31}$$

where p is the pressure, ζ has been discussed in Sec. 8.2 (Eq. (8-35)) and $u(y)$ is the interaction energy between neighbouring molecules. The above assumes that each side has $N/2$ molecules. In (13-30) ϕ_k^m is contributed by the molecules on the left. The above also assumes that $k \neq 0$.

The sum in (13-30) can be calculated if we regard N as infinite. We get

$$C_k = n\left(\frac{1}{1-\phi_k} + \frac{1}{1-\phi_{-k}} - 1\right)$$

$$= n\,\frac{1-|\phi_k|^2}{|1-\phi_k|^2}. \tag{13-32}$$

Now we look at a special form of $u(y)$, i.e. the hard sphere interaction

$$u(y) = \infty, \qquad y < b,$$
$$= 0, \qquad y > b, \tag{13-33}$$

where b is the diameter of the hard sphere.

Substitute (13-33) into (13-31) and we get

$$\phi_k = \frac{p/T}{(p/T + ik)} e^{-ika}$$

$$= \frac{1}{1 + ik(a-b)} e^{-ika} \ ,$$

$$a \equiv 1/n \ . \tag{13-34}$$

Here we employed $p(a - b) = T$. This gas is just like the ideal gas, but the effective volume is $L - Nb = N(a - b)$. From (13-34) and (13-32) we get

$$C_k = \frac{nk^2 (a-b)^2}{|1 - e^{-ikb} + ik(a-b)|^2} \ . \tag{13-35}$$

Notice that this result is independent of the temperature, because $u(y)$ does not have an energy scale. That is to say, for a collection of moving hard spheres, no matter what the temperature is, the density correlation function will not be affected. If $a \gg b$, i.e. at low density, (13-35) is about

$$C_k \simeq n\left(1 - \frac{2b}{a}\right) \ . \tag{13-36}$$

This equation is appropriate for $kb \ll 1$. If the density is high, i.e. $a - b$ small, then when k is not very large, the denominator of (13-35) is very small for $k \approx 2\pi/b$ where m is an integer:

$$1 - e^{-ikb} + ik(a-b)$$

$$\simeq \frac{1}{2}(k'b)^2 + i\left[2\pi m\left(\frac{a}{b} - 1\right) + k'a\right] \ ,$$

$$k \equiv 2\pi m/b + k' \ . \tag{13-37}$$

In the above, $k'a$, $(1 - a/b)$ are regarded as small quantities. Therefore, the denominator of (13-35) is zero when $k = k_m$:

$$k_m \simeq \frac{2\pi m}{a} - i 4\pi^2 m^2 (a-b)^2/a^3 \ ,$$

$$m = \pm 1, \pm 2, \pm 3, \ldots \ . \tag{13-38}$$

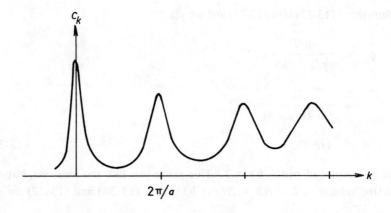

Fig. 13-1 Plot of C_k as a function of k.

Hence C_k has a group of peaks at $2\pi m/a$ with width $4\pi^2 m^2 a^{-3}(a-b)^2$. (See Fig. 13-1.) If m is very large, then (13-38) is not valid. The meaning of these peaks is:

$$C(x) \equiv \langle \rho(x)\rho(0) \rangle_c = \frac{1}{2\pi} \int_{-\infty}^{\infty} dk\, C_k\, e^{ikx} \quad , \tag{13-39}$$

is approximately a function of period a, but gradually decreasing with $|x|$. The imaginary part of (13-38) is the rate of decrease. Hence the correlation length is about

$$\xi = \frac{a^3}{4\pi^2(a-b)^2} \quad ,$$

since

$$C(x) \sim e^{-|x|/\xi} \quad . \tag{13-40}$$

The aim of this example is to point out that the strong repulsion between the molecules, such as the hard sphere interaction can produce a roughly periodic density distribution. In liquids, this phenomenon was observed experimentally long ago. (See Fig. 13-2.)

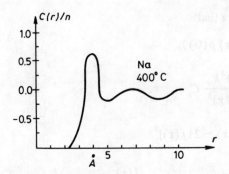

Fig. 13-2 The density correlation function of liquid sodium at 400°C.

13.3. The Fermion Gas

Next consider the density correlation function in a quantum gas. We first examine fermions.

The simplest method in determining the density correlation function is to start from its definition:

$$C_\mathbf{k} = \int d^3r \, e^{-i\mathbf{k}\cdot\mathbf{r}} \langle \rho(\mathbf{r}) \rho(0) \rangle_c \quad , \tag{13-41}$$

and is proportional to the scattering rate. The scattering process is as follows. An external particle comes in and transfers a momentum $\mathbf{k}$ to the body and goes out. Hence, it changes the momentum of a particle from $\mathbf{p}$ to $\mathbf{p}+\mathbf{k}$. We know that the scattering rate must be proportional to $f_\mathbf{p}(1 - f_{\mathbf{p}+\mathbf{k}})$ (see Chapter 3)

$$C_\mathbf{k} \propto \frac{2}{V} \sum_\mathbf{p} f_\mathbf{p}(1 - f_{\mathbf{p}+\mathbf{k}}) \quad . \tag{13-42}$$

The factor 2 on the right hand side comes from the two states of spin $\tfrac{1}{2}$ and $f_\mathbf{p}$ is the average particle number in the $\mathbf{p}$ state, given by

$$f_\mathbf{p} = \frac{1}{e^{(\epsilon-\mu)/T} + 1} \quad , \tag{13-43}$$

while $1 - f_{\mathbf{p}+\mathbf{k}}$ is the probability that the state $\mathbf{p}+\mathbf{k}$ is empty. The proportionality constant in (13-42) is in fact 1 because when f is very small (high temperature, low density case), (13-42) becomes the well known ideal gas result (13-23).

From (13-42) we find

$$C(\mathbf{r}) = \langle \rho(\mathbf{r}) \rho(0) \rangle_c$$

$$= \int \frac{d^3k}{(2\pi)^3} C_\mathbf{k}$$

$$= n\delta(\mathbf{r}) - 2|f(\mathbf{r})|^2 \quad , \tag{13-44}$$

$$f(\mathbf{r}) \equiv \frac{1}{V} \sum_\mathbf{p} f_\mathbf{p} e^{i\mathbf{p}\cdot\mathbf{r}} = \frac{I(r)}{2\pi^2 r} \quad ,$$

$$I(r) \equiv \int_0^\infty dp \, \frac{p \sin pr}{e^{(\epsilon-\mu)/T} + 1} \quad . \tag{13-45}$$

The integral $I(r)$ is known[a]

$$I(\mathbf{r}) = \frac{\partial}{\partial r} \left[\frac{1/2\xi}{\sinh(r/2\xi)} \sin(p_F r) \right] \quad ,$$

$$2\xi \equiv p_F/mT \quad ,$$

$$\mu = p_F^2/2m \quad . \tag{13-46}$$

If $r \ll \xi$ or $T \to 0$, then

$$I(\mathbf{r}) \simeq \frac{\partial}{\partial r} \left[\frac{1}{r} \sin p_F r \right] \quad ,$$

$$C(\mathbf{r}) \simeq n\delta(\mathbf{r}) - \frac{p_F^4}{2\pi^4 r^2} \left[\frac{\cos p_F r}{p_F r} - \frac{\sin p_F r}{(p_F r)^2} \right]^2 \quad . \tag{13-47}$$

The special feature of this result is the appearance of periodic change of period $2\pi/2p_F$. This is caused by the sudden cutoff of the particle distribution at the Fermi surface. The exclusion principle causes the density to form a layered structure around the origin.

---

[a] See Landau and Lifshitz (1980) p. 358.

When $T \to 0$, there is no correlation length, i.e. $C(\mathbf{r})$ does not decrease as of $e^{-r/\xi}$. Rather $C(\mathbf{r}) \propto 1/r^4$, which is still a rapid decrease when r is large. If $T > 0$, there is a correlation length ξ (See Eq. (13-46)). When r is very large,

$$C(\mathbf{r}) \simeq \frac{p_F^2 \, e^{-r/\xi}}{2\pi^4 \xi^2 r^2} \sin^2 p_F r \quad ,$$

$$\xi = \frac{1}{p_F} \left(\frac{\epsilon_F}{T} \right) , \qquad (13\text{-}48)$$

where $\epsilon_F = p_F^2/2m$. At low temperatures, ξ is much larger than the distance between the molecules. In deriving (13-48), we have assumed that $\xi \gg 1/p_F$.

Let us now look at the response function $G(\mathbf{r})$ and its relation with $C(\mathbf{r})$. From (13-14) we get $G = C/T$. But this is valid only classically. Quantum mechanically this is incorrect. This will be discussed a little later. Let us directly examine G.

The meaning of $G(\mathbf{r})$ is this: if we add a potential $-\lambda \delta(\mathbf{r})$, then the change in the density is $\lambda G(\mathbf{r}) + O(\lambda^2)$. Now we add this potential and calculate the density. The effect of this potential is to change the wavefunction of the particle. Let the volume of the gas be a sphere centred at the region. Only the $l = 0$ part of the wavefunction is affected, because $\delta(\mathbf{r})$ only affects the wavefunction at the origin. This wavefunction is

$$\varphi_0(\mathbf{r}) = \frac{1}{\sqrt{V}} \frac{\sin(pr + \delta_0)}{pr}$$

$$\simeq \frac{1}{\sqrt{V}} \left[\frac{\sin pr}{pr} + \frac{\delta_0 \cos pr}{pr} \right] ,$$

$$\delta_0 = -\lambda m p / 2\pi \quad . \qquad (13\text{-}49)$$

The reader can evaluate δ_0 for himself. The change of density is

$$\lambda G(\mathbf{r}) = 2 \sum_{\mathbf{p}} f_{\mathbf{p}} \left[|\varphi_0(\mathbf{r})|^2 - |\varphi_0(\mathbf{r})|^2_{\lambda=0} \right] , \qquad (13\text{-}50)$$

the factor of 2 on the right comes from the two spin states.

Hence from (13-50) and (13-49), and after re-arrangement, we get

$$G(\mathbf{r}) = \frac{-m}{2\pi^2 r^2} I(2\mathbf{r}) \quad . \tag{13-51}$$

$I(2\mathbf{r})$ is just (13-45) but we need to change $\mathbf{r}$ to $2\mathbf{r}$. When $r \ll \xi$ or $T \to 0$, from (13-46) we get

$$G(\mathbf{r}) \simeq -\frac{2m p_F^4}{\pi^2} \left[\frac{\cos 2p_F r}{(2p_F r)^3} - \frac{\sin 2p_F r}{(2p_F r)^4} \right] \quad . \tag{13-52}$$

When $T > 0$, ξ becomes very important. When $r \gtrsim \xi$,

$$G(\mathbf{r}) \simeq -\frac{p_F m \, e^{-r/\xi}}{2\pi^3 r^2 \xi} \cos(2p_F r) \quad . \tag{13-53}$$

Therefore, although $C(\mathbf{r})$ and $G(\mathbf{r})$ have many properties in common, e.g. the periodic change and the correlation length, the relation between them is not trivial.

13.4. The Response and Correlation Functions in Quantum Mechanics

Now let us mention the relation of response and fluctuation in quantum mechanics. We again start from (13-2) where $\hat{B}, \hat{A}$ are operators which may not commute. Therefore, (13-3) may not be correct. Now let us find the first term of the expansion of (13-2) with respect to λ. Let

$$U(\beta) \equiv e^{\beta H} e^{-\beta(H - \lambda \hat{A})} \quad . \tag{13-54}$$

Then its differential is

$$\frac{\partial U}{\partial \beta} = e^{\beta H} (H - H + \lambda \hat{A}) e^{-\beta(H - \lambda \hat{A})}$$

$$= \lambda \hat{A}(\beta) U(\beta) \quad ,$$

$$\hat{A}(\beta) \equiv e^{\beta H} \hat{A} e^{-\beta H} \quad . \tag{13-55}$$

Integrating, we get

$$U(\beta) = 1 + \lambda \int_0^\beta d\tau \, \hat{A}(\tau) + O(\lambda^2) \quad . \tag{13-56}$$

Now (13-2) can be written as

$$B = \frac{\text{Tr } e^{-H/T} U(1/T) \hat{B}}{\text{Tr } e^{-H/T} U(1/T)} . \qquad (13\text{-}57)$$

The statistical sum $\sum_s$ in the phase space is now written as a trace. Substitute (13-56) into (13-57) and we get

$$\chi = \frac{\partial B}{\partial \lambda} = \int_0^{1/T} d\tau \left[\langle \hat{A}(\tau) \hat{B} \rangle - \langle \hat{A} \rangle \langle \hat{B} \rangle \right] , \qquad (13\text{-}58)$$

If $[\hat{A}, H] = 0$, i.e. A is a conserved quantity, then (13-58) is the same as (13-3), and the conclusion following (13-3) is then valid. Of course $[\hat{A}, H] \to 0$ as $\hbar \to 0$, i.e. in classical mechanics they commute.

Let $|s\rangle$ be a stationary state i.e. an eigenstate of H, then

$$\int_0^{1/T} d\tau \langle A(\tau) B \rangle = - \int \frac{d\omega}{2\pi} \frac{S(\omega)}{\omega} , \qquad (13\text{-}59)$$

$$S(\omega) \equiv 2\pi (1 - e^{-\omega/T}) \frac{1}{Z} \sum_{s,s'} e^{-E_s/T} A_{ss'} B_{s's} \delta(\omega - E_{s'} + E_s)$$

$$\equiv (1 - e^{-\omega/T}) S'(\omega) ,$$

$$Z \equiv \sum_s e^{-E_s/T} . \qquad (13\text{-}60)$$

E_s is the eigenvalue of the energy, and $A_{ss'} \equiv \langle s|A|s' \rangle$. The reader can prove

$$\langle AB \rangle = \int \frac{d\omega}{2\pi} S'(\omega)$$

$$= \int \frac{d\omega}{2\pi} \frac{S(\omega)}{1 - e^{-\omega/T}} . \qquad (13\text{-}61)$$

The above result can be summarised as follows: $S(\omega)$ is the frequency spectrum of the response function and $S'(\omega)$ is the frequency spectrum of the correlation function. Only at low frequencies, i.e. when $\omega \ll T$, is it true that

$$\frac{1}{\omega} S(\omega) \simeq \frac{1}{T} S'(\omega) \quad . \tag{13-62}$$

The Hamiltonian H not only represents the total energy but also governs the motion. $S'(\omega)$ is in fact the frequency distribution of the fluctuation:

$$S'(\omega) = \int dt\, e^{i\omega t} \langle \hat{A}(t)\hat{B} \rangle \quad ,$$

$$\hat{A}(t) \equiv e^{iHt} \hat{A} e^{-iHt} \quad , \tag{13-63}$$

where $\hat{A}(t)$ represents the behaviour of $\hat{A}$ with time. Hence from the analysis of the response function, we can at the same time analyse the motion[b]. Here ω is the energy difference, and on the other hand it also represents a time scale of the motion of $\hat{A}$. In classical mechanics $\hbar\omega$ can be neglected, i.e. $\hbar\omega/T \to 0$.

Therefore, the relation between the response function and the correlation function is relatively complicated in quantum mechanics. For the same reason the differential of the thermodynamic potential cannot simply be written as the correlation value of the fluctuation as in (13-12).

Now let us look at a simple example. Consider again the density fluctuation of a fermion gas. Let

$$\hat{A} = \frac{1}{\sqrt{V}} \rho_k \quad ,$$

$$\hat{B} = \frac{1}{\sqrt{V}} \rho_{-k} = \frac{1}{\sqrt{V}} \rho_k^* \quad . \tag{13-64}$$

Equation (13-60) can be used to calculate $S(\omega)$ and $S'(\omega)$.

It looks complicated, but in practice it is not hard to perform the calculation. The action of ρ_k^* is to increase the momentum of a particle by $\mathbf{k}$. This has been discussed in the analysis of the scattering experiments more explicitly,

$$\rho_k^* = \sum_{\mathbf{p},\sigma} a_{\mathbf{p}+\mathbf{k},\sigma}^* a_{\mathbf{p}\sigma} \quad , \tag{13-65}$$

---

[b] These techniques were developed by Matsubara (1955) and by the Russian School. See Abrikosov et al. (1963).

CORRELATION FUNCTIONS 241

where $\sigma = \pm 1$ is the spin direction and $a_{p\sigma}$ is the annihilation operator of a particle (of momentum **p** and spin σ), and $a^*_{p+k,\sigma}$ is the creation operator. Equation (13-65) implies a propagation of the state (p, σ) to state $(p + k, \sigma)$. Hence the $|s'\rangle$ in (13-60) is different from $|s\rangle$ in that the momentum of a particle is increased by **k**. The sum of (13-60) over s is

$$\frac{1}{Z} \sum_{s'} e^{-E_s/T} |\langle s' | \rho^*_k | s \rangle|^2 \, \delta(\omega - E_{s'} + E_s) \quad , \tag{13-66}$$

i.e. the average over $|s\rangle$. Let us now consider the term $a^*_{p+k\sigma} a_{p\sigma}$ in (13-65). If the state $|s\rangle$ has a particle in state (p, σ) and no particle in $(p + k, \sigma)$ then $\langle s | a^*_{p+k\sigma} a_{p\sigma} | s \rangle$ is 1; otherwise it is zero. The probability of being 1 is $f_p(1 - f_{p+k})$. The sum in (13-66) is to calculate this probability. Hence (13-66) is

$$f_p(1 - f_{p+k}) \, \delta(\omega - \epsilon_{p+k} - \epsilon_p) \quad . \tag{13-67}$$

This is just the result of one term in (13-65). To calculate (13-66), we have to sum over **p** and σ, and notice that $E_s - E_{s'} = \epsilon_{p+k} - \epsilon_p$, because only the momentum of a particle is changed. Substituting in (13-60) we get

$$S'(\omega) = \frac{2\pi}{V} \sum_p 2 f_p (1 - f_{p+k}) \, \delta(\omega - \epsilon_{p+k} + \epsilon_p) \quad ,$$

$$S(\omega) = \frac{2\pi}{V} \sum_p 2 (f_p - f_{p+k}) \, \delta(\omega - \epsilon_{p+k} + \epsilon_p) \quad . \tag{13-68}$$

Because of the δ-function, the particle numbers in (13-68) have the following relation

$$f_p - f_{p+k} = f_p (1 - f_{p+k})(1 - e^{-\omega/T}) \quad . \tag{13-69}$$

The reader may verify this for himself.

Substitute (13-68) into (13-61) and (13-59) and we get (13-42) and the response function

$$G_k = 2 \int \frac{d^3 p}{(2\pi)^3} \frac{f_p - f_{p+k}}{\epsilon_p - \epsilon_{p+k}} \quad . \tag{13-70}$$

This integral is not easy to do. We only consider the case when **k** is very small. In this case

$$f_p - f_{p+k} \simeq -\mathbf{k}\cdot\mathbf{v}\,\frac{\partial f_p}{\partial \epsilon_p},$$

$$\epsilon_p - \epsilon_{p+k} \simeq -\mathbf{k}\cdot\mathbf{v},$$

$$\mathbf{v} \equiv \frac{\partial \epsilon_p}{\partial \mathbf{p}},$$

$$G_k \simeq 2\int \frac{d^3p}{(2\pi)^3}\,\frac{\partial f_p}{\partial \epsilon_p} \simeq \frac{mp_F}{\pi^2}. \tag{13-71}$$

This result can also be obtained by integrating (13-51):

$$\lim_{k\to 0} G_k = \int d^3r\, G(r).$$

We use (13-45) to get $I(2r)$. This integration is easy and we immediately get (13-71).

Problems

1. In the last two chapters, we have talked about the independence of the parts of a system, the central limit theorem, scattering experiments, and the correlation function. The reader should go over these again. This chapter emphasises the problems of calculation. The reader should derive every equation for himself.

2. The expansion (13-11) of the thermodynamic potential is a series in the cumulants. This expansion formula had been dicussed in problem 4 of the last chapter. Let the interaction in that problem be

$$u(r) = e^2/r,$$

i.e. a particle with charge e.

(a) Calculate the first two terms of the expansion of the model (in powers of e^2).

(b) Is the result of (a) reasonable? That is, is the thermodynamic potential proportional to the volume and particle number? If not, why? Notice that in many cases of calculation, various kinds of infinite integrals appear. Such infinite integrals are usually regarded as a mathematical problems.

The usual practice is to "regulate" these integrals or simply discard them. But there is a reason for the appearance of these integrals. Either the model is not reasonable or there is problem with the limiting process, or it is due to the use of a certain approximation. Usually these seemingly unreasonable result will bring out new discoveries. Hence, unreasonable results must be understood. An infinite integral is not zero and cannot just be discarded.

3. The response function describes the response of a body to an external force. Of course, any molecule in the body can be regarded as an external agent on the other molecules. Hence, the response function can also be used to analyse the structure of the body itself. The following is an example. Suppose we have a hard sphere gas. Let G_{0k} be the density response function $C_{0k} = G_{0k}/T$, and C_{0k} is the density correlation function discussed in Sec. 13.2. Now assume that there is in addition an attractive force between the hard spheres, with an interaction $-v(r)$ where r is the distance between two spheres, and $v > 0$.

(a) Prove that the response function G_k is approximately,

$$G_k = G_{0k}/(1 - v_k G_{0k}) \quad , \tag{13-72}$$

where v_k is the Fourier transform of $v(r)$. This approximate result is similar to the mean field approximation in Chapter 27.

(b) If $v_k G_{0k}$ tends to 1, then (13-72) shows that G_{0k} becomes very large. Notice that $G_{0k} = C_{0k}/T$ and C_{0k} do not change with temperature (a property of the hard sphere gas; see the example in Sec. 13.2). Therefore, when

$$T \to v_k C_{0k} \quad ;$$

this system is very sensitive to the external force and undergoes large density fluctuations. It is not a stable condition. This can be regarded as the initial stage of a phase transition.

Let the radius of the hard sphere be a. Let

$$-v(r) = -\epsilon \, \theta(b - r) \quad . \tag{13-73}$$

Calculate $T_0 \equiv v_0 C_{00}$ and discuss the meaning of this temperature. Equation (13-26) can be used to calculate C_{00}. This is of course only a crude approximation.

4. Prove that one-dimensional model (13-29) − (13-32) does not exhibit instability, i.e. G_k and C_k are finite. Hence no phase transition can occur.

5. Expand the results of (13-29) − (13-32) in terms of the density, and use this to dicuss the low density approximation.

6. The last section on the quantum mechanical calculation can be found in some books on solid state. However, the reader should derive it for himself.

PART IV
APPLICATIONS

These six chapters discuss some slightly more complicated examples, and provide a crude analysis of a number of well-known phenomena. Most of the previous examples concern the ideal gas, but the examples below focus upon the interaction between molecules and introduce the phenomena of magnetism, impurities and phase equilibrium. The analysis of mutual interaction is the main theme of modern statistical mechanics, while the ideal gas model, without interactions, belongs to the realm of elementary statistical mechanics. The inclusion of interactions is a highly complicated problem, beyond the machinery of present day mathematics. We have to resort to some highly simplified models as well as various approximations.

Chapter 14 discusses the corrections to the ideal gas law; that is, we regard the interaction between molecules as a small quantity and make an expansion in this quantity to first order. In this chapter, we emphasise the influence of the interaction and the collision time on the thermodynamic potential. Chapter 15 gives a simple review of phase equilibrium, possibly the most remarkable consequence of mutual interaction. Chapter 16 reviews the different types of magnetism. Chapter 17 introduces the Ising model; this is the simplest model which can be used to analyse different phenomena, and is a very important tool. Chapter 18 discusses impurities, emphasising especially the difference between mobile and frozen impurities. Chapter 19 discusses the electrostatic interaction, which cannot be neglected even at large distances, leading to many interesting phenomena.

Chapter 14

CORRECTIONS TO THE IDEAL GAS LAW

When we discuss the ideal gas, we regard the interaction between molecules as causing collisions, but we neglect the duration of collision and other details. As a result, collisions only lead to changes of the momenta of the molecules. However, the effect of the interaction is more than this. In this chapter we shall discuss this and its influence on the ideal gas law $p = TN/V$. We only discuss the first correction term in the low density limit, i.e. the coefficient B of $T(N/V)^2$, the so-called second virial coefficient. Let us first review the interaction energy between molecules and point out its relation with B. Then we introduce the concept of time delay in collisions, and describe in greater depth the role played by collisions. We also touch upon some elementary concepts in the theory of scattering.

14.1. Intermolecular Interactions

The interaction between molecules naturally depends on the structure of the molecules. For simplicity, we shall only consider the interaction between simple molecules. If two molecules are so close that their electronic shells touch, a strong repulsion is produced because of the fermionic character of the electrons. If the two molecules are very far apart, there is a weak attractive force. This is mainly due to the electric dipole interaction. An atom alone has no electric dipole (i.e. the centre-of-mass of the electrons coincides with the nucleus), but this is only a time averaged property — the dipole changes rapidly, averaging to zero. If there are two atoms, the electrons will mutually interact. The interaction energy is proportional to r^{-6}, where r is the distance between the atoms or molecules. This r^{-6} interaction can be understood as follows:

CORRECTIONS TO THE IDEAL GAS LAW

Consider an atom at the origin. It possesses a fast rotating dipole **d**. The electric field at **r** is

$$\mathbf{E} \propto \mathbf{d}/r^3 \quad . \tag{14-1}$$

If we place another atom at **r**, then **E** will distort its orbit and produce a dipole moment

$$\mathbf{d}' \propto \mathbf{E} \quad . \tag{14-2}$$

So these two atoms produce an interaction energy

$$U(\mathbf{r}) = -\mathbf{d}' \cdot \mathbf{E} \propto -\mathbf{d}' \cdot \mathbf{d}/r^3$$

$$\propto -d^2/r^6 \quad . \tag{14-3}$$

Although **d** and **d**' each has a zero average, the average of d^2 is nonzero, and this is the reason for the r^{-6} attractive potential. The short-distance repulsion and the long-distance attraction result in an interaction potential as shown in Fig. 14-1, with a minimum $-\epsilon$ at $r = a$. The "6–12 potential",[a]

$$U(\mathbf{r}) = \epsilon \left[\left(\frac{a}{r}\right)^{12} - 2\left(\frac{a}{r}\right)^6 \right] \quad , \tag{14-4}$$

is commonly used as an approximation. Some examples are listed in Table 14-1. Even this crude interaction model has extensive applications. Many properties of gases, solids and liquids can be explained quite well by this model.

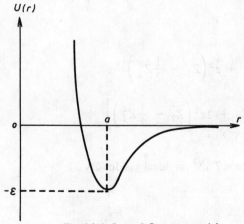

Fig. 14-1 Lenard-Jones potential.

[a] or Lenard-Jones potential.

Table 14-1 Minimum interaction energy and its distance

	$a(A)$	ϵ(Joule)
He	2.2	1×10^{-22}
H_2	2.7	4
Ar	3.2	15
N_2	3.7	13
CO_2	4.5	40

Tabor, (1979) p. 29

14.2. Corrections to the Ideal Gas Law

Now let us consider the corrections to the ideal gas law $p = (N/V)T$. We only discuss the $(N/V)^2$ term. We can start from the method of open systems (grand canonical ensemble). In Chapter 9 we have the expansion (see Eqs. (9-3) to (9-7)). Here, we keep up to order z^2:

$$p = \frac{T}{V} \ln \Sigma ,$$

$$\frac{TN}{V} = z \frac{\partial p}{\partial z} , \tag{14-5}$$

$$\ln \Sigma = z Z_1 + z^2 \left(Z_2 - \tfrac{1}{2} Z_1^2 \right) ,$$

$$\frac{N}{V} = \frac{1}{V} \left[z Z_1 + 2z^2 \left(Z_2 - \tfrac{1}{2} Z_1^2 \right) \right] . \tag{14-6}$$

$$z \equiv e^{\mu/T} (2\pi m T)^{3/2} = \text{ideal gas density} ,$$

$$Z_1 = V , \tag{14-7}$$

$$Z_2 = \frac{1}{2} \int d^3 r_1 \, d^3 r_2 \, \exp(-U(|\mathbf{r}_1 - \mathbf{r}_2|)/T) . \tag{14-8}$$

CORRECTIONS TO THE IDEAL GAS LAW

Rearranging the above equations, and after eliminating z^2, we obtain

$$p = \frac{N}{V}T\left(1 + B\frac{N}{V}\right) \qquad (14\text{-}9)$$

$$B \equiv -\frac{1}{V}\left(Z_2 - \frac{1}{2}Z_1^2\right)$$

$$= -\frac{1}{2V}\int d^3r_1\, d^3r_2\,\left[e^{-U(|\mathbf{r}_1-\mathbf{r}_2|)/T} - 1\right]$$

$$= -\frac{1}{2}\int d^3r\,\left[e^{-U(r)/T} - 1\right]\,. \qquad (14\text{-}10)$$

Therefore, to get B we only need to know the pairwise interaction of the particles. The value of the integral (14-10) is approximately equal to b^3, where b is the effective range of the potential $U(r)$. As seen from above, the expansion is a series in

$$zb^3 \sim \frac{N}{V}b^3\,. \qquad (14\text{-}11)$$

The low density limit applies when there is seldom any particle within the interaction range, i.e. the interaction is seldom effective.

Because $U(r)$ consists of a strongly repulsive core and a weakly attractive tail (see Fig. 14-1),

$$B \simeq \frac{2\pi}{3}a^3 + \frac{1}{2T}\int_a^\infty U(r)\,d^3r$$

$$\equiv v_0 - y/T\,. \qquad (14\text{-}12)$$

Substituting into (14-9) we obtain

$$p = \frac{N}{V}T\left(1 + \frac{Nv_0}{V}\right) - y\left(\frac{N}{V}\right)^2$$

$$\simeq \frac{NT}{(V - Nv_0)} - y\left(\frac{N}{V}\right)^2\,. \qquad (14\text{-}13)$$

250 STATISTICAL MECHANICS

So the correction to the ideal gas law can be seen as (1) due to repulsion, which reduces the available volume for the molecules to $(V - Nv_0)$; and (2) due to attraction, which reduces the pressure. The latter arises because as a molecule approaches the wall, it is pulled back by inter-molecular attraction, so that the collision with the wall is less violent.

Equation (14-13) is the famous Van der Waals equation. We shall encounter it later when we discuss phase transition.

Density expansion is usually called virial expansion, and B is the so-called second virial coefficient. Equation (14-9) can be directly obtained from the virial theorem. This is left as an exercise. (See Problem 2.)

We can also obtain B from measurements of the pressure of the gaseous state. Together with Eq. (14-10), it gives a good knowledge on the interaction energy of Fig. 14-1.

14.3. Time Delay in Collisions

The above analysis is quite simple. Now we can look at the problem from a different viewpoint. This is more complicated, but exhibits more clearly the role of the motion of the molecules.

As in our earlier discussion on detailed balance, we emphasise the importance of collisions. To say that a gas is ideal does not imply that collisions never occur, but that the duration of collision is extremely short. So the corrections to the ideal gas law is caused by the finite collision time. That is to say, the time during which each molecule can move freely is influenced by the interaction. The following analysis shows that B in Eq. (14-10) is directly proportional to the "time delay" in a collision.

We first rewrite $z^2 Z_2$ as follows:

$$z^2 Z_2 = \frac{e^{-2\mu/T}}{2h^6} \int d^3p_1 \, d^3p_2 \, d^3r_1 \, d^3r_2 \, e^{-H/T} \quad , \tag{14-14}$$

$$H = \frac{p_1^2}{2m} + \frac{p_2^2}{2m} + U(|\mathbf{r}_1 - \mathbf{r}_2|)$$

$$= \frac{P^2}{2M} + \frac{p^2}{2m'} + U(r) \quad ,$$

$$d^3p_1 \, d^3r_1 \, d^3p_2 \, d^3r_2 = d^3R \, d^3P \, d^3r \, d^3p \quad , \tag{14-15}$$

where P is the total momentum of the two particles, $M = 2m$ is the total mass, $\mathbf{r}$, $\mathbf{p}$ are the relative position and momentum in the centre-of-mass system respectively, and $m' = m/2$,

$$\mathbf{P} = \mathbf{p}_1 + \mathbf{p}_2 \quad , \qquad \mathbf{p} = \tfrac{1}{2}(\mathbf{p}_1 - \mathbf{p}_2) \quad ,$$

$$\mathbf{R} = \tfrac{1}{2}(\mathbf{r}_1 + \mathbf{r}_2) \quad , \qquad \mathbf{r} = \mathbf{r}_1 - \mathbf{r}_2 \quad . \tag{14-16}$$

We assume that $U(r)$ is independent of the direction of $\mathbf{r}$. Substitute (14-15) into (14-14), and as the integrals over d^3R and d^3P do not involve $U(r)$, they can be performed immediately to give

$$z^2 Z_2 = \frac{e^{2\mu/T}}{h^3}(2\pi MT)^{3/2} V \times V \int d\epsilon\, g(\epsilon)\, e^{-\epsilon/T} \quad ,$$

$$Vg(\epsilon) \equiv \frac{1}{h^3}\int d^3r\, d^3p\, \delta\!\left(\epsilon - \frac{p^2}{2m'} - U(r)\right) \quad . \tag{14-17}$$

In this equation, $g(\epsilon)$ is obviously the energy distribution of the particle in its trajectory, and it can be directly computed from the knowledge of the trajectories. Trajectories can be classified by their constants of motion, which, besides the energy ϵ, also include the angular momentum (see Fig. 14-2)

$$\mathbf{l} = (\mathbf{r} \times \mathbf{p})/\hbar \quad . \tag{14-18}$$

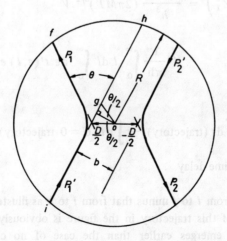

Fig. 14-2 The trajectory of the particle in the centre-of-mass system.

252 STATISTICAL MECHANICS

Equation (14-17) can now be written as

$$g(\epsilon) = \frac{1}{8\pi^3 \hbar} \int 2\pi l \, dl \, \frac{dp_r}{d\epsilon} \, dr \, d\Omega \quad , \tag{14-19}$$

where p_r is the radial component of p and the integration $d\Omega$ is over all incoming directions. Because $d\epsilon/dp_r = v_r = dr/dt$, we have

$$\frac{dp_r}{d\epsilon} \, dr = \frac{dt}{dr} \, dr = dt \quad , \tag{14-20}$$

which is the time differential, and (14-19) becomes

$$V g(\epsilon) = \frac{1}{8\pi^3 \hbar} \int 2\pi l \, dl \, 4\pi \int dt \quad . \tag{14-21}$$

If $U(r) = 0$, then

$$z^2 Z_2 = \frac{1}{2} Z_1^2 \quad . \tag{14-22}$$

Therefore, putting (14-21) into (14-17) we get

$$z^2 \left(Z_2 - \frac{1}{2} Z_1^2 \right) = \frac{e^{2\mu/T}}{h^3} (2\pi MT)^{3/2} V$$

$$\times \frac{1}{\pi \hbar} \int_0^\infty l \, dl \int_0^\infty d\epsilon \, t'(\epsilon, l) \, e^{-\epsilon/T} \quad , \tag{14-23}$$

$$t'(\epsilon, l) \equiv \int dt \, (\text{trajectory}) - \int dt \, (U=0 \text{ trajectory})$$

$$= \text{time delay} \quad . \tag{14-24}$$

This is the time from i to f minus that from i to h as illustrated in Fig. 14-2. The time delay of this trajectory in the figure is obviously negative, i.e. the outgoing particle emerges earlier than the case of no collision. However, $t'(\epsilon, l)$ as depicted in Fig. 14-3 would be positive.

CORRECTIONS TO THE IDEAL GAS LAW

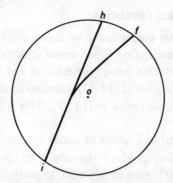

Fig. 14-3 Same as Fig. 14-2 except that the interaction is attractive and the trajectory of particle 2 is not shown.

To conclude, the above analysis shows that

$$B = \frac{2^{3/2} \lambda^3}{\pi \hbar} \int dl \, l \int d\epsilon \, t'(\epsilon, l) \, e^{-\epsilon/T} \quad ,$$

$$\lambda \equiv h/(2\pi mT)^{1/2} \quad . \tag{14-25}$$

This result shows that B is related to the trajectory only through the time delay. This time delay can be defined from the free moving parts of the trajectory. In Fig. 14-2 we only need to measure the times the particle enters and leaves the circle. The details of the trajectory within the circle are not required at all, i.e. to calculate B we only need information on the incoming and outgoing particle, while information during the interaction is not essential.

From Sec. 14-3 to the present analysis we notice an essential point — the coefficient B is not necessarily connected with the details of the interaction between the molecules. It can be directly expressed in terms of quantities before and after the interaction. Although B is a concept in statistical mechanics, and from Eq. (14-10) it is directly linked to the interaction $U(r)$, the above analysis shows that its calculation is not related to the short-distance details, and the information outside the effective range is sufficient. This is more clearly demonstrated in the quantum mechanical case, where details of the wavefunction $\psi(r)$ in the small r region is not essential. What is important is the $\psi(r)$ in the free region, where r is very large. A bound state can then be treated as a free particle.

The above analysis can be further generalised.[b]

[b] Dashen, Ma and Bernstein (1969); Dashen and Ma (1971).

14.4. Quantum Mechanical Calculation

To calculate B quantum mechanically we follow steps similar to those above. There are two modifications. One is concerned with the bosonic or fermionic character of the particles. This is not very difficult and we shall not discuss it. The second is to calculate $g(\epsilon)$ in (14-17) quantum mechanically without using the short wavelength approximation in (14-17). This is an interesting problem, as discussed below.

We restrict r to be within a sphere of radius L. The potential $U(r)$ is non-zero only in the neighbourhood of the origin. In quantum mechanics, the distribution function $g(\epsilon)$ for the energy ϵ can be written as

$$g(\epsilon) = \sum_{l=0}^{\infty} (2l+1) \sum_{n} \delta(\epsilon - \epsilon_{nl}) \quad, \tag{14-26}$$

where ϵ_{nl} are the eigenvalues of the wave equation

$$\left(-\frac{1}{2m'} \nabla^2 + U(r) \right) \psi_{nlm} = \epsilon_{nl} \psi_{nlm} \quad,$$

$$\psi_{nlm} = Y_{lm}(\mathbf{r}) \phi_{ln}(r) \quad, \tag{14-27}$$

with the boundary condition that ϕ_{ln} vanishes on the boundary $r = L$, which is far away from the origin. Outside the effective range of $U(r)$,

$$\phi_{ln}(r) \sim \frac{1}{r} \sin\left(kr + \frac{l\pi}{2} + \delta_l \right) . \tag{14-28}$$

The value of k is determined by the condition $\phi(L) = 0$, i.e.

$$kL + \frac{l\pi}{2} + \delta_l = n\pi \quad, \qquad n = \text{integer} \quad,$$

$$\epsilon_{nl} = \frac{k^2}{2m} . \tag{14-29}$$

In these equations, δ_l is the quantum mechanical phase shift. Its meaning can be simply explained as follows.

We are familiar with the concept of "image" in elementary optics. Mirrors and lenses cause light from a source to be reflected or refracted. The most convenient analysis is to say that the mirror or the lens creates an image. Light

CORRECTIONS TO THE IDEAL GAS LAW   255

then appears to radiate from the image. Once the position of the image is determined, the whole problem is solved. The phase shift in scattering is analogous. If $U(r) = 0$, i.e. the particle is not influenced by any interaction, then $\delta_l = 0$. In this case $\phi_{ln}(r)$ can be written as the sum of the incoming and outgoing waves and (14-28) becomes

$$\phi_{ln}(r) \propto -(-1)^l e^{-ikr} + e^{ikr} \quad , \qquad (14\text{-}30)$$

The last term is the outgoing wave. The first term on the right is the incoming wave and the $-(-1)^l$ factor is connected with the geometry of the 3-dimensional space, which we need not worry about at present. If $U(r) \neq 0$, the outgoing wave is altered and (14-28) becomes

$$\phi_{ln}(r) \propto -(-1)^l e^{-ikr} + e^{2i\delta_l} e^{ikr} \quad . \qquad (14\text{-}31)$$

The second term in (14-30) is now multiplied by a factor $e^{2i\delta_l}$. So the result of the collision is to add a phase of $2\delta_l$ to the outgoing wave. To see how the wave is scattered, we superpose waves with momenta near k, resulting in a spherical wave packet of finite thickness. We then consider how this wave moves outwards:

$$\psi(r, t) \propto \int dq\, A(q)\, e^{2i\delta_l(k+q)}\, e^{i(k+q)r}\, e^{-i\epsilon(k+q)t} \quad . \qquad (14\text{-}32)$$

In the above we consider only the outgoing wave and put back the time factor,

$$A(q) = \int dr\, e^{-iqr}\, a(r) \quad ,$$

$$a(r) = \frac{1}{2\pi} \int dq\, e^{iqr}\, A(q) \quad . \qquad (14\text{-}33)$$

The envelope $a(r)$ is much wider than $1/k$ (see Fig. 14-4),

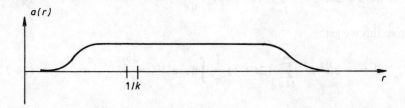

Fig. 14-4 Envelope $a(r)$ as a function of r.

and because of this, its Fourier transform $A(q)$ is very narrow, and the exponent in (14-32) can be expanded:

$$\delta_l(k+q) \simeq \delta_l(k) + q \frac{\partial \delta_l}{\partial k} ,$$

$$\epsilon(k+q) \simeq \frac{1}{2m'}(k+q)^2 \simeq \epsilon(k) + \frac{k}{m'} q . \quad (14\text{-}34)$$

Using the approximation in (14-34), we can write (14-32) as

$$\psi(r,t) \propto e^{2i\delta_l(k)} e^{ikr - i\epsilon(k)t} \int dq\, A(q)\, e^{iq(r - vt - r_0)}$$

$$\propto e^{2i\delta_l(k)} e^{ikr - i\epsilon(k)t} a(r - vt + r_0) , \quad (14\text{-}35)$$

$$v \equiv k/m' ,$$

$$-r_0 \equiv 2\partial \delta_l / \partial k . \quad (14\text{-}36)$$

The interpretation of (14-35) is very clear: the outgoing wave moves outwards with the velocity v and its source is situated at r_0. That is to say, the position of the "image" is at $r = r_0$. If $\delta_l = 0$ the source is at the origin. The introduction of $U(r)$ is to produce an "image" at r_0.

Let us return to (14-29) and (14-26). We want to find $g(\epsilon) - g_0(\epsilon)$ where $g_0(\epsilon)$ is the energy distribution when $U(r) = 0$. From (14-29) we get

$$\frac{dn}{d\epsilon} - \left(\frac{dn}{d\epsilon}\right)_0 = \frac{1}{\pi} \frac{d\delta_l}{d\epsilon} . \quad (14\text{-}37)$$

where $dn/d\epsilon$ indicates the number of states with angular momentum l within an energy interval $d\epsilon$. Therefore from (14-29) and (14-37) we get

$$g(\epsilon) - g_0(\epsilon) = \sum_{l=0}^{\infty} (2l+1) \frac{1}{\pi} \frac{d\delta_l}{d\epsilon} . \quad (14\text{-}38)$$

From this we get[c]

$$B = -2^{3/2} \lambda^3 \sum_{l=0}^{\infty} (2l+1) \frac{1}{\pi} \int d\epsilon\, e^{-\epsilon/T} \frac{d\delta_l}{d\epsilon} , \quad (14\text{-}39)$$

---

[c] This result appears in Bethe and Uhlenbeck (1937).

CORRELATIONS TO THE IDEAL GAS LAW 257

which is the second virial coefficient calculated quantum mechanically. This formula has not yet taken into account the statistics of the particles. If the particles are bosons, then l must be even. If the particles are fermions, we must consider the spin of the particles first. For spin $\frac{1}{2}$ fermions, when the spins are parallel, l must be odd and the whole expression must be multiplied by 3. When the two spins are antiparallel, then l must be even. The sum of the expression for parallel and antiparallel spins gives the coefficient B.

Now we can compare (14-39) with (14-25). From (14-36) we get

$$2 \frac{d\delta_l}{d\epsilon} = 2 \frac{d\delta_l}{dk} \bigg/ \frac{d\epsilon}{dk}$$

$$= -r_0/v$$

$$= t'(\epsilon, l) \quad . \tag{14-40}$$

Since r_0 is the displacement of the image, t' is just the time delay. That is to say, the outgoing wave is delayed by t'. Hence except for changing l to $l + \frac{1}{2}$ and replacing the integration by summation, (14-25) and (14-39) are exactly the same.

14.5. Bound State and the Levinson Theorem

The above analysis does not include the bound states of two particles. If the two particles can combine to form a new molecule, this is a chemical reaction

$$A + A \leftrightarrow A_2 \quad . \tag{14-41}$$

The chemical potential of A_2 is 2μ, its mass is $M = 2m$ and the total energy is the kinetic energy minus the binding energy W:

$$\frac{P^2}{2M} - W \quad . \tag{14-42}$$

The pressure of this A_2 gas is

$$p_2 = T \frac{e^{(2\mu + W)/T}}{\lambda_2^3} \simeq 2^{3/2} T \lambda^3 \left(\frac{N}{V}\right)^2 e^{W/T} \quad ,$$

$$\lambda_2 \equiv h/(2\pi MT)^{1/2} = \lambda/\sqrt{2} \quad . \tag{14-43}$$

To make one A_2, we need two A's. So the reaction (14-41) reduces the total pressure by p_2

$$p = \frac{N}{V} T - p_2$$

$$\simeq \frac{N}{V} T \left(1 - 2^{3/2} \left(\frac{N}{V}\right) \lambda^3 e^{W/T}\right) . \qquad (14\text{-}44)$$

Comparing with (14-9), we see that the effect of the bound state A_2 gives

$$B = -2^{3/2} \lambda^3 e^{W/T} . \qquad (14\text{-}45)$$

Combining the effects of collision and bound state we can write

$$g(\epsilon) - g_0(\epsilon) = \sum_{l=0}^{\infty} (2l+1) \left[\sum_\alpha \delta(\epsilon + W_{l\alpha}) + \frac{1}{\pi} \frac{d\delta_l}{d\epsilon} \right] , \qquad (14\text{-}46)$$

which is just Eq. (14-38) with the bound state $l\alpha$ added. Of course, if a certain l does not possess a bound state, then the delta function term $\delta(\epsilon + W_{l\alpha})$ does not exist. Equation (14-25) also does not include the effect of the bound state, and must be corrected.

By the way we mention here an interesting theorem in the theory of scattering, i.e. Levinson's theorem[d]

$$n_l = \frac{1}{\pi} \left[\delta_l(0) - \delta_l(\infty) \right] . \qquad (14\text{-}47)$$

Here n_l is the number of bound states of angular momentum l, which is shown to be related to the difference between the low energy and high energy phase shifts. This theorem, relating the results of scattering experiments, i.e. δ_l, to the bound states, seems very profound. But, from the viewpoint of statistics the meaning of this theorem is very obvious. The explanation follows below.

Let $g_l(\epsilon)$ be the energy distribution of different states with angular momentum l then

$$g_l(\epsilon) - g_{0l}(\epsilon) = (2l+1) \left[\sum_\alpha \delta(\epsilon + W_{l\alpha}) + \frac{1}{\pi} \frac{d\delta_l}{d\epsilon} \right] , \qquad (14\text{-}48)$$

---

[d] See books on quantum mechanics.

is the total number of states with angular momentum l. But the volume of the phase space with total angular momentum l should not be affected by the interaction $U(r)$, which only shifts the energy of different states and modifies the wavefunctions, without altering the total number of states. So

$$\int_{-\infty}^{\infty} d\epsilon \left[g_l(\epsilon) - g_{0l}(\epsilon) \right] = 0 \quad . \tag{14-49}$$

The integral on the right of (14-48) is

$$(2l + 1) \left[n_l + \frac{1}{\pi} \left(\delta(\infty) - \delta(0) \right) \right] . \tag{14-50}$$

From (14-49) and (14-50) we immediately get (14-47). (The reader should be aware that although (14-49) is true in general, it may not hold when terms with different l values are summed over, since the series is not guaranteed to converge.)

Problems

1. Prove the virial theorem starting from Newton's law

$$\sum_{i=1}^{N} \langle \mathbf{r}_i, \mathbf{F}_i \rangle + 2K = 0 \quad , \tag{14-51}$$

where $\mathbf{r}_i$ is the position of the i-th particle, $\mathbf{F}_i$ is the force acting on it, K is the total kinetic energy, $\langle \ldots \rangle$ is the long-time average and there are N particles.

2. The volume of a gas is V and the total energy is

$$H = \sum_{i=1}^{N} \frac{p_i^2}{2m} + \frac{1}{2} \sum_{i,j} U(|\mathbf{r}_i - \mathbf{r}_j|) \quad . \tag{14-52}$$

(a) Use the virial theorem to prove:

If $U = 0$, then $pV = \frac{2}{3} K$. $\tag{14-53}$

(b) Let a be the effective range of U, i.e.

$$U(r) = 0, \quad r > a \quad . \tag{14-54}$$

Prove: if $a \to 0$, even if $U(r)$ is very large for $r < a$, pV is still approximately equal to $(2/3)K$.

(c) If $\alpha = Na^3/V \ll 1$, derive the second virial coefficient B:

$$pV = NT\left(1 + B\left(\frac{N}{V}\right) + O(\alpha^2)\right). \qquad (14\text{-}55)$$

Starting from (14-51), use the correlation function in the previous chapter to obtain $\sum_i \langle \mathbf{r}_i, \mathbf{F}_i \rangle$.

3. A metal contains some fixed impurity atoms. The interaction potential between each electron and the impurity atom is $U(r)$, where r is the distance between the electron and an impurity atom. Let the scattering phase shift of the electron be δ_l.

(a) Obtain the change of specific heat due to the impurities.
(Hint: specific heat is only related to $g(\epsilon)$, assuming that the density of the impurity atoms is very low.)

(b) If the impurity atoms are hard spheres with radius a, evaluate the result of (a). Consider the two special uses of $k_F a \ll 1$ and $k_F a \gg 1$ where k_F is the momentum of the electrons at the Fermi surface. The case $k_F a \gg 1$ is very easy. We need not consider δ_l and only the volume occupied by the impurities is needed.

4. The calculation of higher virial coefficients is relatively complicated. The expansion can be started from Problem 4 of Chapter 12. This aspect is discussed in many books, e.g. Brout and Carruthers (1963), Chapter 1, which the reader can consult.

Chapter 15

PHASE EQUILIBRIUM

Generally speaking, under fixed conditions (e.g., at fixed temperature, pressure and magnetic field) the properties of a substance (e.g., its density, entropy and magnetic moment) have unique values. (We disregard the small thermal fluctuations.) However, in some special situations, some properties may not have a unique value and may take on two or more different values. For example, at 100°C and one atmospheric pressure, water can be in the gas phase (low density) or the liquid phase (high density). For a magnet, if there is no external magnetic field, its magnetic moment is not definite and can point in any direction. Under these special situations, vapour and water can coexist, or one-half of the spins of the magnet can point upwards and one-half downwards. This phenomenon is called phase equilibrium. This is a quite general, but extremely difficult problem, and is also one of the most fascinating topics in modern statistical mechanics. This chapter merely introduces some basic concepts and well-known facts about the topic.

15.1. Gaseous and Liquid Phases

The interaction between atoms is a strong repulsion at short distances and a weak attraction at large distances (Fig. 15-1). Thus each atom is like a hard sphere slightly attracting other atoms. When the temperature is low the minimum of $F = E - TS$ is determined by E. To minimize E, the system assumes the crystalline form. At high temperatures, F is dominated by $-TS$, and the system is in the gas phase because S is large for the gaseous state. Liquid appears for intermediate temperatures. On one hand, the interatomic distance is close to a (Fig. 15-1), while on the other hand, the position is not fixed and

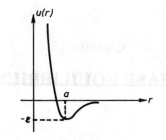

Fig. 15-1 Interaction potential between atoms.

S is not small. The detailed motion of the atoms in a liquid is very complicated — there is crystalline arrangement on a short time scale, as well as sliding, rotation and vibration, etc. The details depend on the structure and other properties of individual atoms or molecules, which will not be discussed here.

Now we discuss some elementary facts about the coexistence of two phases. When two phases coexist, e.g., water in a container with vapour above, the substance is not homogeneous because the densities of water and vapour are different. But the water phase is homogeneous; so is the vapour phase. These two phases can then be treated as two separate bodies under fixed temperature and pressure. As molecules can move freely from one system to the other, the chemical potential of water and that of the vapour must be the same to be in equilibrium:

$$\mu_1(T,p) = \mu_2(T,p) \quad . \tag{15-1}$$

Initially, the meanings of μ_1 and μ_2 seem obvious. However, further scrutiny reveals problems which require some explanation. We shall discuss some simple conclusions from (15-1) and defer the complex ones to the end of this chapter.

Equation (15-1) defines a curve on the (T,p) plane. Only points on this curve satisfy condition (15-1). This curve is the demarcation between phase 1 and phase 2 (see Fig. 15-2). The end point of the line separating the gas phase and the liquid phase is called the critical point. Above this point there is no distinction between the gas and the liquid phases. For the same reason, if there are three phases in equilibrium, then (T,p) must satisfy

$$\mu_1(T,p) = \mu_2(T,p) = \mu_3(T,p) \quad . \tag{15-2}$$

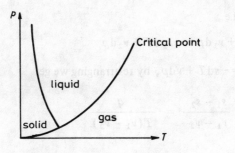

Fig. 15-2

(T, p) can be regarded as the intersection of the three transition lines, i.e., the triple point.

The gas phase is called the vapour. The equilibrium pressure is then called the saturated vapour pressure. Now let us review some terms and basic knowledge.

A. Latent Heat

The latent heat q is the energy required to move a molecule from one phase to another. Let

$$\epsilon_1 = E_1/N_1 \quad , \qquad \epsilon_2 = E_2/N_2 \quad ,$$
$$v_1 = V_1/N_1 \quad , \qquad v_2 = V_2/N_2 \quad , \qquad (15\text{-}3)$$
$$s_1 = S_1/N_1 \quad , \qquad s_2 = S_2/N_2 \quad .$$

To move a molecule from phase 2 to phase 1, we require not only the energy $\epsilon_1 - \epsilon_2$, but we also have to do work $p(v_1 - v_2)$ against pressure. Therefore

$$q = (\epsilon_1 - \epsilon_2) + p(v_1 - v_2) \quad . \qquad (15\text{-}4)$$

Since $\mu_1 - \mu_2 = 0$ and $\mu_{1,2} = \epsilon_{1,2} - Ts_{1,2} + pv_{1,2}$, so

$$(\epsilon_1 - \epsilon_2) - T(s_1 - s_2) + p(v_1 - v_2) = 0 \quad , \qquad (15\text{-}5)$$

i.e., $q = T(s_1 - s_2)$. The latent heat can be measured directly. Therefore the difference in entropy of the two phases is directly measurable.

B. Equation of the transition line

Along the transition line, $\mu_1(T, p) = \mu_2(T, p)$. Differentiating, we get

$d\mu_1 = d\mu_2$, i.e.,

$$-s_1 dT + v_1 dp = -s_2 dT + v_2 dp \quad . \tag{15-6}$$

Since $d\mu = -s dT + v dp$, by re-arranging we get

$$\frac{dp}{dT} = \frac{s_1 - s_2}{v_1 - v_2} = \frac{q}{T(v_1 - v_2)} \quad , \tag{15-7}$$

where q is introduced from (15-5). This is the equation for the transition line of phases 1 and 2. This equation is very important, because q, t, N, and v_2 are directly measurable and the whole transition line can be calculated by integration. If one of the two phases is a gas, the transition line is called the vapour pressure curve.[a]

Now we use (15-7) to examine the transition line between solid and gas at extremely low temperatures. This serves as a simple application. Let 1 be the gas phase and 2 the solid phase. When the temperature is low, the pressure is also low. So we can treat the vapour as an ideal gas. The density of the solid is much greater than that of the vapour. Therefore,

$$v_1 - v_2 \approx v_1 \quad . \tag{15-8}$$

Since $pv_1 = T$,

$$\frac{dp}{dT} = \frac{qp}{T^2} \quad . \tag{15-9}$$

Now

$$q = \epsilon_1 - \epsilon_2 + p(v_1 - v_2)$$

$$\approx \epsilon_1 - \epsilon_2 + T \approx \epsilon_1 - \epsilon_2 \quad , \tag{15-10}$$

where $\epsilon_1 - \epsilon_2$ is the energy required to move one molecule out of the solid. The pressure outside the solid is very low and T is negligible. Hence if $\epsilon_1 - \epsilon_2$ does not change with T, (15-9) can be immediately solved and we get the saturated vapour pressure of the solid,

$$p \propto e^{-(\epsilon_1 - \epsilon_2)/T} \quad . \tag{15-11}$$

This is as expected.

---

[a]Equation (15-7) is also called the Clausius-Clapeyron equation.

C. Rate of evaporation

Let us estimate how many molecules per second evaporate from a unit area of the liquid surface. Let the liquid be in equilibrium with its vapour. Then the evaporation rate is equal to the rate at which the gas molecules enter the liquid. Suppose the vapour is so dilute that it can be regarded as an ideal gas, then the number of molecules hitting the liquid surface per unit time per unit area is

$$n\bar{v}_z = n \frac{\int_0^\infty dp_z (p_z/m) e^{-\beta p_z^2/2m}}{\int_{-\infty}^\infty dp_z\, e^{-\beta p_z^2/2m}} = \frac{nT}{\sqrt{2\pi m T}} \quad , \qquad (15\text{-}12)$$

where n is the density of the gas and $\bar{v}_z$ is the average velocity normal to the liquid surface. Some of the molecules will rebound from the surface, and thus the actual number of molecules entering the liquid is only a fraction α of (15-12), with $0 < \alpha < 1$. The parameter α may be called the "sticking coefficient". As the number entering is equal to that leaving,

$$\text{evaporation rate} = \alpha \frac{p}{\sqrt{2\pi m T}} \quad . \qquad (15\text{-}13)$$

The value of α depends on the details of the motion and the structure of the liquid surface, etc., and $p = nT$ is the saturated vapour pressure.

Now even if the vapour is not saturated, (15-13) is still correct because the evaporation rate is not strongly dependent on the number of molecules entering. We only need to take T as the temperature of the liquid, and the saturated vapour pressure may be regarded as a known function of temperature. Of course, if the vapour is not saturated, more molecules will leave the liquid. After a while all the liquid will evaporate.

Evaporation is different from boiling. If a bubble appears in the liquid and can maintain itself from collapse, this temperature is called the boiling point. The pressure in the bubble is the saturated vapour pressure. Therefore when the saturated vapour pressure equals the liquid pressure (i.e. the external pressure, usually one atmosphere), further heating produces boiling.

15.2. The Growth of Water Droplets

The above discussion focuses on the coexistence of two phases and does not consider the boundary between the two phases. If the number of molecules in the boundary is much smaller than the number in the liquid and the gas, then it is indeed right to neglect them. However, sometimes the number is not negligible. For a water droplet, the molecules near the surface strongly influence the properties of the droplet. The growth of a water droplet is the first step of condensation. This is a major problem and is unavoidable in the discussion of the formation of rain, dew, frost and fog. Meteorologists have done much research on this topic. We introduce the basic concepts of the growth of water droplets here.

The water droplet is not only subjected to constant temperature, constant pressure, but also to an open environment. We define the thermodynamical potential Λ as

$$\Lambda(p, T, N) = G(p, T, N) - \mu N \quad . \tag{15-14}$$

Here N is the number of water molecules inside the droplet, μ is the chemical potential of the environment, i.e. the chemical potential of the vapour, and G is the Gibbs' free energy of the water droplet. In equilibrium, the minimum of Λ occurs when

$$\partial \Lambda / \partial N = 0 \tag{15-15}$$

i.e., $\mu = \partial G / \partial N$. If N is a large number, then $G(N)$ should be

$$G(N) = Ng + N^{2/3} \eta + \ldots , \tag{15-16}$$

where we have neglected terms smaller than $N^{2/3}$. If $N \gg N^{2/3}$, we can neglect the $N^{2/3}$ term, i.e., the boundary can be neglected. Now G must be equal to $\mu' N$, where $\mu'(T, p)$ is the chemical potential of the liquid. Therefore $g = \mu'$. The second term comes from the surface tension of the liquid. The number of molecules on the surface is proportional to $N^{2/3}$, provided the surface is sufficiently flat. These molecules have no neighboring molecules on one side and consequently experience less attractive force, and therefore have smaller negative energy. The energy increases with surface area and η is proportional to the surface tension α. (Note: this α is different from that in (15-13).) $\eta N^{2/3} = \alpha 4 \pi r^2$. Therefore,

$$\alpha = \frac{\eta}{4\pi} \left(\frac{4\pi}{3} n' \right)^{2/3} , \tag{15-17}$$

where n' is the density of the liquid and r is the radius of the water droplet. Notice that we have not yet given a definition of the water droplet. The surface of the water droplet is not a definite geometric shape. Which molecules are inside and which are outside the water droplet is not well-defined and thus the so-called surface of the water droplet is only an approximate boundary. In (15-16) the neglected term has no great significance because the value of N has only an approximate meaning. (Generally speaking, N is defined within $\pm\sqrt{N}$. Note that $\sqrt{N} < N^{2/3}$.) Substituting (15-16) in (15-14) we get

$$\Lambda(N) \simeq (\mu' - \mu)N + N^{2/3}\eta \quad . \tag{15-18}$$

Figure 15-13 shows $\Lambda(N)$. The last term always increases with N, but the behavior of $(\mu' - \mu)N$ is determined by whether μ' or μ is larger. If $\mu' < \mu$, then $\Lambda(N)$ always increases with N. If $\mu' < \mu$, $\Lambda(N)$ first increases and then decreases. Recall that μ is the chemical potential of the vapour around the water droplet, and μ' is the chemical potential of the liquid at the same temperature and pressure. Both μ and μ' can be expressed in terms of the pressure p of the vapour and the saturated vapour pressure p'. If we treat the vapour as an ideal gas, then

$$p = e^{\beta\mu} \frac{z}{\lambda^3} \quad , \tag{15-19}$$

$$\frac{z}{\lambda^3} \equiv \int \frac{d^3p}{(2\pi)^3} e^{-\beta p^2/2m} \sum_\alpha e^{-\beta\epsilon_\alpha} \quad . \tag{15-20}$$

The subscript α in (15-20) denotes the internal states of the water molecule. From (15-19) we get

$$\mu = T[\ln p - \ln(z/\lambda^3)] \quad ,$$
$$\mu' = T[\ln p' - \ln(z/\lambda^3)] \quad , \tag{15-21}$$
$$\mu - \mu' = T\ln(p/p')$$
$$\equiv T \ln s \quad ,$$

where $s = p/p'$ is called the degree of saturation. Note that p' is the vapour pressure when water and vapour are in equilibrium, i.e., the saturated vapour pressure. If the pressure p around the water droplet is smaller than p', i.e. $s < 1$, we call the vapour unsaturated. If $s > 1$, we call it supersaturated. In supersaturation, $\mu' < \mu$. When N is large enough, $\Lambda(\mu)$ will drop (see Fig. 15-3). As N will change in direction to lower $\Lambda(N)$, N will increase and

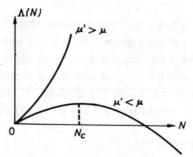

Fig. 15-3 Water droplet thermodynamical potential Λ as a function of number of molecules N.

the water droplet will grow indefinitely. If the vapour is unsaturated, N will not increase and the water droplet will disappear. Of course, the most important point is that in a supersaturated vapour, if the volume of the water droplet is smaller than the critical volume N_c (Fig. 15-3), it is bound to disappear. So, if there is no water droplet with volume greater than the critical volume, the vapour will not condense to water. The value of N_c can be obtained by differentiation.

$$N_c^{1/3} = \frac{2\eta}{3T\ln s}. \qquad (15\text{-}22)$$

Since the degree of saturation s appears in the formula as $\ln s$, its effect on N_c is slight. An increase of s by several times will not affect N_c much. For $T \sim 300\,\text{K}$, $s \sim 250\%$ and $N_c \sim 300$ the water droplet is then of radius 13 Å. Although this critical volume is not large, in practice it is not easy to make 300 molecules come together by collision. Experimentally, even supersaturated vapour with $s \sim 500\%$ will not condense. This, of course, refers only to pure water vapour. If there are impurities in the vapour (e.g., dust, water droplets, and various charged particles such as ions, etc.), condensation can occur very quickly for s slightly greater than 1. Impurities form condensation nuclei and water droplets start to grow on the particles. The reader can consult books on meteorology for discussions on this topic. Many interesting problems such as the motion of fog, dew, frost, snow and raindrops and the techniques of inducing artificial rain, are all in the domain of meteorology.

15.3. Latent Heat and the Interaction Energy of Molecules

As mentioned in Sec. 15.1, the latent heat between solid and gas, i.e. the heat of sublimation, is approximately the same as the energy required to move an atom out of the solid (see (15-10)). This energy, of course, is related to the curve in Fig. 15-1. In the solid phase the interatomic distance is approximately equal to a. The interaction between a pair of atoms is about $-\epsilon$. Therefore, to move an atom out of the solid, we must break up n pairs of atoms if we assume each atom has n neighbors. So the heat of sublimation is

$$q = \frac{1}{2} n \epsilon \ . \tag{15-23}$$

The factor of $\frac{1}{2}$ arises because N atoms can have only $\frac{1}{2} Nn$ pairs of neighbors. For fcc (face-centered cube) and hcp (hexagonal close-packed) crystals, $n = 12$, i.e. $q = 6\epsilon$. Table 15-1 lists the heat of sublimation and the values of some of these types of crystals. The value of ϵ can be calculated or deduced from the properties of the gas (see Chapter 15). All the molecules in Table 15-1 are spherical for which the interaction energy in Fig. 15-1 is the most reliable. The so-called experimental values in the table are actually the sum of the heat of melting and the heat of evaporation. The heat of sublimation is not easily obtained by direct measurement. The heat of melting plus the heat of evaporation, being the energy required to change a solid to a liquid and then to a gas, should differ only slightly from the heat of sublimation. The numbers in Table 15-1 are approximate; nevertheless they show the relation between the intermolecular interaction and the heat of sublimation. Helium is an exception because helium is very light, and there is a large zero-point vibration. Its interatomic distance in a crystal is not near to that of the value of Fig. 15-1, but is shifted to a larger value.

The density of a crystal is very different from that of a gas, but is nearly the same as that of a liquid. So one may expect that the heat of melting is much

Table 15-1

Crystal	Heat of sublimation $q(10^3$ J/mole)		$\epsilon(J) \times 10^{22}$
	Experimental value	Calculation (15-23)	
He	0.08	0.33	0.9×10^{-22}
Ne	1.3	1.7	4.8
Ar	7.3	5.9	16.5
Kr	10.0	9.0	25

Tabor (1979), p. 1952.

less than that of sublimation, and the heat of evaporation is comparable to the heat of sublimation. This is indeed the case. In general, the heat of melting is around 3 – 10% of the heat sublimation.

15.4. Melting and the Lindemann Formula

Melting is a very complicated phenomenon. The atoms in a solid are arranged in a definite manner. As the temperature rises, the amplitude of vibration of the atoms increases and will produce defects in the structure. When the temperature reaches the melting point, the whole structure collapses, becoming a liquid. The melting point is related to the pressure and can be determined by the transition curve (15-7). If the volume of the solid is less than that of the liquid, then an increase in pressure will raise the melting point (i.e., it does not melt so easily). If the volume of the solid is larger than that of the liquid, an increase in pressure will lower the melting point. Ice and water is the most important example of the latter.

Since the density of the solid differs very little from that of the liquid, the interaction energy curve in Fig. 15-1 is not very helpful for understanding melting. The values of r before and after melting are both close to a. However, from very simple assumptions, one can obtain a surprisingly accurate relation between the melting point and the interaction energy. This is the Lindemann formula[b]. We shall briefly mention it here.

Let us assume that when the molecules in a crystal vibrate to a certain extent, the crystal melts. The amplitude of vibration is related to the temperature by the law of equipartition

$$\frac{1}{2} K \langle (r-a)^2 \rangle = \frac{1}{2} T \tag{15-24}$$

where K is the second derivative of the interaction potential at its minimum. Assume that when the amplitude reaches αa, the crystal melts; then

$$K a^2 \alpha^2 = T_0 \tag{15-25}$$

where T_0 is the melting point. The value of K can be directly determined from the elastic constants. The relation of Young's modulus E with K in a cubic crystal is $K = Ea$. Therefore

$$T_0 = E \frac{\alpha^2}{nk} , \tag{15-26}$$

---

[b] See books on solid state physics.

where $n = 1/a^3$ is the density of the crystal and k is Boltzmann's constant (we had set $k = 1$ previously). Both E and n can be measured directly, and we must choose an α value to fit the measured melting points. Table 15-2 shows some results. The good agreement of (15-26) is surprising. The fitted value of α is 0.1, which shows that the amplitude of vibration is not too large even when the crystal melts. Hence to understand the process of melting, we have to understand first why (15-26) works so well.

15.5. Definitions of μ_1 and μ_2

The preceding discussion mainly reviews some phenomena of the coexistence of liquid and gas. We now want to clarify some basic conceptual problems posed in (15-1). In (15-1), the definitions of μ_1 and μ_2 were not clear. Initially, μ_2 was the chemical potential of the gas and μ_1 that of the liquid, but the basic assumptions in statistical mechanics allow only one μ, i.e. $G(T, p)$ divided by N,

$$e^{-G(T,p)/T} = \sum_s e^{-[H(s)+pV(s)]/T} ,$$

$$\mu = G/N . \qquad (15\text{-}27)$$

From (15-27) the calculation of G does not mention the gas phase or the liquid phase or its equilibrium. At a fixed temperature and pressure, there

Table 15-2 Melting point calculated from Lindemann formula with $\alpha^2 = 0.01$.

	T_0 from (15-26) (K)	T_0 from experiment (K)
Lead	400	600
Silver	1100	1270
Iron	1800	1800
Tungsten	4200	3650
Common salt	1200	1070
Fused quartz	1900	2000

Taken from Tabor (1979), p. 246.

is only one G and only one μ. Then where do the two chemical potentials μ_1 and μ_2 come from? Now let us examine carefully the meanings of μ_1 and μ_2.

Under a fixed pressure, Eq. (15-1) states that we can find the intersection point of two curves $\mu_1(T)$ and $\mu_2(T)$. This intersection point fixes the equilibrium temperature. This implies that $\mu_2(T)$ is still defined as below $T(p)$ (at least in the neighborhood of $T(p)$) and $\mu_1(T)$ is defined as above $T(p)$. Otherwise we do not know the intersection point. But what is the meaning of $\mu_1(T)$ above $T(p)$ and $\mu_2(T)$ below $T(p)$? From the basic assumptions of statistical mechanics, they are meaningless because (15-27) clearly indicates that there is one and only one $\mu(T)$ curve. We should really think of (15-1) and Fig. 15-4 as the situation depicted in Fig. 15-5, where the unwanted parts have been removed and there is only one curve. At $T(p)$ there is a discontinuity in the slope of this curve.

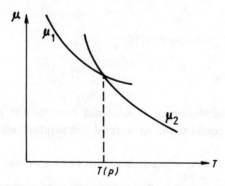

Fig. 15-4 Plots of the chemical potentials for 2 phases.

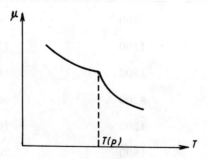

Fig. 15-5 Discontinuity of $\mu(T)$ at $T(p)$.

Now the problem is: (15-27) looks like a continuous function of T and p, and cannot develop any discontinuity in its slope. Each term in the sum, i.e., each statistical weight

$$e^{-(H+pV)/T}, \qquad (15\text{-}28)$$

is smooth in T and p and infinitely differentiable. Where does the discontinuity in the slope come from? This mathematical problem is of course extremely important. If the sum in (15-27) contains a finite number of terms there would be no discontinuity. However the number of the molecules, N, is large. When N tends to infinity, continuity becomes a problem. Yang and Lee pointed out that when $N \to \infty$, a discontinuity can occur, and if N is sufficiently large, the slope of the curve $\mu(T)$ appears to be discontinuous. There is considerable literature on this subject.[c]

Now that the meaning of the curve in Fig. 15-5 and of $T(p)$ is clarified, we can see that the extrapolated portions of μ_1 and μ_2 in Fig. 15-5 are not entirely meaningless. Below $T(p)$, the gas phase can still exist i.e., as a supercooled gas (i.e., supersaturated vapour). Above $T(p)$, the liquid phase can also exist, i.e., as superheated liquid. However these supercooled and superheated states are metastable. A slight disturbance will vaporise or condense the particles. The cloud chamber and bubble chamber for detection of high energy charged particles utilise the principles of supercooled vapour and superheated liquid. When a charged particle passes through, vapour condenses to form tiny water droplets, or liquid forms small bubbles. In Sec. 15.2, where we discussed the water droplet, we encountered no difficulty; only the duration of time is an important problem. Direct calculation from (15-27) often misses the properties of metastability, which are often important characteristics of a phase transition.

Finally we mention some problems in terminology. Since in Fig. 15-5, the thermodynamical potential is discontinuous in its first derivative, transitions such as that from liquid to gas are called first-order transitions. If in a certain transition the second derivative, e.g., the specific heat

$$C = T \frac{\partial S}{\partial T} = -T \frac{\partial^2 G}{\partial T^2}, \qquad (15\text{-}29)$$

[c]Yang and Lee (1952), Griffiths (1972).

is discontinuous, then this transition is called second-order. In fact, some commonly known second-order transitions such as the transition between the paramagnetic and ferromagnetic phases and the superfluid transition of liquid helium, should not be called second-order, since at the transition temperature (i.e., the critical temperature), the specific heat approaches infinity. Only under certain approximations (the mean field calculation) is C finite and discontinuous. We should not attach too much significance to the terms "first-order" and "second-order".

Problems

1. The vapour pressure curve of a liquid also follows (15-9). Experiments show that at 99°C and 101°C the water vapour pressure is 733.7 and 788.0 mm Hg respectively. Use (15-9) to obtain the latent heat evaporation of water. (The latent heat of evaporation of waters is 9717 cal per mole at 100°C.)

2. Experimental results show that the entropy of evaporation (latent heat of evaporation divided by temperature) at the boiling point (at one atmosphere) is approximately the same for many liquids. See Table 15-3 (taken from Zemansky (1957), p. 322, Trouton's rule). How do we explain these results?

3. From the results of Table 15-3 and (15-9), prove that

$$\left(\frac{dT}{T}\right) \Big/ \left(\frac{dp}{p}\right) \approx 0.1 \quad,$$

where dp is the change in pressure, and dT is the change in the boiling point due to dp.

4. The boiling point of a certain liquid is 95°C at the top of a mountain and 105°C at the bottom. Its latent heat is 1 000 cal/mole. Calculate the height of mountain.

5. Discuss the evaporation and growth of water droplets in two-dimensional and in one-dimensional space (i.e., extend the discussion of Sec. 15.2).

6. When a gas and liquid are in equilibrium, $\partial p/\partial V = 0$, i.e., the total volume can change but the pressure is constant. Under these conditions, what is the meaning of the formula (13-19) in Chapter 13?

7. Above the critical temperature T_c, there is no distinction between gas and liquid. Also, above T_c, $\partial p/\partial V < 0$. At temperatures close to T_c, $-\partial p/\partial V$ is very small. Discuss the scattering of neutrons or photons in the neighborhood of T_c.

Table 15-3 Experimental value for boiling points and latent heat for some liquids

	T_b (boiling point K)	L Latent heat (cal/mole)	ΔS L/T
Ne	27.2	415	15.3
Ar	87.3	1 560	17.9
Kr	120	2 158	18.0
Xe	165	3 021	18.3
Rn	211	3 920	18.6
F_2	85.0	1 562	18.4
Cl_2	239	4 878	20.4
HCl	188	3 860	20.5
HBr	206	4 210	20.4
HI	238	4 724	19.9
N_2	77.3	1 330	17.2
O_2	90.2	1 630	18.1
NO	121	3 293	27.1
N_2O	373	9 717	26.0
CS_2	319	6 400	20.0
SO_2	263	5 955	22.6
H_2S	216	4 463	21.0
N_2O	185	3 956	21.4
NH_3	240	5 581	23.3
CH_4	112	1 955	17.5
CF_4	145	3 010	20.7
CCl_4	350	7 140	20.4
CH_2O	254	5 850	23.0
CF_2Cl_2	243	4 850	20.0
$CHCl_3$	334	6 970	20.8
C_2N_2	252	5 576	22.1

Table 15-3 Cont'd

	T_b (boiling point K)	L Latent heat (cal/mole)	ΔS L/T
C_2H_6	185	3 517	19.1
C_6H_6	353	7 350	20.8
$(C_2H_5)_2O$	307	6 470	21.1
Li	1 599	30 800	19.3
Na	1 156	21 800	18.8
K	1 030	18 600	18.1
Rb	985	17 100	17.4
Cs	986	15 700	15.9
Hg	630	14 100	22.4
Ga	2 540	62 000	24.4
In	2 273	51 600	22.7
Cd	1 040	23 900	22.9

Chapter 16

MAGNETISM

From the time of our ancestors, magnetism has always been a subject for research and application. In modern statistical mechanics and many domains of natural science, magnetism has played a very important role. Magnetic interactions can be classified into two types: the interaction of the external magnetic field with individual electrons or atoms, and the interaction among electrons or atoms. If the second type of interaction is relatively weak, the problem is simple. In general, paramagnetism and diamagnetism belong to this class. If the second type of interaction is very strong, the problem becomes very complicated because we have to consider the collective behavior of many spins: ferromagnetism and antiferromagnetism, etc. belong to this second class of problems. Of course, in this second class of problems, we also have to consider the external magnetic field. This chapter discusses paramagnetism and diamagnetism caused by electron orbits. Finally, we review the cause of the mutual interaction.

16.1. Paramagnetism

Let $s_1, s_2 \ldots s_N$ be N spins. Each spin comes from an electron or an atom or a nucleus, etc. Let the total energy be

$$H = H_0 - \mathbf{h} \cdot \mathbf{M} \quad ,$$

$$\mathbf{M} = g\mu_B \sum_i \mathbf{s}_i \quad . \tag{16-1}$$

The product $g\mu_B$ is the ratio of the magnetic moment to the spin. If it is the electron spin, then

$$g = 2,$$
$$\mu_B = e/(2mc) = 9.27 \times 10^{-18} \text{ erg/kG}$$
$$= 6.72 \times 10^{-2} \text{ K/kG} , \qquad (16\text{-}2)$$

where m is the mass of the electron and $kG = 10^3$ Gauss. The term H_0 in (16-1) includes all energy which does not involve the external field $\mathbf{h}$. Now assume that H_0 does not depend on s_i.

Suppose $s = \frac{1}{2}$. Each spin can exist in two states, i.e. parallel or antiparallel to $\mathbf{h}$. The free energy can be calculated very easily:

$$e^{-F/T} = \sum_s e^{-H/T}$$
$$= (e^{h'/T} + e^{-h'/T})^N e^{-F_0/T} ,$$
$$F = F_0 - NT \ln(2 \cosh h'/T) . \qquad (16\text{-}3)$$

The term F_0 is the free energy due to H_0 and

$$h' \equiv \frac{1}{2} g\mu_B h . \qquad (16\text{-}4)$$

By differentiation, we obtain the entropy and other quantities

$$S = -\left(\frac{\partial F}{\partial T}\right)_h = S_0 + N\left[\ln\left(2\cosh\frac{h'}{T}\right) - \frac{h'}{T}\tanh\frac{h'}{T}\right] , \qquad (16\text{-}5)$$

$$M = -\left(\frac{\partial F}{\partial h}\right)_T = \frac{1}{2} g\mu_B N \tanh\frac{h'}{T} \qquad (16\text{-}6)$$

$$E = E_0 - hM , \qquad (16\text{-}7)$$

$$F' \equiv E_0 - TS \equiv F + hM , \qquad (16\text{-}8)$$

where M is the total magnetic moment along $\mathbf{h}$. We shall discuss the meaning of F' below. E_0 and S_0 are the energy and entropy, respectively, calculated from H_0.

Note that here the free energy F is that under a fixed magnetic field, including the potential energy of the spins in the magnetic field, i.e. $-\mathbf{h} \cdot \mathbf{M}$. The magnetic moment $\mathbf{M}$ is treated as a variable and not a constant. To understand this point we use a gas as an illustration. In Chapter 7, when we discussed the thermodynamic potential, we pointed out that the energy E is the external energy and does not include the potential energy due to the external pressure. If we want to include this, then E and F should be replaced by

$$E \to A = E + pV ,$$
$$F \to G = F + pV , \qquad (16\text{-}9)$$

where pV is the potential energy due to the external pressure p.

The work in the first law of thermodynamics is defined in terms of the internal energy:

$$dW = -p\,dV ,$$
$$p = -\left(\frac{\partial E}{\partial V}\right)_S = -\left(\frac{\partial F}{\partial V}\right)_T . \qquad (16\text{-}10)$$

But in A and G, p is a constant and V is the variable whose equilibrium value is

$$V = \left(\frac{\partial A}{\partial p}\right)_S = \left(\frac{\partial G}{\partial p}\right)_T . \qquad (16\text{-}11)$$

In (16-1), (16-3) and (16-7), we have included the potential energy due to h. Therefore E_0 and F', like E and F in (16-9), do not include the potential energy $-hM$, ($-hM$ is like pV):

$$dW = h\,dM$$
$$h = \left(\frac{\partial E_0}{\partial M}\right)_S = \left(\frac{\partial F'}{\partial M}\right)_T . \qquad (16\text{-}12)$$

On the other hand, A and G in (16-10) are like E and F in (16-7) and (16-3).

$$M = -\left(\frac{\partial E}{\partial h}\right)_S = -\left(\frac{\partial F}{\partial h}\right)_T . \qquad (16\text{-}13)$$

In general, the free energy F' of any magnetic substance can be regarded as a function of M and T, while $F = F' - hM$ is a function of h and T:

$$dF' = -S dT + h dM \quad ,$$
$$dF = -S dT - M dh \quad . \tag{16-14}$$

This is not limited to paramagnetism.

Now let us discuss cooling by adiabatic demagnetisation or "nuclear cooling", a major achievement in experimental low temperature physics. In low temperature experiments, we attempt to absorb as much heat as possible from the substance under observation and thus we must lower the temperature of its environment. A low temperature environment is simply one which absorbs heat readily. To absorb heat requires freedom of molecular, atomic or nuclear motion. At low temperatures, all substances become solids. Vibrations essentially stop below 10^{-2} K. The only remaining mode of motion is that of the nuclear spins, whose mutual interaction is extremely small.[a] At 1K they can flip freely. Adiabatic demagnetisation utilises this kind of motion to absorb heat. Figure 16-1 is the curve (16-5) of entropy versus T for different magnetic fields h_1 and h_2, (S_0 is discarded.). If $h = 0$, then $S/N = \ln 2$. In the figure, $h_2 < h_1$. The entropy S is lower for higher magnetic fields because the magnetic field restricts the motion of the spin.

Cooling proceeds as follows: we first apply a strong magnetic field to align the spins (see Fig. 16-1, from A to B). Then we decrease the magnetic field

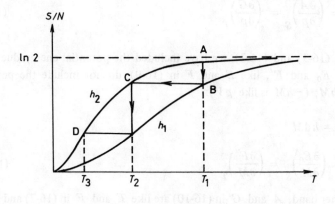

Fig. 16-1

[a] The magnetic interaction energy of two nuclear spins at a distance of 10^{-8} cm is approximately 10^{-30} K.

adiabatically. In adiabatic processes, entropy is unchanged. Therefore as the magnetic field is decreased the temperature is lowered along BC. This is cooling by adiabatic demagnetisation.

Note that when the magnetic field is lowered, the total magnetic moment M is unchanged. Entropy is a function of M alone and (16-5) can be written as

$$S = \frac{1}{2} N \left[(1+m) \ln \frac{2}{1+m} + (1-m) \ln \frac{2}{1-m} \right],$$

$$m \equiv M/N \quad . \tag{16-15}$$

Since the entropy is not changed, M is also unchanged. In small magnetic fields, the spins can flip more easily. This allows the absorption of more energy and hence a lowering of the temperature.

The process $A \to B \to C$ in Fig. 16-1 can be continued to D and then repeated. Of course, we can use magnetic fields of different strengths.

Although the above theory is very simple, the experiment is much more complex. Equilibrium is defined only on sufficiently long time scales, so that during the observation time, every spin must undergo many changes. These changes are brought about by the interaction of the spin with its surrounding environment. On one hand, we require the interaction to be strong so that the spins can change continually to reach equilibrium, thus keeping the observation time reasonably short. On the other hand, we want the interaction to be sufficiently weak so that this interaction will not hinder the free rotation of the spins. This poses quite a difficult task for the experimentalist. Some fifteen years ago, the experiment was performed using nuclear spins in metals and a temperature below 10^{-4} K was reached.[b] The use of metals is due to the fact that free electrons can still move at extremely low temperatures. At low temperatures, the number of mobile electrons is proportional to T, i.e., the specific heat is proportional to T. Although this is small, it is nevertheless larger than the T^3 specific heat due to phonons. The nuclear spin, through a slight interaction with the electrons, is then able to reach equilibrium. Even so, the observation time may require a week.

16.2. Magnetic Susceptibility

The magnetic susceptibility is defined as

$$\chi = \frac{\partial M}{\partial h} \quad , \tag{16-16}$$

[b] Osgood and Goodkind (1967); Dundon and Goodkind (1974).

usually under the condition $h \to 0$, i.e. $\chi = M/h$. The magnetic susceptibility for the model of the preceding section can be calculated from (16-6):

$$\chi = \tfrac{1}{4} N(g\mu_B)^2/T \ . \tag{16-17}$$

This inverse proportionality to T is a common phenomenon, called Curie's law.[c] In Chapter 13 (see(13-6)), we found that

$$\chi = \frac{1}{T} \langle M^2 \rangle_c \ . \tag{16-18}$$

If every spin is independent, then

$$\langle M^2 \rangle_c = \tfrac{1}{3} \langle (\Sigma s_i)^2 \rangle_c (\mu_B g)^2$$
$$= \tfrac{1}{3} N s(s+1)(\mu_B g)^2 \ . \tag{16-19}$$

The factor $\tfrac{1}{3}$ comes from the definition that M is the magnetic moment along the direction of h, i.e., $\langle M^2 \rangle = \tfrac{1}{3}\langle \mathbf{M} \cdot \mathbf{M} \rangle$. If $s = \tfrac{1}{2}$, we then get (16-17). Note that (16-18) assumes $[\mathbf{M}, H] = 0$, which is valid if $\mathbf{M}$ is conserved, or if we neglect quantum effects.

Now let us examine the magnetic susceptibility of a collection of fermions, e.g., electrons in a metal, or ^{3}He. We consider only the spin of the electrons and disregard magnetism due to orbital motion.

At $T = 0$, the energy levels are filled up to the Fermi energy ϵ_F. Each spin has two states — up spin and down spin. Therefore at $T = 0$, all the spins are fixed — one-half upwards, and one-half downwards. At $T > 0$, there are some particles outside the Fermi surface and some holes inside. The spins of these particles and holes are free to orient, contributing to the magnetic susceptibility. The number of these particles and holes is proportional to T. Therefore, at very low temperatures χ is a constant, because the denominator of (16-18) is also T. The calculation of this constant value of χ is very simple. Let $N_\pm$ be the number of up and down spins respectively. Then

$$M = \tfrac{1}{2} g\mu_B (N_+ - N_-) \ ,$$

$$\langle M^2 \rangle_c = \tfrac{1}{4}(g\mu_B)^2 [\langle N_+^2 \rangle_c - \langle N_-^2 \rangle_c] \ , \tag{16-20}$$

$\langle N_+ N_- \rangle = 0$ because N_+ and N_- are uncorrelated.

---

[c] Discovered by Mme. Curie.

Now use the relation in Chapter 13 (see (13-17)):

$$\langle N_\pm^2 \rangle_c = T \frac{\partial N_\pm}{\partial \mu} = \frac{1}{2} T \frac{\partial N}{\partial \mu} \quad . \tag{16-21}$$

At low temperatures $\partial N/\partial \mu$ is the density of states (see (4-14)):

$$Vg(0) = V \frac{m p_F}{\pi^2} \quad , \tag{16-22}$$

and thus for an increase in $d\mu$ of the Fermi energy μ, the number of occupied states will increase by $dN = Vg(0)d\mu$. From (16-18), (16-20) and (16-21) we finally obtain

$$\chi = \frac{1}{4} (g\mu_B)^2 \frac{\partial N}{\partial \mu} = \frac{V}{4} (g\mu_B)^2 \frac{m p_F}{\pi^2} \quad . \tag{16-23}$$

Note that at high temperatures, $\langle N_\pm^2 \rangle_c = N_\pm$, and χ reverts to the form (16-17). Equation (16-23) is the Pauli susceptibility.[d]

The experimental value of χ for ^{3}He is far greater than that given by (16-23). (The magnetic moment $g\mu_B$ of ^{3}He must be taken to be its nuclear moment.) The reason is that due to the interatomic interaction, liquid ^{3}He is far from being an ideal gas. Repulsion causes $\langle N_+ N_- \rangle_c < 0$, which, in turn, increases $\langle M^2 \rangle_c$ considerably.

16.3. Diamagnetism of Charged Particles

Charged particles move in circular orbits in a magnetic field. This can produce a magnetic moment opposite to the external magnetic field, resulting in diamagnetism. However, this argument is not entirely correct. Figure 16-2 shows some orbits for positive charges moving in a magnetic field pointing into the

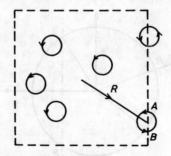

Fig. 16-2 Orbits of charged particles in an external magnetic field.

[d] After Pauli, who discovered the exclusion principle.

paper. The magnetic moments of the circular orbits point upward. Now let us choose any one region in the gas (inside the dotted line in Fig. 16-2). Orbits totally within the region contribute an upward magnetic moment but orbits near the boundary produce anti-clockwise current, forming a downward magnetic moment.

We can prove that the total magnetic moment within the region is zero. The orbital magnetic moment is proportional to

$$\sigma \equiv \frac{1}{2} \oint \mathbf{r} \times d\mathbf{r} \quad , \tag{16-24}$$

the integral being along the orbit. If the orbit is circular, then σ is the area of the circle. Let us suppose that the magnetic field points in the $-z$ direction and the orbit lies in the xy plane. The motion along z is neglected. If an orbit lies within the region in Fig. 16-2, then

$$\sigma = \pi a^2 \quad . \tag{16-25}$$

The direction of σ is along $+z$ and a is the radius of the circle. If the orbit is partly outside the region, then the calculation is slightly more complicated. Let $\mathbf{R}$ be the position of the centre of the circle.

$$\mathbf{r} \equiv \mathbf{R} + \mathbf{r}' \quad ,$$

$$\sigma = \tfrac{1}{2} \mathbf{R} \times \int d\mathbf{r}' + \tfrac{1}{2} \int \mathbf{r}' \times d\mathbf{r}' \quad . \tag{16-26}$$

Figure 16-3 is an enlarged view of the orbit on the right of Fig. 16-2. The first

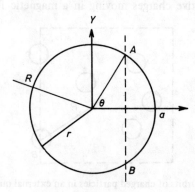

Fig. 16-3

integral on the right-hand side of (16-26) is

$$\int d\mathbf{r}' = -2a \sin\theta \, \mathbf{y} , \qquad (16\text{-}27)$$

where $\mathbf{y}$ is the unit vector in the direction of the y-axis. All orbits within $\pm a$ of the boundary lie partially within the region. There are

$$n(2a \, dR) \qquad (16\text{-}28)$$

such orbits along a segment dR of the boundary, where n is the density of the particles (number per unit area). Therefore, the total magnetic moment of those orbits along the boundary is, from (16-27) and (16-28)

$$2an \, dR \, \frac{1}{2} \mathbf{R} \times (-2a \langle \sin\theta \rangle) \mathbf{y}$$

$$= -\pi a^2 \, \frac{n}{2} \mathbf{R} \times d\mathbf{R} , \qquad (16\text{-}29)$$

$$\langle \sin\theta \rangle = \frac{1}{2} \int_0^\pi \sin\theta \, d(\cos\theta) = \frac{\pi}{4} . \qquad (16\text{-}30)$$

The differential dR is in the anti-clockwise direction and we have neglected the last term in (16-26) because its contribution to the magnetic moment is proportional to the perimeter while that of (16-29) is proportional to the area. Integrating (16-29) along the boundary, we get

$$-\pi a^2 \, \frac{n}{2} \oint \mathbf{R} \times d\mathbf{R} = -\pi a^2 n \mathbf{A} , \qquad (16\text{-}31)$$

where $\mathbf{A}$ is the area of the region pointing towards z. The magnitude of (16-31) is the same as that of the magnetic moment due to those orbits wholly within the region, but with opposite sign. So the magnetic moment of the whole region is zero.

The above analysis suffices to prove that orbital motion does not produce a net magnetic moment. This is a result of classical mechanics. Indeed, a magnetic field can only change the direction of motion of the particles, but not the energy. The thermodynamic potential is related only to energy and hence is not a function of the magnetic field. Since it is not a function of h, its derivative with respect to h must vanish, i.e., the total magnetic moment is zero:

$$M = -\frac{\partial F}{\partial h} = 0 . \qquad (16\text{-}32)$$

The situation will be slightly different if we take quantum mechanics into account. The result of the quantum mechanical calculation is: the energy density of states of the particles and hence the thermodynamic potential are changed by the magnetic field, so the magnetic moment is non-zero. This result is extremely important.[e] We shall only highlight the essential points here. For simplicity, we consider motion in the xy plane with the magnetic field in the z direction.

From the wave equation in quantum mechanics, we can solve for the energy of each particle

$$\epsilon_n = (n + \tfrac{1}{2})\omega, \quad n = 1, 2, 3, \ldots,$$

$$\omega \equiv eh/mc, \qquad (16\text{-}33)$$

where h is the magnetic field, c is the velocity of light, $\hbar = 1$, and ω is the rotational frequency of the particle. Therefore, the original kinetic energy $p^2/2m$ now takes the form of energy characterising simple harmonic motion. The density of states of the particle becomes

$$L^2 g(\epsilon) = \sum_n W_n \delta(\epsilon - \epsilon_n). \qquad (16\text{-}34)$$

L^2 is the area of the object and W_n is the statistical weight of the energy ϵ_n. In the absence of the magnetic field, the number of states between ϵ_n and ϵ_{n+1} is

$$L^2 \int_{\epsilon_n}^{\epsilon_{n+1}} d\epsilon \, g_0(\epsilon) \approx L^2 \omega g_0(\epsilon_n), \qquad (16\text{-}35)$$

where $g_0(\epsilon)$ is the density of states at zero magnetic field and $\epsilon = 0$ is the lowest energy state. The above expression assumes that $g_0(\epsilon)$ does not change considerably over this energy interval. The effect of the magnetic field is to squeeze all the states between ϵ_n and ϵ_{n+1} into the same energy state, so

$$g(\epsilon) \approx g_0(\epsilon) \omega \sum_{n=0}^{\infty} \delta(\epsilon - (n + \tfrac{1}{2})\omega). \qquad (16\text{-}36)$$

With $g(\epsilon)$, we can calculate the thermodynamic potential and the other equilibrium quantities. The thermodynamic potential of free electrons is

$$\frac{1}{L^2} \Omega = -T \int_0^\infty d\epsilon \, g(\epsilon) \ln(1 + e^{-(\epsilon - \mu)/T}). \qquad (16\text{-}37)$$

---

[e] Landau's theory of diamagnetism; see any textbook on solid state.

MAGNETISM 287

Note that the only difference between the integral of $g(\epsilon)$ and that of $g_0(\epsilon)$ is that one is a discrete sum with step size ω while the other is a continuous integral. If ω is very small, the difference is negligible. Figure 16-4 shows the difference between integration and summation. As the reader can verify, any integral can be written as

$$\int_0^\infty R(\epsilon)\, d\epsilon \approx \sum_{n=0}^\infty R((n+\tfrac{1}{2})\omega) - \frac{\omega^2}{24}\left(\frac{dR}{d\epsilon}\right)_{\epsilon=0} + O(\omega^3), \qquad (16\text{-}38)$$

as long as $R(\epsilon)$ is sufficiently smooth and the integral is finite. The sum in (16-38) is the area of the rectangular strips in Fig. 16-4, each with height $R((n+\tfrac{1}{2})\omega)$. Therefore, only the value at $\epsilon = 0$ is related to ω and (16-37) becomes

$$\frac{1}{L^2}\Omega - \frac{1}{L^2}\Omega_0 = \frac{-\omega^2}{24} T \frac{d}{d\epsilon}\left[g_0(\epsilon)\ln(1+e^{-(\epsilon-\mu)/T})\right]_{\epsilon=0}$$

$$= \frac{\omega^2 m}{24\pi} \frac{e^{\mu/T}}{1+e^{\mu/T}}. \qquad (16\text{-}39)$$

Note $g_0(\epsilon) = m/\pi$ is a constant and Ω_0 is the thermodynamic potential for $\omega = 0$.

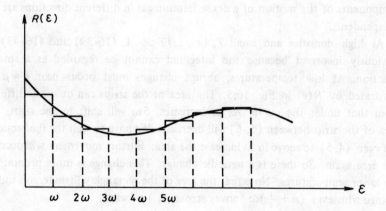

Fig. 16-4 Area approximated as a sum of elementary rectangular areas.

If the density of the gas is low, then $e^{\mu/T}$ is small, and (16-39) is well approximated by

$$\frac{1}{L^2}(\Omega - \Omega_0) \simeq \frac{\omega^2 m}{24\pi} e^{\mu/T} \;. \tag{16-40}$$

The density of an ideal gas is

$$n = 2e^{\mu/T} \int \frac{d^2 p}{(2\pi)^2} e^{-p^2/2mT} = \frac{mT}{\pi} e^{\mu/T} \;. \tag{16-41}$$

From (16-40) and (16-41), we get the total magnetic moment

$$M = -\frac{\partial \Omega}{\partial h} = -N \frac{\mu_B^2 h}{3T} \;,$$

$$\omega = 2\mu_B h \;. \tag{16-42}$$

Therefore, the orbit of the particle still produces a small effect but purely as a quantum effect. This result is also valid in three dimensions because the motion of an ideal gas in the z-direction is independent of the field.

If the density of the gas is very high and $\mu = \epsilon_F \gg T$, then (16-39) becomes

$$\frac{1}{L^2}(\Omega - \Omega_0) \simeq \frac{(\mu_B h)^2 m}{6\pi} \;,$$

$$M \simeq -N \frac{\mu_B^2 h}{3\epsilon_F} \;. \tag{16-43}$$

This result is valid only in two dimensions because, in three dimensions, the components of the motion of a dense fermion gas in different directions are not independent.

At high densities and small T, i.e., $\omega/T \gg 1$, (16-39) and (16-43) are obviously incorrect because the integrand cannot be regarded as a smooth function. At low temperatures, abrupt changes must occur near $\epsilon = \mu$, as illustrated by $R(\epsilon)$ in Fig. 16-5. The area of the strips can be quite different from that under the curve. As ω increases, 5ω will shift to the right. The area of the strip between (5, 6) will decrease. We have to wait for the rectangle between (4, 5) to move in to increase the area. Further movement will decrease the area again. So there is a periodic change. This change is more pronounced for lower temperatures. Note that the area of the rectangle will undergo a sudden change whenever $(n + \frac{1}{2})\omega$ moves across μ, i.e., when

$$n + \frac{1}{2} = \frac{\mu}{\omega} \;. \tag{16-44}$$

MAGNETISM 289

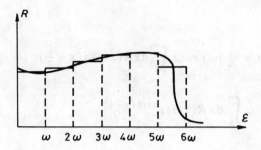

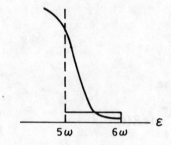

Fig. 16-5 Abrupt changes can occur near $\epsilon = \mu$. As ω increases, 5ω will shift to the right. The area of the square between (5, 6) will decrease.

Therefore, the change is a periodic function of $1/\omega$. This is a most important area of research in solid state physics, but we shall not discuss the details here. Note that the expansion in powers of ω is not permissible: no matter how small ω is, this periodic change always exists. No power series in ω can explain this change. Therefore, we have to re-examine (16-38) which is a mathematical problem.

How much does the integral of a function $R(\epsilon)$ differ from the area of the rectangles in Fig. 16-4? Equation (16-38) is a direct expansion in terms of ω, but the results of (16-44) are disturbing. Now write (16-36) as

$$\frac{g(\epsilon)}{g_0(\epsilon)} = \sum_{n=0}^{\infty} \delta\left(\frac{\epsilon}{\omega} - \frac{1}{2} - n\right) \equiv q\left(\frac{\epsilon}{\omega} - \frac{1}{2}\right), \qquad (16\text{-}45)$$

$$q(x) \equiv \sum_{k=-\infty}^{\infty} e^{2\pi i k x} . \qquad (16\text{-}46)$$

The expression in (16-46) is a δ-function with period 1. We consider $\epsilon > 0$. So the sum over equally spaced discrete arguments of any function can be written as

$$\omega \sum_{n=0}^{\infty} R((n + \tfrac{1}{2})\omega) = \omega \int_0^{\infty} dx\, R(x\omega) q(x - \tfrac{1}{2})$$

$$= \sum_{k=-\infty}^{\infty} \omega \int_0^{\infty} dx\, R(x\omega) e^{2\pi i k (x - \tfrac{1}{2})}$$

$$= \sum_{k=-\infty}^{\infty} R_k (-1)^k \quad , \tag{16-47}$$

$$R_k \equiv \int_0^{\infty} d\epsilon\, R(\epsilon) e^{2\pi i k \epsilon / \omega} \quad . \tag{16-48}$$

The integral of $R(\epsilon)$ is just R, so

$$\omega \sum_{n=0}^{\infty} R((n + \tfrac{1}{2})\omega) - \int_0^{\infty} d\epsilon\, R(\epsilon)$$

$$= \sum_{k=1}^{\infty} (R_k + R_{-k})(-1)^k \quad . \tag{16-49}$$

This is the so-called "Poisson sum rule". If the k series converges rapidly, this formula is very useful. Let us look at two examples.

Example 1:
$R(\epsilon) = e^{-\epsilon/T}$, then

$$R_k + R_{-k} = \frac{2\omega^2/T}{(\omega/T)^2 + (2\pi k)^2} \quad . \tag{16-50}$$

The series is convergent and there is nothing peculiar about it.

Example 2:
$R(\epsilon) = e^{-\epsilon^2/2\sigma^2}/\sqrt{2\pi\sigma}$, then

$$R_k + R_{-k} = e^{-\tfrac{1}{2}(2\pi k \sigma)^2/\omega^2} \tag{16-51}$$

This function is quite peculiar. It cannot be expanded in terms of ω. When $\omega = 0$, every derivative is zero, so according to (16-38) we would obtain zero. But the k series in (16-49) converges rapidly. This example demonstrates that expansion in a power series is not always reliable.

Note that if there is any discontinuity or rapid change of $R(\epsilon)$ near $\epsilon = \mu$, in Eq. (16-48) there will appear terms of the type,

$$e^{2\pi i k \mu/\omega} ,$$

i.e. a periodic function of period $1/\omega$ will be obtained.

16.4. Spin-Spin Interaction

Now let us review the mutual interactions between spins, the most important being the short distance exchange force and the long distance dipole interaction. In solids, because of the crystal structures and impurities, etc. there are numerous types of interactions, resulting in a variety of magnetic properties. Here we discuss mainly the exchange interaction.

A. Short distance exchange interaction

This is the strongest interaction among electron spins. Generally speaking, it is caused by the electric repulsion and the exclusion principle of the electrons. The wavefunction of fermions must be anti-symmetric. The wavefunction of two electrons can be written as

$$\phi(1,2) = \phi(\mathbf{r}_1, \mathbf{r}_2) \chi(s_1, s_2) , \qquad (16\text{-}52)$$

where $\mathbf{r}_1$ and $\mathbf{r}_2$ are the positions of the electrons and s_1 and s_2 are their spins. Electrons are fermions, so $\phi(1,2)$ satisfies

$$\phi(1,2) = -\phi(2,1) . \qquad (16\text{-}53)$$

Now we know that if the spins are parallel, then $\chi(s_1, s_2) = \chi(s_2, s_1)$; but if they are anti-parallel $\chi(s_1, s_2) = -\chi(s_2, s_1)$. Therefore

Parallel spins: $\phi(\mathbf{r}_1, \mathbf{r}_2) = -\phi(\mathbf{r}_2, \mathbf{r}_1)$

Anti-parallel spins: $\phi(\mathbf{r}_1, \mathbf{r}_2) = \phi(\mathbf{r}_2, \mathbf{r}_1)$. $\qquad (16\text{-}54)$

Thus when $\mathbf{r}_1 = \mathbf{r}_2$, ϕ for parallel spins is zero, but ϕ for anti-parallel spins will not vanish. The condition $\mathbf{r}_1 = \mathbf{r}_2$ means that the two electrons are at the same position. Therefore, if the spins are parallel, the two electrons will seldom be close. This implies a lower average potential energy due to electrostatic repulsion. Hence, electrons with parallel spins have lower energy. If the spins are anti-parallel, the electrons can be closer and the electric energy is higher.

This can be summed up in the following formula

$$\text{effective interaction} = -J\mathbf{s}_1 \cdot \mathbf{s}_2 \tag{16-55}$$

where J is a constant.

We shall make a crude estimate of J. Suppose two hydrogen nuclei are fixed at 0 and R. Let

$$\phi_1(\mathbf{r}) = \phi(\mathbf{r}) ,$$
$$\phi_2(\mathbf{r}) = \phi(\mathbf{r} - \mathbf{R}) , \tag{16-56}$$

where $\phi(\mathbf{r})$ is the 1s wavefunction of the hydrogen atom. If the electrostatic interaction between the electrons is neglected, then the wavefunction describing the two electrons is (16-52), with

$$\phi_\pm(\mathbf{r}_1, \mathbf{r}_2) = \frac{1}{\sqrt{2}} \Big(\phi_1(\mathbf{r}_1)\phi_2(\mathbf{r}_2) \pm \phi_1(\mathbf{r}_2)\phi_2(\mathbf{r}_1) \Big) , \tag{16-57}$$

where ϕ_- is for parallel spins and ϕ_+ is for anti-parallel spins. Under this approximation, the energies ϕ_+ and ϕ_- are degenerate. Now we include the electrostatic interaction

$$U(\mathbf{r}_1 - \mathbf{r}_2) = \frac{e^2}{|\mathbf{r}_1 - \mathbf{r}_2|} . \tag{16-58}$$

Using first-order perturbation theory, we get

$$E_\pm = \langle \phi_\pm | U | \phi_\pm \rangle = \int d^3\mathbf{r}_1 d^3\mathbf{r}_2 |\phi_1(\mathbf{r}_1)\phi_2(\mathbf{r}_2)|^2 \frac{e^2}{|\mathbf{r}_1 - \mathbf{r}_2|} \pm \frac{J}{2} ,$$
$$\frac{J}{2} \equiv \int d^3\mathbf{r}_1 d^3\mathbf{r}_2 \, \phi_1(\mathbf{r}_2)\phi_2(\mathbf{r}_1)\phi_1(\mathbf{r}_1)\phi_2(\mathbf{r}_2) \frac{e^2}{|\mathbf{r}_1 - \mathbf{r}_2|} . \tag{16-59}$$

The difference in the energies of ϕ_+ and ϕ_- is $E_+ - E_- = J$, i.e., electrons with parallel spins have an energy lower by the amount J. The meaning of (16-55) is precisely this. Note that

$$\mathbf{s}_1 \cdot \mathbf{s}_2 = \tfrac{1}{4} \quad \text{parallel}$$
$$\phantom{\mathbf{s}_1 \cdot \mathbf{s}_2} = -\tfrac{3}{4} \quad \text{anti-parallel} . \quad \text{for } \mathbf{s}_1, \mathbf{s}_2 \tag{16-60}$$

Note also that J is a short distance interaction because the product $\phi_1(\mathbf{r}_1) \phi_2(\mathbf{r}_2)$ in the integrand of (16-59) is non-zero only when both ϕ_1 and ϕ_2 are non-zero, i.e., only in the overlap of ϕ_1 and ϕ_2. But ϕ_1 is centered at $r = 0$ while ϕ_2 is centred at $\mathbf{r} = \mathbf{R}$.

$$\phi_1(\mathbf{r}) \sim \frac{1}{r} e^{-r/a} ,$$

$$\phi_2(\mathbf{r}) \sim \frac{1}{|\mathbf{r} - \mathbf{R}|} e^{-|\mathbf{r} - \mathbf{R}|/a} , \qquad (16\text{-}61)$$

where a is the Bohr radius of the hydrogen atom. Therefore $J \to 0$ if $R \gg a$, while J is very large if $R \lesssim a$.

The above crude analysis shows how (16-55) comes about. The expression (16-55) is the energy difference between the two states of a pair of electrons. It is surprising that J is totally unrelated to the magnetism of the electrons: the spins only enter via the symmetry of the wavefunction, and the energy difference comes from the electrostatic interaction. The response of the whole system to an external magnetic field is now determined by J. Although every electron is a magnetic dipole, the macroscopic magnetic property is not directly determined by the magnetic interaction among these dipoles.

The calculation of the effective interaction between spins is, of course, not really so simple. Even the basis of (16-55) is not entirely clear. Our aim is merely to discuss the dynamics of the spins and to summarise all the effects of the other degrees of freedom into the effective interaction strength J. This J carries a heavy burden. An interesting question is that if there are three spins, can we write the effective interaction as

$$-J_{12}\, \mathbf{s}_1 \cdot \mathbf{s}_2 - J_{23}\, \mathbf{s}_2 \cdot \mathbf{s}_3 - J_{31}\, \mathbf{s}_3 \cdot \mathbf{s}_1 \ ? \qquad (16\text{-}62)$$

The answer is no; it is not so simple.[f] We can only hope that (16-62) is a reasonable approximation. Of course, the safest way is to solve for the motion of all the electrons and the problem of the spins is obtained consequently. There would be no need to discuss the effective interaction but this method is too cumbersome. If we are interested in the spins only, then the method involving an effective interaction is still a convenient way of handling the problem.

One of the most common models of magnetism, based on (16-55), is to

---

[f] For the theory of J, see Mattis (1965).

write the total energy of the spins as

$$H = -\frac{1}{2} \sum_{i,j} J_{ij}\, \mathbf{s}_i \cdot \mathbf{s}_j - \mathbf{h} \cdot \sum_i \mathbf{m}_i \quad,$$

$$\mathbf{m}_i = g\mu_e \mathbf{s}_i \quad. \tag{16-63}$$

J_{ij} is the exchange energy between spins i and j and (16-63) is the so-called vector model or Heisenberg model. This is a quantum model: $\mathbf{s}_i$ is an operator and so is H. We assume (16-63) to be correct and calculate some results from it to compare with experiments. We can either fix J by experiment or we can calculate J and predict the result of experiments. In any case, (16-63) is a tool for analysis. In certain situations, (16-63) can be very useful. In other cases, corrections may be needed; e.g., terms like

$$J_{ijk}\, \mathbf{s}_i \cdot \mathbf{s}_j \times \mathbf{s}_k \tag{16-64}$$

may be included in the Hamiltonian. Moreover, $\mathbf{s}_i$ need not be the spin of an electron, but may be the total spin of the atom. J_{ij} may also be negative.

B. Long distance dipole interaction

At a separation r, the interaction energy of two magnetic dipoles $\mathbf{m}_1$ and $\mathbf{m}_2$ is

$$\frac{\mathbf{m}_1 \cdot \mathbf{m}_2}{r^3} - \frac{3(\mathbf{m}_1 \cdot \mathbf{r})(\mathbf{m}_2 \cdot \mathbf{r})}{r^5} \quad. \tag{16-65}$$

This interaction, though much weaker than the exchange interaction, is long-ranged. If the spins exhibit collective alignment, then this interaction cannot be neglected.

The result of a mutual interaction is not easy to analyse, be it the interaction between spins, or the attraction and repulsion between molecules. If we wish to discuss the correlation among many degrees of freedom, then the usual mathematical analysis is inadequate, so we must try to simplify the models. However, even the simplest model presents a very difficult problem. In the next chapter we shall introduce the Ising model, with the purpose of obtaining a crude understanding of magnetism.

Problems

1. A container of volume V contains a dilute gas of N particles. On the wall of the container there are N' spins ($N' \ll N$). Each spin has two states of energy $\pm h$ where h is proportional to the external magnetic field. The initial temperature is T. Assume that there is energy exchange between the gas molecules and the spins,

   (a) If the magnetic field is turned off slowly, calculate the change of temperature ΔT, assuming that there is no heat loss.

   (b) If the magnetic field is suddenly switched off, calculated ΔT.

2. Cooling by pressurisation.

   If liquid ^{3}He is pressurised adiabatically, it becomes a solid and the temperature drops. This cooling method has been pointed out by Pomeranchuk and is named after him. The following lists the steps of the theory.

   (a) Calculate the low temperature entropy of ^{3}He, using the fermion gas model with the Fermi energy approximately equal to 5 K.

   (b) At low temperatures the entropy of solid ^{3}He comes almost entirely from the spins. Below 10^{-3} K, ^{3}He exhibits diamagnetism and the spins are frozen. Draw the entropies of the liquid and the solid phases on the same diagram as a function of T.

   (c) From the diagram above, explain the method of cooling by pressurisation.

3. Magnetic susceptibility is related to the correlation function of the spins. Equation (16-19) is the result if the spins are uncorrelated. Assume a model with N spins s_i of spin $\frac{1}{2}$, $i = 1, 2, \ldots N$.

   (a) If $s_i, s_{i+1}, i = 1, 3, 5 \ldots$ are always parallel and s_i and s_{i+2} are uncorrelated, calculate the magnetic susceptibility.

   (b) If $s_i, s_{i+1}, i = 1, 3, 5, \ldots$ are always anti-parallel, then what is the magnetic susceptibility?

   (c) Discuss the correlation of the spins among the electrons in metals, and compare with anti-ferromagnetic materials. The spins in an anti-ferromagnetic system are arranged periodically, e.g., in a cubic lattice, each spin is opposite to those of its six nearest neighbors.

   This problem is, of course, related to problem 2.

4. Write the thermodynamic potential (16-37) directly as

$$\Omega = -T \sum_n W_n \ln(1 + e^{-(\epsilon_n - \mu)/T}) \quad . \tag{16-66}$$

Differentiate with respect to h to get

$$M = -\frac{\partial \Omega}{\partial h}$$

$$= \sum_n W_n \left(-\frac{\partial \epsilon_n}{\partial h}\right) f_n$$

$$+ T \sum_n \frac{\partial W_n}{\partial h} \ln(1 + e^{-(\epsilon_n - \mu)/T}) \, ,$$

$$f_n \equiv \frac{1}{e^{(\epsilon_n - \mu)/T} + 1} \, . \tag{16-67}$$

The first term on the right of (16-67) can be interpreted as the magnetic moment due to the circular orbit of the electrons. (Treat each n as one orbit.) What is the interpretation of the second term? First review the discussion from (16-31) to (16-54).

5. The total energy function of an electron in a magnetic field can be written as

$$H = \frac{1}{2m}\left(p + \frac{e}{c}A\right)^2$$

$$= \frac{1}{2m}[p_x^2 + (p_y + m\omega x)^2] \, , \tag{16-68}$$

where the magnetic field h is normal to the xy plane. We neglect the motion along the z direction. The vector potential is given by $A = hx$ and points in the y direction and $\omega = eh/mc$.

(a) From the equations of classical mechanics, prove that the orbit of the electron is a circle centered at (x_0, y_0) and

$$p_y = -m\omega x_0 \tag{16-69}$$

is an invariant.

(b) Let $x - x_0 = x'$, then (16-68) can be written as

$$H = \frac{1}{2m}(p_x^2 + m^2\omega^2 x'^2) \, . \tag{16-70}$$

Quantum mechanically, $[H, p_y] = 0$ and thus p_y is conserved. The expression in (16-70) is the energy operator for a simple harmonic oscillator with eigenvalues given by (16-33). Derive $g(\epsilon)$. This problem is the main content of Landau theory of diamagnetism.

Chapter 17
ISING MODEL

The simplest possible variable is one which takes on two values. (If there is only one value, no change is possible.) This kind of variable is called an Ising variable. The two values can be chosen to be $+1$ and -1. The simplest collective model is a set of Ising variables, and we call this the Ising model. In modern statistical mechanics, the applications of this model are numerous. Although it is quite different from real physical systems, it nevertheless describes certain general features of many phenomena. We shall introduce the simplest Ising ferromagnetic model — its solution in one-dimensional space and the proof of its ferromagnetic nature in two dimensions. We shall then briefly mention some models akin to the Ising model. The models mentioned in this chapter are related to phase transitions. The Ising model has received the greatest attention in this respect, but its application is by no means limited to phase transitions.

17.1. Ising Ferromagnetism in One Dimension

The Ising model in modern physics starts from the ferromagnetic model. Let s_r $(r = 1, 2, \ldots, N)$ be N Ising variables. Their values ± 1 denote the two directions of the spins. Each s_i is an element of motion. The total energy is

$$H(s) = -J \sum_{r=1}^{N-1} s_r s_{r+1} - h \sum_{r=1}^{N} s_r \quad . \tag{17-1}$$

This is a one-dimensional ferromagnetic model. There is interaction between neighbouring elements (or spins) and h is the external magnetic field.[a]

---

[a] This model was first discussed in Ising's thesis. Generalisations are likewise named after him.

This model has no time-dependent dynamics, that is to say, it does not say how each s_r varies with time. In this respect, it is unlike the gas model and quantum mechanical models, in which the Hamiltonian function or operator determines the equation of motion. The evolution of s_r must be determined by other means. We shall only assume that each s_r can be changed at any time, so that the basic assumption of statistical mechanics are still valid.

The equilibrium properties of this model can be derived by the method of fixed temperature. We first look for the solution to the zero-field, $h = 0$ case.

If the temperature $T = 0$, then either all s_r are $+1$ or all s_r are -1, so that H is a minimum, with a value

$$E(T = 0) = -J(N-1) \quad . \tag{17-2}$$

If $T > 0$, then, some s_r will be $+1$, and others -1. The boundary between $+1$ region and the -1 region is called the partition point (Fig. 17-1). At $T = 0$ there is no partition point and at low temperatures the partition points are few. The energy of each partition point is $2J$. This model is then transformed to a model of a gas of partition points. The number of partition points is not constant, so its chemical potential is 0. The possible sites are between any r and $r+1$, with at most one partition point per site, so that the partition points form a fermion gas, with the result

$$N'/N = \frac{1}{e^{2J/T} + 1} ,$$

$$\frac{1}{N}[E(T) - E(0)] = \frac{2J}{e^{2J/T} + 1} , \tag{17-3}$$

where N' is the number of partition points. The above assumes $N, N' \gg 1$. The reader can calculate the other quantities.

Because $N'/N \neq 0$ (unless $T = 0$), the s_r do not point in a unique direction, and the total magnetic moment is zero:

$$\frac{M}{N} = \frac{1}{N} \sum_{r=1}^{N} s_r = 0 , \tag{17-4}$$

i.e. this model has no ferromagnetism (except at $T = 0$). This is a general result: one-dimensional models do not normally exhibit ferromagnetism.

This conclusion holds for macroscopic bodies, i.e.

$$N \gg \xi \equiv e^{2J/T} \; , \tag{17-5}$$

where ξ is the average distance between two partition points. If N is smaller than ξ, then every s_r would be the same and this body can be regarded as a ferromagnet. At a certain temperature, if the total length of the system is large enough, ferromagnetism will disappear due to the appearance of partition points.

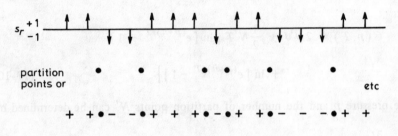

Fig. 17-1

Now let us look at the situation with an external field $h \neq 0$. Figure 17-1 shows how the partition points break up the body into sections with the spins in each section having the same value. Let $y_1, y_2, y_3, \ldots$ be the lengths of the sections, i.e. y_i is the number of spins in the i-th section. Therefore, the total energy (1) can be written as

$$H(y_1, y_2, \ldots) = 2JN' + h(y_1 - y_2 + y_3 - y_4 \ldots - y_{N'}) \; . \tag{17-6}$$

This energy is measured from the zero point and we assumed that the spins in the first section are $+1$, those in the second are -1, etc. and N' is assumed to be even. The term $2JN'$ is the energy of the partition points. The total length is the sum of the lengths of the sections:

$$N = y_1 + y_2 + \ldots + y_{N'} \; . \tag{17-7}$$

This can be regarded as the volume of the gas of partition points. We can determine the various equilibrium quantities by the method of fixed temperature and fixed pressure. This is similar to the last example in Chapter 10. Let

$$G(p, T) = -T \ln Z'$$

$$Z' = \sum_{y_1, \ldots, y_{N'}} \exp\{-(H + pN)/T\} \; . \tag{17-8}$$

Substituting (17-6) into (17-8) we get

$$Z' = e^{-2JN'/T} \left[\sum_{y=1}^{\infty} e^{-y(p+h)/T} \right]^{N'/2}$$

$$\times \left[\sum_{y=1}^{\infty} e^{-y(p-h)/T} \right]^{N'/2}, \qquad (17\text{-}9)$$

$$G(p,T) = 2N'J + \frac{1}{2}N'T \left\{ \ln \left[e^{(p+h)/T} - 1 \right] \right.$$

$$\left. + \ln \left[e^{(p-h)/T} - 1 \right] \right\}. \qquad (17\text{-}10)$$

The pressure p and the number of partition points N' can be determined by

$$\frac{\partial G}{\partial p} = N,$$

$$\frac{\partial G}{\partial N'} = 0. \qquad (17\text{-}11)$$

The results are

$$N = \frac{1}{2}N' \left[\frac{1}{1 - e^{-(p+h)/T}} + \frac{1}{1 - e^{-(p-h)/T}} \right],$$

$$e^{-4J/T} = \left(e^{(p+h)/T} - 1 \right)\left(e^{(p-h)/T} - 1 \right). \qquad (17\text{-}12)$$

From these two equations we can solve for N' and p. Notice that p is the pressure of the partition points, (i.e. the force necessary to make the y's decrease) and increases with temperature. The magnetic field acts as a tension for spins of $+1$, increasing y, and as a pressure for spins of -1. From the second equation of (17-12) we get

$$e^{p/T} = \cosh \frac{h}{T} \pm \sqrt{ \cosh \frac{2h}{T} - 1 + e^{-4J/T} }. \qquad (17\text{-}13)$$

In order for the pressure p to be meaningful, the square root must be positive. Putting (17-13) into the first equation of (17-12), we then get N'.

When h is small, the total magnetic moment and the magnetic susceptibility are

$$M = -\left(\frac{\partial G}{\partial h}\right)_{p,T,N}$$

$$= \frac{hN'}{T} e^{-p/T} (1 - e^{-p/T})^{-2} \quad , \tag{17-14}$$

$$\chi = \frac{M}{h} = \frac{N}{T} e^{2J/T} = \frac{N}{T} \xi \quad . \tag{17-15}$$

If every spin is independent then $\chi = N/T$. Now this is modified by a factor ξ, which is just the average distance over which the spins can maintain the same direction.

17.2. Proof of the Existence of Ising Ferromagnetism in Two Dimensions

The example in one dimension is simplest. The Ising model is not easy to solve for two or more dimensions. But the proof that the Ising model exhibits ferromagnetism (in dimensions ≥ 2) is not difficult. To prove the existence of ferromagnetism is to prove that in zero external magnetic field and at nonzero temperatures, the average value of the magnetic moment is nonzero. We now give such a proof.

The definition of the model in two dimensions is as follows. Consider a square lattice, and define a variable s_i for every lattice point i. The total energy is

$$H(s) = -\frac{1}{2} \sum_{i,j} s_i s_j J_{ij} - h \sum_i s_i \quad , \tag{17-16}$$

$$J_{ij} = J \quad \text{if } i, j \text{ are nearest neighbours}$$

$$= 0 \quad \text{otherwise} \ .$$

Every point has four nearest neighbours. We first give a simple argument to show why ferromagnetism can occur. Then we present a rigorous derivation.

The concept of partition points in the previous section can be extended to two dimensions. Figure 17-2 shows some values of the spins ($+1$ or -1, abbreviated as $+$, $-$). There exists a partition point between every pair of neighbouring spins with opposite signs. These points can be joined to form

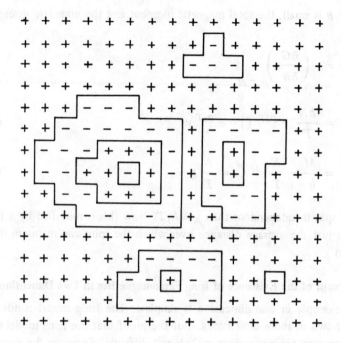

Fig. 17-2 A two-dimensional partition of spins for a square lattice.

partition lines, dividing the whole lattice into regions. The length of the partition line is determined by the number of points on the line. The total magnetic moment is the area of positive regions minus that of negative regions. To create partition lines, energy must be spent. At $T = 0$ all the spins are in the same direction (say positive), and there is no partition line. When the temperature is increased to T, negative regions will appear. If the perimeter of a certain negative region is L, the energy then is $2JL$ because the energy of each partition point is $2J$. So the probability of having this negative region is

$$e^{-2JL/T} \quad . \tag{17-17}$$

Suppose $h = 0$, and L is at least 4. Therefore if the temperature is sufficiently low, negative regions of large area are unlikely. Therefore most of the spins are positive and the total magnetic moment is positive, i.e. there is ferromagnetism.

This is different from the situation in one dimension. In one dimension, if the density of the partition points is nonzero, we have positive and negative regions. The lengths of each region can be changed without expenditure of energy. So on the average the numbers of positive and negative spins are the same. In two dimensions, although the density of partition lines is nonzero, the negative regions cannot increase at low temperature because expansion requires longer boundary lines and hence more energy. The result is similar in three dimensions, where partition lines become partition surfaces. At low temperatures, except for some small regions, all the spins will point to the same direction.

Now let us repeat the above argument rigorously.[b] Take any spin s_1. Let the probabilities that s_1 is $+1$ or -1 be p_+ and p_-. Then

$$\langle s_1 \rangle = p_+ - p_- . \qquad (17\text{-}18)$$

Now we have to prove that if T is sufficiently small, $\langle s_1 \rangle \neq 0$. Let all the spins at the boundary of the square lattice be $+1$. We shall prove that p_- is smaller than p_+.

If s_1 is -1, then it must be surrounded by one loop or three loops or five loops ... of closed partition lines. The number of loops must be odd. (If it is surrounded by two loops or four or six ... loops, s_1 must be $+1$. If there is no loop surrounding it, naturally it is $+1$.)

If we change s_1 and all the spins within the innermost loop to $+1$, we get one loop less and this is the configuration for $s_1 = +1$. So the probability of any configuration $s_1 = -1$ can be written as

$$e^{-2JL/T} \times e^{-2JL'/T}/Z , \qquad (17\text{-}19)$$

where L is the length of the innermost loop and L' that of the sum of the other loops, and Z is the partition function.

Equation (17-19) is only the probability for one configuration. To calculate p_-, we first fix the outer loops and calculate the number of configurations with length L for the innermost loop:

$$g(L) = \text{number of configurations with length } L \text{ for the innermost loop}.$$

$$(17\text{-}20)$$

This $g(L)$ naturally depends on the positions of the other partition lines. We cannot draw a loop intersecting the other loops. But if we neglect this restriction and count all the configurations looping s_1 once, then we over-estimate $g(L)$.

---

[b] This is similar to the proof by Peierls (1936), differing only in some details.

Therefore, fixing the outer loops, the probability that $s_1 = -1$ must be smaller than

$$\left(\sum_L g(L) e^{-2JL/T} \right) e^{-2JL'/T} \bigg/ Z \quad . \tag{17-21}$$

The factor $g(L)$ in this formula is calculated neglecting the existence of the outer loops. Now we sum up the different configurations of the outer loops. The term outside the bracket in (17-21) is the probability for a configuration of $s_1 = +1$, because getting rid of the innermost loop s_1 becomes $+1$. So the sum is p_+. The result is

$$p_- < \left(\sum_L g(L) e^{-2JL/T} \right) p_+ \quad , \tag{17-22}$$

where $g(L)$ is the number of ways of drawing a loop around s_1. Starting from any point, there are 3 ways of drawing the next steps, i.e. except for going back on the previous step, the remaining 3 directions can be chosen. To draw L steps we have $4 \times 3^{L-1}$ ways. The 4 comes from the first step, when all the four directions are permissible. We require L to return to the starting position. So the total number is less than $4 \times 3^{L-1}$. In addition we can start from any point on the loop, clockwise or anticlockwise. Therefore we should divide $4 \times 3^{L-1}$ by $2L$. We can write

$$g(L) < \frac{4}{2L} 3^{L-1} \left(\frac{L}{4} \right)^2 \quad . \tag{17-23}$$

The factor $(L/4)^2$ is just the largest area that can be enclosed by the partition line, i.e. a square with side $L/4$. Because s_1 can be any point within the loop, it is obvious that $g(L)$ must contain this area factor. Substituting (17-23) into (17-22), we get

$$p_- < \sum_{L=4}^{\infty} \frac{L}{8} 3^{L-1} e^{-2JL/T} \quad . \tag{17-24}$$

Notice that the shortest loop is $L = 4$ and L must be even.

The factor p_+ at the end of (17-22) is neglected because $p_+ < 1$. The series in (17-24) is easily computed. For simplicity we add the odd terms and get

$$p_- < \frac{5}{24} e^{-4\alpha}/(1-e^{-\alpha}) \ ,$$

$$\alpha \equiv \frac{2J}{T} - \ln 3 \ , \tag{17-25}$$

$$\langle s_1 \rangle = 1 - 2p_- > 1 - \frac{5}{12} e^{-4\alpha}/(1-e^{-\alpha}) \ . \tag{17-26}$$

When $T = J$, $e^{-\alpha}$ is less than $\frac{1}{2}$. If $T < J$ the above conclusion gives

$$\langle s_1 \rangle > \frac{91}{96} \ . \tag{17-27}$$

In the above proof we assume that the spins at the boundary are +1. If we do not make any such assumption, then $\langle s_1 \rangle$ must be zero because when $h = 0$ this model has a symmetry between + and −, i.e.

$$H(s) = H(-s) \ . \tag{17-28}$$

Ferromagnetism means that a very small applied force (e.g. a force applied on a tiny portion of the system) will cause $\langle s_1 \rangle \neq 0$. The boundary is just such a tiny portion of the total volume. The above proof can be improved by fixing fewer s_i to be +1. We can also use a very small magnetic field to define a preferential direction.

The above proof can be extended to three or more dimensions. This is left as an exercise.

The model in two dimensions has been solved,[c] but the mathematical manipulation is too complicated to be discussed here. The following are some results (see Fig. 17-3)

$$\sinh \frac{2J}{T_c} = 1 \ ,$$

$$T_c = 2.269 J \ . \tag{17-29}$$

[c] Onsager (1944).

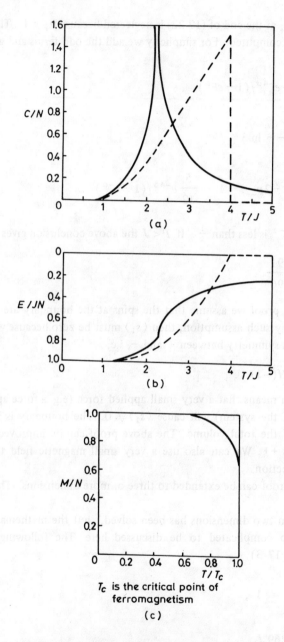

Fig. 17-3 (a) specific heat, (b) energy, (c) magnetic moment.
The dotted lines are results from mean field approximation.
(See Chapter 27, T_c is $4J$ in the mean field approximation).

When $h = 0$ and $T < T_c$ the total magnetic moment is[d]

$$M = N \frac{\cosh^2 2J/T}{\sinh^4 2J/T} [\sinh^2 (2J/T) - 1]^{1/8} . \qquad (17\text{-}30)$$

At low temperatures, M/N is almost 1. Near T_c,

$$\frac{M}{N} \propto (T_c - T)^{1/8} . \qquad (17\text{-}30')$$

17.3. Other Ising Models

The above ferromagnetic model, under slight modifications, becomes models for other substances. Therefore, many different phenomena can be analysed by the same model, enabling us to understand the common features of these different phenomena. Let us give several examples.

A. *Ising antiferromagnetic model*

If we replace J in (17-16) by $-J$ ($J > 0$) then energy is lowered for neighbouring spins with different signs. At low temperatures neighbouring spins are antiparallel. This becomes the antiferromagnetic model.

When $h = 0$, if we write every alternate s_i as $-s_i$, then H is the same as before because the extra minus sign cancels that from $-J$. In other words, if $h = 0$, the sign of J does not affect the thermodynamic potential. In calculating the thermodynamic potential the positive and negative values of each s_i must be taken into account. From the results of the last two sections, we know that in one dimension the Ising model does not exhibit antiferromagnetism but in two dimensions it does.

If $h \neq 0$, then the ferromagnetic and antiferromagnetic models are different and we need to re-analyse the problem.

B. *Model of binary alloy*

Let $s_i = 1$ represent an atom A at point i and $s_i = -1$ represents an atom B at point i. Let ϵ_{AA}, ϵ_{BB}, ϵ_{AB} be the interaction energies between neighbouring atoms. This is then a model of a binary alloy. This model can be transformed

---
[d] Yang (1952).

to the Ising ferromagnetic model. The energy of a pair of neighbouring atoms i, j can be written as

$$\epsilon_{ij} = -J s_i s_j - \frac{1}{z} h (s_i + s_j) + K ,$$

$$H = \frac{1}{2} \sum_{i,j} \epsilon_{ij} , \qquad (17\text{-}31)$$

where z is the number of nearest neighbours for each spin.

To determine the constants J, K, h, first suppose both i and j are occupied by A, then $\epsilon_{ij} = \epsilon_{AA}$, $s_i = s_j = 1$, i.e.

$$-J - \frac{2h}{z} + K = \epsilon_{AA} . \qquad (17\text{-}32)$$

Similarly

$$-J + \frac{2h}{z} + K = \epsilon_{BB} ,$$

$$J + K = \epsilon_{AB} . \qquad (17\text{-}33)$$

From (17-32) and (17-33) we can solve for J, h and K:

$$J = \frac{1}{2} \epsilon_{AB} - \frac{1}{4}(\epsilon_{AA} + \epsilon_{BB}) ,$$

$$K = \frac{1}{2} \epsilon_{AB} + \frac{1}{4}(\epsilon_{AA} + \epsilon_{BB}) ,$$

$$h = \frac{z}{4}(\epsilon_{BB} - \epsilon_{AA}) . \qquad (17\text{-}34)$$

Using (17-34), Eq. (17-16) now describes a binary alloy. At sufficiently low temperatures the atoms A will gather together while atoms B will form another collection. Hence we have an equilibrium of different phases (assuming $J > 0$). If $J < 0$ and the temperature is low, the atoms A and B will arrange alternately, similar to the antiferromagnetic case.

C. *Lattice gas model*

Let $s_i = 1$ represent a molecule at point i, and $s_i = -1$ represent no molecule at point i. We assume that the same point cannot accommodate two or more molecules. Let the interaction energy of neighbouring molecules be $-\epsilon$. Then we may write the energy between two neighbouring points as (17-31)

$$-J - \frac{2h}{z} + K = -\epsilon , \qquad s_i = s_j = 1 ,$$

$$-J + \frac{2h}{z} + K = 0 , \qquad s_i = s_j = -1 ,$$

$$J + K = 0 , \qquad s_i = 1, \quad s_j = -1 . \qquad (17\text{-}35)$$

Solving these, we get

$$J = \frac{1}{4}\epsilon ,$$

$$h = \frac{z}{4}\epsilon . \qquad (17\text{-}36)$$

Therefore, Eq. (17-16) is also a model of gas molecules, where $-\epsilon$ represents the attraction between the gas molecules. The restriction of at most one molecule per site represents short-range repulsion. The high density state, i.e. $\langle s_i \rangle > 0$ corresponds to the liquid phase while the low density state, i.e. $\langle s_i \rangle < 0$ corresponds to the gas phase.

The above examples show that many different phenomena can be analysed by the same model. Other applications include the inhomogeneous interaction, random interactions, long-ranged interaction, etc. Analysis may refer to phase transitions, impurities, boundary, etc. We shall return to some of these later. There are of course many phenomena which cannot be analysed with this model, e.g. the problems of spherical symmetry and spin waves, etc. Spin is a vector while the Ising variable does not contain the concept of angles. So in using the Ising model[e] to analyse various phenomena, we must first understand its limitations.

---

[e] The Ising model is studied extensively by physicists, chemists and mathematicians. The literature is enormous. The recent introduction by Fisher (1981) is a good starting point for references.

Problems

1. Many one-dimensional models can be solved by the method of the transfer matrix if the interaction is limited to nearest neighbours. Let

$$H = \sum_{i=1}^{N} K(s_i, s_{i+1}) ,$$

$$s_{N+1} = s_1 . \tag{17-37}$$

In the above, K can be any function of s_i and s_{i+1}.
(a) Prove that

$$Z = \sum_{s_1, s_2, \ldots, s_N} e^{-H/T}$$

$$= \text{Tr}(Q^N) . \tag{17-38}$$

The definition of the transfer matrix Q is

$$Q_{\alpha\beta} \equiv e^{-K(\alpha, \beta)/T} . \tag{17-39}$$

(b) Prove that for large N,

$$F = -T \ln Z = N\lambda , \tag{17-40}$$

where λ is the smallest eigenvalue of K, i.e. $e^{-\lambda/T}$ is the largest eigenvalue of Q.

(c) Use the method of the transfer matrix to solve model (17-1): First prove that

$$Q = \begin{pmatrix} e^{(J+h)/T} & e^{-J/T} \\ e^{-J/T} & e^{(J-h)/T} \end{pmatrix} . \tag{17-41}$$

2. The spin variable s_i need not be limited to only two values. If s_i has n values, then Q is an $n \times n$ matrix. Solve the $n = 3$ case, i.e. $s_i = 1, 0, -1$, the energy being still given by (17-1).

3. Consider any quantum mechanical model in which there are n states. Then the Hamiltonian is an $n \times n$ matrix K. If $\psi(0)$ is the state vector at $t = 0$, then the state vector at time t is

$$\psi(t) = e^{-iKt} \psi(0) \quad . \tag{17-42}$$

From this, we see that the one-dimensional problem in statistical mechanics is similar to the problem of finding the zero-point energy in quantum mechanics. The matrix K in (17-42) is like K in (17-39); it is equivalent to $-N/T$ and e^{-iKt} is equivalent to Q^N in (17-38). Thus the Ising model (1) is equivalent to the spin $\tfrac{1}{2}$ model in quantum mechanics:

$$K = a + \mathbf{b} \cdot \boldsymbol{\sigma} \quad , \tag{17-43}$$

where $\mathbf{b}$ is the "magnetic field", a is a constant and $\boldsymbol{\sigma}$ is the spin $\tfrac{1}{2}$ operator. Derive the relation between a, $\mathbf{b}$ and J, h.

4. Extend the proof in Sec. 17.2 to the Ising model in three dimensions.

5. If the temperature in a two-dimensional Ising model is very low, i.e. $T \ll T_c$, and there is a small positive magnetic field h, then almost all the spins are $+1$. At a certain moment the magnetic field is suddenly reversed to $-h$. Discuss the subsequent motion. (Hint: refer to the section Growth of Water Droplets in Chapter 15.)

Chapter 18

IMPURITIES AND MOTION

Impurities refer to foreign molecules mixed in a host substance. Examples include solutions and alloys. According to the motion of the impurity atoms, they can be classified into two types: mobile impurities, such as oxygen dissolved in water, and stationary impurities, such as manganese in steel. The state of motion is strongly related to the equilibrium properties of the system. This chapter emphasises the difference between these two situations, and the statistical mechanical analyses are not the same. Solute molecules in a solution have an osmotic pressure, much like a gas. The impurity atoms are stationary and exhibit behaviour quite different from that of a gas. There is no motion, hence no osmosis. Whether the arrangement is random or regular, there is always a difference between stationary impurities and mobile ones. We cannot discuss both in the same way. A random but frozen state is called an amorphous state or simply a glass. The motion of molecules in an amorphous state often have different time scales, and there is no sharp division between stationary and mobile. This chapter gives a few examples to illustrate some characteristics of the amorphous state.

18.1. Solutions and Osmotic Pressure

Osmotic pressure is the pressure produced by the motion of solute molecules. This pressure can be observed by using a semipermeable membrane, which allows molecules of the solvent but not the solute molecules to pass through.

Figure 18-1 shows a semipermeable membrane dividing the liquid into two halves. The right half contains solute particles while the left half contains none. The pressures on the two sides are maintained by pistons. The solute molecules

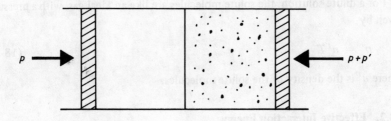

Fig. 18-1 A semipermeable membrane dividing a liquid into two parts.

on the right cannot enter the left while the solvent molecules can move either way, unaffected by the pressure of the membrane. The excess pressure on the right is called the osmotic pressure. The pressures in Fig. 18-1 can also be maintained by suitable weights. As shown in Fig. 18-2, the semipermeable membrane is in the middle of the U tube. The pistons on both sides have the same weight. The right-hand-side requires more weight to maintain equilibrium. The extra weight can be provided by a suitable height of the solution, i.e.

$$p' = \rho g h , \qquad (18\text{-}1)$$

where ρ is the density of the solution. Actually we can take away the pistons without causing any change. Therefore, the liquid on the right is pushed up. From (18-1) the osmotic is easily determined.

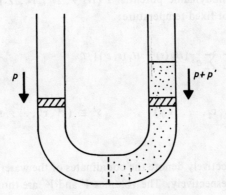

Fig. 18-2 A semipermeable membrane separating right and left halves of a U tube.

For a dilute solution, the solute molecules are like an ideal gas, with a pressure given by

$$p' = n'T , \qquad (18\text{-}2)$$

where n' is the density of the solute molecules.

18.2. Effective Interaction Energy

We now examine carefully the action of the solute molecules in the solvent (referred to as "water" below for the sake of simplicity). Suppose there are N water molecules and N' solute molecules. The total energy is

$$H = H_0 + H_1 , \qquad (18\text{-}3)$$

where H_0 includes the kinetic energy of the water molecultes, the interaction among water molecules and the kinetic energy of the solute molecules and H_1 is the interaction energy of the solute molecules among themselves and with the water molecules:

$$H_1 = \sum_{k=1}^{N} \sum_{i=1}^{N'} u(\mathbf{r}_k - \mathbf{r}'_i) + \frac{1}{2} \sum_{i,j=1}^{N'} v(\mathbf{r}'_i - \mathbf{r}'_j) , \qquad (18\text{-}4)$$

where $\mathbf{r}'_i$ is the position of the solute molecule i and $\mathbf{r}_k$ the position of the water molecule k and u and v are the interaction energies of a pair of molecules.

The total thermodynamic potential $F(N, V, N', V', T)$ can be computed from the method of fixed temperature:

$$e^{-F/T} = \sum_{s'} \sum_{s} e^{-[H_0(s) + H_1(s, s')]/T} ,$$

$$s \equiv (\mathbf{r}_1, \mathbf{r}_2, \ldots, \mathbf{r}_N) , \qquad s' \equiv (\mathbf{r}'_1, \mathbf{r}'_2, \ldots, \mathbf{r}'_{N'}) ,$$

$$(18\text{-}5)$$

where s and s' collectively denote the coordinates of the water molecules and the solute molecules respectively. The volumes V and V' are those occupied by the water molecules and solute molecules respectively, V' being a part of V. The kinetic energy can be immediately integrated out; the result is not related to V

and V', and will be ignored. The sum in (18-5) can be done in two steps: first sum over s, then over s'.

$$e^{-F/T} = e^{-F_0/T} \sum_{s'} e^{-W(s')/T} ,$$

$$e^{-F_0/T} \equiv \sum_{s} e^{-H_0(s)/T} , \qquad (18\text{-}6)$$

$$e^{-F_0/T - W(s', T)/T} \equiv \sum_{s} e^{-[H_0(s) + H_1(s, s')]/T} . \qquad (18\text{-}7)$$

In these expressions F_0 is the thermodynamic potential when H_1 is omitted, while $W(s', T)$ is the effective interaction energy of the solute molecules, and is the thermodynamic potential when every s' is fixed. This is an example of statistics by classification: we fix s' and sum over all the configurations s of the water molecules. The function W represents an interaction energy taking into account the motion of the molecules, the interaction among the solute molecules and the interaction between solute and water. This effective interaction energy is a very important concept. When we isolate a certain part of the system (here the solute molecules) for examination, the interaction of this part with its surrounding may be very strong and must be taken into account. Using this concept of effective interaction energy, we can treat the surrounding as a medium, whose only effect is to make a suitable correction to the interaction energy. Then we can forget about the medium. This method is very common, e.g. in the electrostatic interaction in a medium, the influence of the molecules of the medium can be represented by a dielectric constant. The external charge divided by the dielectric constant is the effective charge. After replacing the charge by the effective charge, we can forget about the medium. This is clearly very convenient.

Notice that the effective interaction energy is defined for a specific use and cannot be used for other purposes. For example, the function $W(s', T)$ defined here is used to analyse equilibrium phenomena and cannot be used directly to analyse the motion of the molecules. Of course, it may help in the analysis of the motion, but we cannot derive the equation of the motion of the molecules by taking its derivative.

Of course we can also define an effective interaction between the water molecules. Because of the introduction of solute molecules, the interaction between water molecules is influenced, and can be replaced by the effective interaction energy.

Equation (18-7) can also be written as an average:

$$e^{-W/T} = \langle e^{-H_1/T} \rangle_0$$

$$\equiv \frac{\sum_s e^{-H_0/T} e^{-H_1/T}}{\sum_s e^{-H_0/T}} , \qquad (18\text{-}8)$$

where the average value is taken over the motions of water molecules with the solute molecules fixed. Note that W is not the average of H_1, but is a more complicated object:

$$W = -T \ln \langle e^{-H_1/T} \rangle_0 \qquad (18\text{-}9)$$

18.3. Low Density Case

Evaluation of the effective interaction energy is not easy. However, in some special cases approximation can be employed. We now look at the low density case.

Low density implies that the solute molecules are always far apart. So the interaction v can be neglected. But the interaction u with the water molecules cannot be neglected (see Eq. (18-4)). The interaction between water molecules is short-ranged. If the distance between solute molecules is large enough they become uncorrelated. The calculation (18-8) can be simplified as follows

$$\langle e^{-H_1/T} \rangle_0 \simeq \prod_{i=1}^{N'} \left\langle \exp\left(-\frac{1}{T} \sum_k u(\mathbf{r}_k - \mathbf{r}'_i)\right) \right\rangle_0 , \qquad (18\text{-}10)$$

i.e. we neglect the last term of (18-4) and each $\mathbf{r}_i$ in the first term is independent.

In addition, the water is homogeneous and the average value of (18-10) does not change with $\mathbf{r}'_i$. So in this approximation

$$W \simeq -TN' \ln \left\langle \exp\left(-\frac{1}{T} \sum_k u(\mathbf{r}_k)\right) \right\rangle_0 \equiv N' f' , \qquad (18\text{-}11)$$

where f' does not depend on $\mathbf{r}'$. Putting (18-11) into (18-6), we get

$$\sum_{s'} e^{-W/T} = e^{-N'f'/T} \frac{V^{N'}}{N'!} ,$$

$$F = F_0 + N'f' - TN'\left(\ln \frac{V'}{N'} + 1\right) . \qquad (18\text{-}12)$$

This result shows that each solute molecule has a thermodynamic potential f'. Except for this, its equilibrium properties are the same as those of an ideal gas. The osmotic pressure is

$$p' = -\frac{\partial F}{\partial V'} = \frac{TN'}{V'} , \qquad (18\text{-}13)$$

because F_0 is not a function of V' and f' is a function of T only. Equations (18-13) and (18-2) say the same thing. We have thus reached the important conclusion: no matter how strong the interaction is between the water molecules and the solute molecules, so long as the density of the solute is sufficiently low, the osmotic pressure is given by (18-13).

At low densities, the effect of the interaction between solute and water is contained in f'. We shall illustrate the importance of f' by an example.

Example 1:

The pressure of the solute above the water surface is p. Calculate the density N'/V' of the solute in the water.

The solute molecules can freely enter or leave the water surface. So the chemical potential is the same above or below the water surface. The gas above the water surface can be treated as an ideal gas, and, therefore, the chemical potential is

$$\mu = T\ln(\lambda^3 p/T) ,$$

$$\lambda \equiv h(2\pi mT)^{-\frac{1}{2}} . \qquad (18\text{-}14)$$

The chemical potential below the water surface can be obtained from (18-12).

$$\mu = \frac{\partial F}{\partial N'} = T\ln\left(\lambda^3 \frac{N'}{V'}\right) + f' . \qquad (18\text{-}15)$$

318 STATISTICAL MECHANICS

Notice that the F_0 in (18-12) includes all contributions due to the motion of the solute. In fact the solute in water is the same as an ideal gas except for the energy f'. From (18-14) and (18-15),

$$\frac{N'}{V'} = \frac{p}{T} e^{-f'/T} . \qquad (18\text{-}16)$$

Therefore, if we know f' and p we can calculate the density of the solute in water.

If the density of the solute is not very low, then (18-10) must be corrected. The result is that besides f' we have the effective interaction energy between the solute molecules. This point will become clearer in the next section.

18.4. Mobile and Stationary Impurities

The solute molecules mentioned above are mobile impurities. During the time of observation solute molecules move to and fro.

If the impurity molecules do not move but remain in a fixed position, then they are stationary or frozen impurities. The impurities in solids are usually stationary. In highly viscous liquids, impurities also may not move. Whether or not there is motion is a question of the time scale. The motion of impurities can be described by diffusion. In time t the diffusion distance is

$$l \sim (Dt)^{\frac{1}{2}} , \qquad (18\text{-}17)$$

(see Sec. 12.9.) where D is the diffusion constant. If n is the number of impurities, then $n^{-1/3}$ is their typical separation. Let $\mathfrak{J}$ be the observation time; the impurities are

$$\text{mobile if} \quad \mathfrak{J} \gg \frac{1}{Dn^{2/3}} ,$$

$$\text{stationary if} \quad \mathfrak{J} \ll \frac{1}{Dn^{2/3}} . \qquad (18\text{-}18)$$

In solids, the diffusion constant of impurity atoms has the empirical dependence

$$D = D_0 e^{-E/T} . \qquad (18\text{-}19)$$

Figure 18-3 shows the diffusion constant of carbon atoms in iron, which is fitted by (18-19) with $D_0 = 0.02 \text{ cm}^2/\text{sec.}, E = 0.87 \text{ eV}$. The exponential in (18-19)

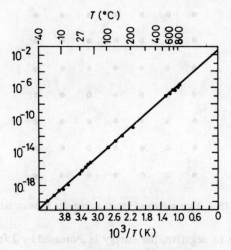

Taken from Kittel (1966) p. 569

Fig. 18-3 Fit of experimental data on carbon atoms in iron against Eq. (18-19).

shows that in order for the impurity atom to move, it must overcome a barrier of energy E, it is surrounded by other atoms and cannot move easily. From (18-19) it can be seen that D is very sensitive to temperature. Usually E is between 1 and 2 eV, i.e. about 10^4 K. At low temperatures, D is negligible. The impurities, apart from small vibrations, are stationary. In many cases, the atoms are not free to move until the solid reaches the melting temperature.

In applying the basic assumptions, mobile impurities and stationary impurities are totally different situations. The former is like a gas in which the coordinates of the impurity atoms change continually. The latter is like a solid for which the mean position of every impurity atom is fixed; whether the positions of the impurity atoms are random or regular, they are constant positions. This is very important. Let us look at an example.

Figure 18-4 shows a square lattice fully occupied by two species of atoms, say positive (black) and negative (white), with numbers, N_+ and N_-. Let the interaction between similar atoms be $-J$ and that between different atoms be J. The interaction is limited to nearest neighbours.

Suppose $N_- \ll N_+$, i.e. the negative atoms are very few in number and can be treated as impurity atoms. If $N_- = 0$, i.e. there are no negative atoms, the total energy is $-\frac{1}{2} N z J$. Here $z = 4$ is the number of nearest neighbours. The total number of atoms N is equal to the number of lattice sites. If we change

320 STATISTICAL MECHANICS

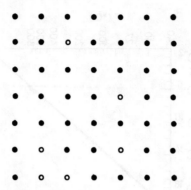

Fig. 18-4 Two species of atoms occupying a square lattice.

one positive atom to negative, the energy is increased by $2Jz$. Therefore, if the density of the negative atoms is very low, the total energy is

$$H_0 = -\frac{1}{2}NzJ + 2JzN_- \ . \tag{18-20}$$

If there are nearest neighbour negative atoms, (18-20) must be modified. Every pair of neighbouring negative atoms increases the energy by $-4J$. So the total energy is

$$H = H_0 + H_1 \ ,$$

$$H_1 = \frac{1}{2} \sum_{i,j=1}^{N_-} u(|\mathbf{R}_i - \mathbf{R}_j|) \ , \tag{18-21}$$

$\mathbf{R}_i$, $i = 1, 2, \ldots, N_-$ is the position of the negative atom i ,

$$u(0) = \infty \ ,$$

$$u(1) = -4J \ ,$$

$$u(r) = 0 \ , \quad \text{if } r > 1 \ .$$

The distance between neighbouring atoms is 1. This is the Ising gas model at the end of Chapter 16. The positive atoms are treated as a medium or the solvent, while the negative atoms are the solute. However, we now want to discuss whether these atoms can move or not.

When an atom moves, it must exchange its position with the nearest neighbour. Exchange of position needs a "push". In crystals this is a rather complicated process. Every atom is continually making a small vibration. If the neighbour on the left moves down a bit, and the atom in question moves up a bit, while the neighbour on the right gives a push to the left, the probability that the atom in question will move to the left will be enhanced (see Fig. 18-5). This process is a special topic in itself. Here we only suppose the existence of such processes and assume that such processes occur very quickly. We only consider the situations before and after the process and do not ask what happens during the process.

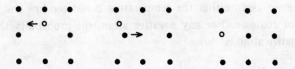

Fig. 18-5

If the negative atoms move frequently, this problem is similar to the problem of the solution. The only difference is that the position of the atoms are restricted to the lattice sites. In this simple model, the term $2Jz$ in (18-20) is just f' in (18-12). In the above section we did not discuss the interaction between the impurity atoms. Hence, H_1 in (18-21) has an important effect. From the discussion on the Ising model in the last chapter, we know that at low temperatures the negative atoms will coalesce together, segregating from the positive atoms in order to reduce the energy. Now let us make a simple calculation. Let F be the thermodynamic potential at constant temperature, then

$$F = -T \ln Z ,$$

$$Z \equiv \frac{1}{N_-!} \sum_{R_1} \cdots \sum_{R_{N_-}} e^{-H/T} .$$
(18-23)

If we neglect J/T, the result is that of an ideal gas

$$F \simeq F_0 = H_0 - T S_0 ,$$

$$S_0 = N_- \ln\left(\frac{N}{N_-} + 1\right) .$$
(18-24)

If the concentration of the negative atoms is low, we can use the low temperature expansion (see Sec. 14.1). We simply write down the results, leaving the details to the reader.

$$F = F_0 + F_1 \;,$$

$$F_1 = -\frac{TN_-^2 \, v}{2N}(e^{3J/T} - 1) + \frac{TN_-^2}{2N} \;. \tag{18-25}$$

In the above we assume that the temperature is not too low and the negative atoms do not coalesce. For any negative atom, the probability that its neighbour is a negative atom is

$$p_-(T) = \frac{N_-}{N} e^{3J/T} \;. \tag{18-26}$$

Equations (18-23) to (18-26) are the results of the motion of the negative atoms.

If the negative atoms do not move, the situation is totally different. Each $\mathbf{R}_i$ is a fixed quantity and there is no change in configuration. So the thermodynamic potential is just

$$F'(\mathbf{R}_1, \ldots, \mathbf{R}_{N_-}) = H \;. \tag{18-27}$$

The function H is defined from (18-21) and (18-20), and the value of F' is controlled by the predetermined positions of the negative atoms, e.g. predetermined by cooling from high temperatures. But these positions are related to the process of cooling. Different processes of cooling will result in different distributions of the positions. Once the $\mathbf{R}_i$ are frozen in, they will not be influenced by the changes of temperature (provided they are not heated to the temperature of motion again). So the temperature does not influence the entropy and the specific heat. Notice that no matter how randomly the negative atoms are distributed, the entropy is not affected, because there is no motion. Equation (18-27) is quite different from (18-24) and (18-25), showing that the influence of the impurities depends very much on the mobility of the impurity atoms.

Under special conditions, F' in (18-27) can be simplified, e.g. when a body is cooled suddenly (so-called quenching) from a high temperature T_0 at which the negative atoms can move. In this case, the distribution of the negative atoms reflects the distribution before quenching, at $T = T_0$. At that time, among the neighbours of each negative atoms there are on the average $zp_-(T_0)$ negative atoms. (See Eq. (18-26) for p_-). Therefore, F', i.e. the value of H, is

$$F'(\mathbf{R}_1, \ldots, \mathbf{R}_{N_-}) = -\frac{1}{2}NzJ + 2JzN_- - \frac{1}{2}N_- z(3J)\frac{N_-}{N}e^{3J/T} \,. \tag{18-28}$$

Notice that (18-28) is not just the average of F', but its actual value. There is no fluctuation in F' and hence no average. The average number of negative neighbours of a negative atom is statistical, obtained from N_- negative atoms. We assume this average value to be $zp_-(T_0)$.

To summarise, if the impurity atoms are allowed to move, the trajectory occupies a certain region in configuration space. If the range of $\mathbf{R}_i$ is large, the extent of this region becomes larger. If they are stationary, then the trajectory is just a point.

The extent of motion is determined by the motion. This has already been emphasised in Chapter 8 when we discussed the hydrogen molecules.

The above discusses two cases: mobile and stationary. What happens if the time scale falls between these two? The impurities do not move very far, but neither are they absolutely stationary. This case is very difficult. We have to take into account the duration of observation as well as the details of the motion. That is to say the problem is not an equilibrium problem but that of non-equilibrium.

In many substances there is no well defined unique time scale. Rather, there is a range of time scales, both long and short. No matter how long the observation time, we cannot apply the simple basic assumptions. This situation is the most troublesome and occurs in the amorphous state. In the next section we shall give an introduction to the amorphous state.

18.5. The Amorphous State

The main point of the discussion in the above section is on the distinction between motion and immobility. The amount of impurity atoms is not important. In the above example, if the numbers of the positive and negative atoms are comparable, the result will be more complicated. But the basic concept is the same. If the numbers are comparable, the term "impurity" will not be appropriate, and we must call this model a "mixture". If each atom is stationary but randomly distributed, we call this the amorphous state.

The concept of an amorphous state does not require the mixing of different molecules. The same type of molecules can have random but frozen structure. Any rándom but frozen structure can be called amorphous. Glass is a good example; ice is another.

Most solids in nature are amorphous, either containing frozen impurities or possessing a random structure. In modern materials science, the amorphous state is an important topic. The use of statistical mechanics to analyse amorphous materials started about ten years ago. The theory is still in its infancy.

The following is an example that illustrates some concepts related to the amorphous state.

The problem of amorphous magnets (or spin glasses) has received much recent attention. The most widely studied material is manganese-copper alloy. Some manganese atoms are dissolved in copper and then cooled to low temperature. The electrons in the outermost shell of the manganese atom mix with the free electrons of copper and the remaining manganese ion is then fixed with spin $\frac{5}{2}$. This spin interacts with the spins of the electrons in the metal and peculiarly influences the motion of the electrons in the metal; this is the so-called Kondo problem, which we shall not discuss. Because of the interaction of the manganese ion with electrons in the metal, the spins of two manganese ions (simply called the spin below) mutually interact. That is, spin 1 influences the electrons, and the electrons in turn influence spin 2. This mutual interaction is approximately

$$J(\mathbf{R}_{12})\, \mathbf{s}_1 \cdot \mathbf{s}_2 = A \left[\frac{\cos 2p_F R}{(2p_F R)^3} - \frac{\sin 2p_F R}{(2p_F R)^4} \right] . \qquad (18\text{-}29)$$

This formula is correct for low densities of the manganese atoms and at low temperatures. In this formula p_F is the Fermi momentum of the free electrons of copper and R_{12} is the distance between spins 1 and 2; A is a constant given by

$$A \sim \frac{K^2}{\epsilon_F} \left(\frac{p_F a}{2} \right)^6 \frac{1}{2\pi^3} , \qquad (18\text{-}30)$$

where K is the interaction energy between the spin and the spin of the electron, a is the lattice distance and ϵ_F is the Fermi energy of the free electrons. This is the so-called RKKY interaction.[a] The most important characteristic of this interaction is that it changes sign with R. This characteristic comes from the correlation function of the spins of the electrons, which is approximately proportional to $J(R)$, because the interaction between $\mathbf{s}_1$ and $\mathbf{s}_2$ is mediated by the free electrons. The fluctuation with period π/p_F is due to the sharp cut-off of the free electrons distribution at the Fermi surface (see Chapter 13).

---

[a] Ruderman, Kittel, Kasuya, Yoshida (RKKY) interaction. For detailed derivation see Mattis (1965) p. 193–198.

Because p_F is very large and the distribution of the positions of the manganese atoms is random, the mutual interaction $J(R_{ij})$ between the spins is positive for some pairs and negative for others, causing a random interaction. Note that the position of every manganese atom is fixed and so every $J(R_{ij})$ is fixed too. Experiments are usually performed for $T \leqslant 20\,\text{K}$ and the concentration of manganese below 10%.

The experimenters are interested not in the random distribution of manganese, but in the random interaction of the spins due to this distribution. At high temperatures, these spins form a paramagnetic body and each spin can freely rotate. When the temperature is lowered below a certain T_0, some spins will become frozen. If the temperature decreases further, more spins will be frozen out. The frozen directions of the spins are random, and the total magnetic moment is zero. This is the amorphous phase of the spins, also called the spin glass phase.

This amorphous phase has many interesting properties, the most important being its "remanence", i.e. if we magnetise with a magnetic field, the magnetic moment M does not vanish immediately when the magnetic field is removed, but decays gradually

$$M(t) - M(0) \sim -aT \ln(t/\tau) \quad , \tag{18-31}$$

where a is a constant and τ is a microscopic time scale. Not only the magnetic moment, but other quantities also follow this logarithmic decay. Such changes show the existence of different time scales of motion: some are fast, some change so slowly that they may be regarded as frozen during the time of observation, while some are in the intermediate case. This is the characteristic of not only the amorphous magnets but also of other amorphous systems. To understand the amorphous state we cannot avoid the problem of nonequilibrium. Therefore, we shall analyse the problem again in the discussion of dynamics.

Problems

1. In a certain solution, the osmotic pressure of the solute is p'. Prove that the vapour pressure of the solution $p + \Delta p$ is lower than that of the pure solvent p by an amount

$$\Delta p = -\frac{\rho_g}{\rho} p' \quad , \tag{18-32}$$

where ρ is the density of the solution and ρ_g is the density of the vapour of the solvent. The above assumes that the vapour pressure of the solute is very low.

326 STATISTICAL MECHANICS

Hint: Figure 18-6 shows an O-shaped tube separated into two parts by a semipermeable membrane. From the pressure difference on both sides and the heights of the liquid surfaces we can get (18-32). The vapour pressure of the solute is so low that there is not enough time for the solute molecules to be transferred to the left via the upper tube.

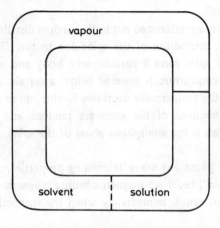

Fig. 18-6

2. Calculate the osmotic pressure of salt in sea water. If we use a semipermeable membrane, how high can we raise the sea water? The reader will find that this is quite a considerable number.

3. An estuary generator: Figure 18-7 shows two dams built across an estuary. The fresh water generates electricity by flowing under the dam and then flows to sea via the semipermeable membrane. (This idea was learnt from oceanographer J. Isaacs.) Discuss the feasibility of this construction.

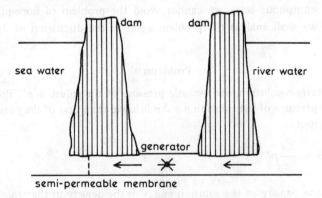

Fig. 18-7

4. The one-dimensional Ising model

$$H = -J \sum_{i=1}^{N-1} s_i s_{i+1} - h \sum_{i=1}^{N} s_i \quad, \tag{18-33}$$

can be transformed to another similar model. The steps are as follows:

(a) Let $s_1, s_3, s_5 \ldots$ be part (1) of the body and $s_2, s_4, s_6 \ldots$ be part (2) of the body. Write an equation for H_1' where H_1' is the effective interaction energy of part (1).

(b) Let

$$s_1' = s_1, \quad s_2' = s_3, \quad s_3' = s_5 \ldots \quad. \tag{18-34}$$

Write H_1' as

$$H_1' = -J' \sum_{i=1}^{N'-1} s_i' s_{i+1}' - h \sum_{i=1}^{N} s_i' + F_0 \quad,$$

$$N' = \frac{1}{2} N \quad. \tag{18-35}$$

(Assume N is even.) Obtain J', h' and F_0. Assume that $h = 0$.

Answer: $h' = 0$,

$$J' = \frac{1}{2} T \ln \left(\cosh \frac{2J}{T} \right) \quad,$$

$$F_0 = -\frac{1}{4} NT \ln \left(4 \cosh \frac{2J}{T} \right) \quad. \tag{18-36}$$

This model is very simple. H_1' and H have the same structure.

5. As in the above problem, but take $h \neq 0$. Calculate h', J' and F_0.

6. Let $N = 2^m$. The function H in the above model becomes a constant after m transformations, i.e. the thermodynamic potential F. Derive F under the assumption that $2J/T \ll 1$.

7. Assume a one-dimensional Ising model of ferromagnetism to be described by:

$$H = -J \sum_{i=1}^{N-1} s_i s_{i+1} (1 - \lambda n_i) \quad, \tag{18-37}$$

where $n_i = 0$ or 1, and λ is a constant. The numbers n_i can be regarded as the number of impurity atoms, i.e. if there is one impurity atom between s_i and

s_{i+1} the interaction is weakened. Let $N' = \sum_{i=1}^{N} n_i$.

(a) If the position of the impurity atoms are not fixed, i.e. they can move to and fro, calculate the thermodynamic potential, entropy, specific heat and magnetic susceptibility.

(b) If the impurity atoms are stationary and distributed randomly, calculate the same quantities.

Chapter 19

ELECTROSTATIC INTERACTION

The electrostatic interaction is very strong at short distances and is weaker at large distances. However, it can interact with many particles at the same time and, therefore, cannot be analysed by a simple expansion. This chapter uses some examples to illustrate the characteristics of the electrostatic interaction, emphasising the long distance effect. The method of analysis is to examine the response of the particles to a stationary charge and introduce the screening effect. The examples include the plasma gas, the ionic solution, electrons in metals; and we introduce in a simple way the experiment on electron crystals in two dimensions. Lastly we discuss an abstract Coulomb gas model in two dimensions, in which the interaction energy between the charges is $q^2 \ln r$, where q is the charge and r is the distance between the two charges. This gas is a plasma above a certain temperature T_0. Below T_0 the positive and negative ions coalesce into pairs forming dipoles. This model is quite important in current work on statistical mechanics. Its main characteristics can be explained by quite simple analysis.

19.1. Short Distance and Long Distance are Both Important

The electrostatic interaction energy is inversely proportional to the distance between the charges and does not have a characteristic length scale. Let e_i be the electric charge of the i-th particle, then the electrostatic potential of a collection of particles is

$$U = \frac{1}{2} \sum_{i,j} \frac{e_i e_j}{r_{ij}} \quad . \tag{19-1}$$

Since there is no length scale, we can enlarge or contract the body. Let $r_{ij} = \lambda r'_{ij}$, then U will acquire a factor $1/\lambda$ in the new scale. If U is negative then all particles will coalesce into one point and $U \to -\infty$. If U is positive then the charges will fly-off to infinity to lower the energy. Obviously other effects must exist; otherwise the electrostatic interaction alone cannot describe stable matter.

In Chapter 9 we mentioned that for the thermodynamic limit to be established, the necessary conditions are

(a) the total charge is zero, i.e. $\sum_i e_i = 0$,

(b) kinetic energy is quantum mechanical, and

(c) at least one type of charge (positive or negative) is fermionic.

Condition (a) means that uncompensated charges will be dispersed (to the surface of the body if possible). Condition (b) avoids the coalescing of all the particles into one point. The requirement of quantum mechanics is this: If the position of a particle is restricted, then its kinetic energy is increased, i.e.

$$\text{kinetic energy} \sim \hbar^2/mr^2 , \qquad (19\text{-}2)$$

where r = extent of the position. So, if r is too small, (19-2) will be larger than the potential energy e^2/r. Thus condition (b) is not enough. It has to be supplemented by the exclusion principle, i.e. condition (c). (See Sec. 19.2.)

Therefore the discussion of the electrostatic interaction must either be coupled with that of other effects, otherwise we must use approximations to avoid the instability mentioned above. In statistical mechanics, there are many problems directly related to the electrostatic interaction. In these problems, the main characteristic of the electrostatic interaction is its long-range effect. Because $1/r$ is still significant even when r is large, a particle can interact simultaneously with many other particles. The discussion below stresses this long-range interaction.

19.2. The Plasma Gas and Ionic Solutions

Suppose we have N particles with positive charge e and N with negative charge $-e$. Also suppose that there is a repulsion at short distances preventing the charges from approaching each other indefinitely.

Now consider the density correlation function of the particles and the thermodynamic potential. We shall follow the method in Chapter 13 (see Sec. 13.2). First put a positive charge e at the origin and see how the other particles are distributed.

Let $n_\pm(\mathbf{r})$ be the densities of positive and negative charges. Let $\phi(\mathbf{r})$ be the electric potential. Then

$$n_\pm(\mathbf{r}) = n e^{\mp e\phi(\mathbf{r})/T}$$
$$\simeq \mp n e \phi(\mathbf{r})/T + n \quad. \tag{19-3}$$

where n is the average density of the positive or negative charges.

The relation of $\phi(\mathbf{r})$ with the density of the total charge is

$$-\nabla^2 \phi(\mathbf{r}) = 4\pi e(n_+(\mathbf{r}) - n_-(\mathbf{r})) \quad. \tag{19-4}$$

Substitute (19-3) into (19-4), and we get

$$\nabla^2 \phi = \phi/b^2 \quad,$$
$$1/b^2 \equiv 8\pi e^2 n/T \quad, \tag{19-5}$$

with the solution

$$\phi(\mathbf{r}) = \frac{e}{r} e^{-r/b} \quad.$$

If there are no other particles, a charge at the origin produces a potential e/r. Hence the effect of the particles is to reduce the electric potential far away. Now we have a new length scale b, the so-called screening length. Around the origin there is a cloud of charges of density

$$e(n_+ - n_-) = -\nabla^2 \phi/4\pi$$
$$= -\frac{e}{4\pi b^2 r} e^{-r/b} \quad. \tag{19-6}$$

This is the screening layer. The integral of (19-6) over $4\pi r^2 dr$ is $-e$. Hence the charge at the origin is screened and its effect outside b tends to zero. An increase in temperature increases b and (19-5) is more accurate. Notice that from (19-5) we have

$$(8\pi e^2 n^{1/3})/T = 1/(b^3 n)^{2/3} \quad. \tag{19-7}$$

The left-hand-side is approximately the ratio of the interaction energy e^2/r to the kinetic energy T. ($r \sim$ distance between the particles $\sim n^{-1/3}$.) The denominator on the right is the number of particles in the screening layer raised to the $\frac{2}{3}$ power. Therefore, the above approximation is valid when $b^3 n$ is large, but remains as a low density approximation.

The functions $n_\pm(\mathbf{r})$ are conditional distributions. The condition is that there is a charge at the origin. As in Chapter 13, it should be denoted by $n_\pm(\mathbf{r}|0+)$ (see Eq. (13-20)). From $n_\pm(\mathbf{r})$ the density correlation function can be calculated

$$\langle \rho_\pm(\mathbf{r}) \rho_+(0) \rangle_c = \mp \frac{e^2 n^2}{rT} e^{-r/b} \qquad (19\text{-}8)$$

It can be seen that the correlation length of this gas is b. Originally the electrostatic interaction had no length scale; and the new scale b is a function of density, temperature and electric charge (see Eq. (19-5)).

From (19-8) we can derive the interaction energy (19-1):

$$U = \frac{V}{2} \int d^3 r \, \frac{e^2}{r} \langle (\rho_+(\mathbf{r}) - \rho_-(\mathbf{r}))(\rho_+(0) - \rho_-(0)) \rangle$$

$$= -V \frac{T}{8\pi b^3} = -\frac{e^2}{b} N$$

$$= -V(8\pi)^{1/2} e^3 n^{3/2} T^{-1/2} \qquad (19\text{-}9)$$

From this result we can calculate the specific heat and thermodynamic potential, etc. Notice that

$$U \propto -N^{3/2} V^{-1/2} T^{-1/2} \quad , \qquad E = E_0 + U \quad . \qquad (19\text{-}10)$$

Readers are invited to prove the following:

$$\frac{\partial E}{\partial T} = C_V = C_{0V} - \frac{1}{2} U/T \quad ,$$

$$S = S_0 + \frac{1}{3} U/T \quad ,$$

$$F = F_0 + \frac{2}{3} U \quad ,$$

$$p = p_0 + \frac{1}{3} U/V \quad ,$$

$$\mu = \mu_0 + U/N \quad . \qquad (19\text{-}11)$$

The subscript 0 means quantities evaluated when $e = 0$.

If these charged particles are the solute in a solution, e.g. Na^+ and Cl^- in a saline solution, the above results can be directly applied. The only correction is the electric charge. In water, because the water molecules have an electric dipole moment, the charge of the ions has already been screened partly, so e^2 must be divided by the dielectric constant ϵ. In water $\epsilon = 88$ and the quantities in (19-11) now refer to the ions. The pressure p becomes the osmotic pressure (see Chapter 18).

In water, because $\epsilon = 88$ is quite large, the electrostatic interaction between the ions is reduced considerably. The ratio of the interaction energy to the kinetic energy is (19-7). If (19-7) is very small, then

$$n \ll (kT)^3/[8\pi(e^2/\epsilon)^3] \quad . \tag{19-12}$$

In the above formula, we have written out the Boltzmann constant k explicitly. Putting in numerical values $kT = 1.38 \times 10^{-16} \times 300$ erg, $e^2 = (4.8 \times 10^{-12})^2$ erg cm, $\epsilon = 88$, we get

$$n \ll 1.6 \times 10^{20}/cm^3 \quad . \tag{19-13}$$

This limitation is not too harsh, since each mole of solute contains 6.23×10^{23} molecules.

The analysis in this section is the so-called Debye-Huckel theory[a] for ionic solutions.

19.3. Electrons in Metals

The analysis of the above section can be applied to electrons in a metal. The major differences are:
(1) the positive charges (nuclei) in a metal are more or less stationary;
(2) the density of electrons is very high, $\epsilon_F \gg T$, and the electrostatic interaction energy $e^2/r \sim e^2 p_F$ is much larger than T. Therefore in discussing the electrostatic interaction, we can ignore the temperature.

The ratio of the interaction energy to the kinetic energy is

$$\alpha \equiv (e^2/r)/\epsilon_F = \frac{2me^2}{p_F} \sim \frac{n^{-1/3}}{a} \quad ,$$

$$a \equiv \hbar^2/me^2 \quad . \tag{19-14}$$

---

[a] For a more detailed discussion on Debye-Huckel theory, see books on physical chemistry.

Therefore, this ratio is approximately equal to $n^{-1/3}$, the distance between the electrons (n being the density), divided by the Bohr radius a (i.e. the radius of the hydrogen atom is 0.53 Å). This ratio is not small, and hence the electrostatic interaction of the electrons in a metal is not weak. In the following analysis we still assume it to be small for the convenience of calculation. The result obtained is therefore not very accurate.

We place a charge e at the origin and see how the electrons respond. The nuclei are stationary and hence do not respond.

Let $\phi(\mathbf{r})$ be the electric potential. The equation for the electrostatic potential is

$$-\nabla^2 \phi = -4\pi e n'(\mathbf{r}) \quad , \tag{19-15}$$

where n' is the excess electron density, i.e. the density of the electron minus N/V, N being the total number of electrons and V the volume. We treat the stationary positive charges as homogeneously distributed, so their charge just cancels the mean charge of the electrons, and the remaining density $n'(\mathbf{r})$ is the density distribution within the potential $-e\phi(\mathbf{r})$. Let ϵ_F be the kinetic energy of the electrons at the Fermi surface. Then

$$\epsilon_F + (-e\phi(\mathbf{r})) = \mu \quad , \tag{19-16}$$

i.e. the energy at the top of the Fermi sea is the sum of the potential energy plus the kinetic energy. At $T = 0$ this is the chemical potential. The above assumes $\phi(\mathbf{r})$ to be a slowly varying function of $\mathbf{r}$. To calculate the energy distribution of the electronic states, i.e. the density of states, we can use the short wavelength approximation (see Eq. (3-12))

$$g(\epsilon) = \frac{1}{Vh^3} \int d^3r\, d^3p\, \delta\left(\epsilon - \frac{p^2}{2m} + e\phi(\mathbf{r})\right)$$

$$= \frac{1}{V} \int d^3r\, g_0(\epsilon + e\phi(\mathbf{r})) \quad . \tag{19-17}$$

Here $g_0(\epsilon)$ is the density of states when $\phi = 0$ and $g_0(\epsilon + e\phi)$ can be interpreted as the density of states at $\mathbf{r}$. The energy level is lowered by $e\phi$ because the potential is now $-e\phi$.

Now assume $e\phi(\mathbf{r})$ to be very small, then the density of electrons at $\mathbf{r}$ is

$$n'(\mathbf{r}) + n = \int_0^\mu d\epsilon\, g_0(\epsilon + e\phi) ,$$

$$\simeq \int_0^\mu d\epsilon \left[g_0(\epsilon) + e\phi \frac{\partial g_0}{\partial \epsilon} \right] ,$$

$$n'(\mathbf{r}) = e\phi\, g_0(\mu) = e\phi \frac{m p_F}{\pi^2} . \tag{19-18}$$

The factor $g_0(\mu)$ is just the $g(0)$ in Chapter 4. Substituting (19-18) into (19-15), we get

$$\nabla^2 \phi = \phi/b^2 ,$$

$$1/b^2 \equiv 4e^2 m p_F/\pi = 2\alpha p_F^2/\pi$$

$$= 6\pi e^2 n/\epsilon_F ,$$

$$\phi(\mathbf{r}) = \frac{e}{r} e^{-r/b} . \tag{19-19}$$

This result is somewhat similar to (19-5), and the ratio α is defined in (19-14).

The above analysis assumes α to be very small. However, in most metals α is not small. Therefore the above result only tells us that the electrostatic screening is very strong and the effective distance of the electrostatic interaction is very short.

19.4. Electron Crystals

If the density of the electrons is so small that

$$\epsilon_F \ll T \ll e^2/r , \tag{19-20}$$

then we need not use quantum mechanics, because this ensemble should become a crystal at low temperature. Notice that we have assumed that the positive charges form a homogeneous background. (If the distribution is not homogeneous, but regular as in a crystal, the problem would be entirely different.) This homogeneous arrangement of the positive charges with mobile electrons is not realisable in three dimensions. However, in the laboratory we can use other effects to provide this homogeneous positive background. This has been achieved in two dimensions. Now we present the relevant experiment.

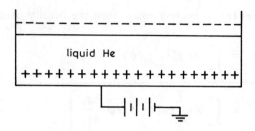

Fig. 19-1

Figure 19-1 is a schematic diagram of the experimental set-up. A flat container holds some low temperature liquid helium (about 0.5 K). Some electrons are sprayed onto the liquid surface; these electrons are the objects under observation. Positive charges are introduced into the bottom of the container. They serve two functions: on the one hand they pull the electrons downwards; on the other hand they provide a uniform positive background.

The electrons cannot enter the liquid, because they are blocked by the helium atoms. In fact the liquid surface attracts the electrons slightly, so that the electrons adhere to the surface. However, the electrons are very light and their wave nature keeps them slightly above the liquid surface.[b] Because the temperature is very low, the vapour pressure of helium can be neglected and the space above the liquid surface is more or less a vacuum, while the electrons on the liquid surface form a two-dimensional system. The electrons can move on the liquid surface and interact via their electrostatic repulsion. There are no impurities. The density of the electrons can be controlled (from 10^5 to $10^9 /\text{cm}^2$). This is theoretically very close to a collective body. Of course, there is some oscillation in the liquid surface, with a slight effect on the motion of the electrons.

If the temperature is sufficiently low, these electrons will be arranged into a crystal. How can we prove experimentally that this crystal exists? The answer comes from the vibration of the crystal. The vibrational frequencies of a crystal are related to its structure. The experimental procedure is to apply a weak oscillating electric field in the perpendicular direction, i.e. periodically pulling the electrons against the liquid surface. This pulling can cause oscillation in the liquid surface and shake the electrons. If the frequency and wavelength of the vibrating surface are the same as those of one of the characteristic vibrations of

---

[b] The binding energy of electron and the liquid surface is 0.7×10^{-3} eV. The average position above the liquid surface is 100 Å.

the crystal, resonance will occur and energy will be absorbed from the electric field. From the frequency of the resonance we can deduce the existence of the crystal structure.

There is extensive experimental and theoretical work on this problem. The reader can consult the literature.[c]

19.5. Two-Dimensional Coulomb Gas Model

Now let us discuss a collection of charges in two dimensions, or vortex lines. In electrostatics we encounter the two-dimensional formula

$$-\nabla^2 \phi = 2\pi q \delta(\mathbf{r}) \quad , \tag{19-21}$$

where ϕ is the potential due to a charge at the origin:

$$\phi(\mathbf{r}) = -q \ln(r/a) \quad , \tag{19-22}$$

and a is a constant. In fluid mechanics if we have a vortex of strength g at the origin, then $\phi(r)$ is the velocity potential and $\nabla \phi$ is the velocity.

Now we consider a collection of such charges. The interaction between them is

$$H = -\frac{1}{2} \sum_{i,j} q_i q_j \ln\left(\frac{r_{ij}}{a}\right) ,$$

$$\sum_i q_i = 0 , \tag{19-23}$$

and we assume that the distance between the i-th and j-th charges, r_{ij}, to be greater than a. The charges q_i can be $+q$ or $-q$. This collection is of course not the same as that in the last section. The preceding section discussed an interaction e^2/r. If we call (19-21) the electrostatic equation, then (19-23) can rightly be called a collection of charges. The model (19-23) has received much attention recently, because it bears a close relation to many two-dimensional models involving magnetism, superfluidity and melting, etc. The analysis of these models was started by Kosterlitz and Thouless, and now these models are named after them.[d]

The $\ln(r/a)$ interaction is not only a long range interaction but increases with r. Such strong interaction makes the analysis of Secs. 19.2 and 19.3 inapplicable.

---

[c] Grimes and Adams (1979) is a good place to start.

[d] Kosterlitz and Thouless (1973), Kosterlitz (1973).

Kosterlitz and Thouless were able to pinpoint the main characteristics of this model, namely: if the temperature is low, the positive charges (q) and the negative charges $(-q)$ will combine in pairs forming "molecules" or an "electric dipole". If the temperature is higher than some T_0, the molecules will be broken up. This feature can be understood as follows.

Starting from the method of Sec. 19.2, we put a positive charge at the origin. This positive charge produces an electric potential $\phi(\mathbf{r})$. We now ask: Will there be screening? (Screening means total screening, like that in Secs. 19.2 and 19.3.) If, when r is very large,

$$\phi(\mathbf{r}) \longrightarrow -\frac{q}{\epsilon} \ln(r/a) \quad , \tag{19-24}$$

and ϵ (the dielectric constant) is unchanging, then we have no free charges but only dipoles, i.e. the positive and negative charges all combine in pairs. If $1/\epsilon \to 0$ when $r \to \infty$, then we have screening. (Note that q/ϵ is the unscreened charge.) Figure 19-2a shows a circle of radius r, centred at the origin where q resides. The total charge within the circle is q/ϵ. If the molecule is small (the dipole distance is small), no screening occurs because the dipole is totally in the circle and cannot change the charge within the circle. The dipoles near the boundary may cross the boundary, but this only causes a finite ϵ. To create screening, we must have many molecules bigger than r, with one end within the circle and the other end outside. If there are many such molecules, then within the circle it will be like a collection of many single positive and negative charges, with their other halves too far away to have much effect. From the experience gained in Secs. 19.2 and 19.3, these single charges will create screening. Let us now estimate the number of these large molecules.

The probability distribution of the dipole separation r' in a molecule is

$$f(r') \propto e^{+q\phi(r')/T} \sim e^{-(q^2/T)\ln(r'/a)}$$

$$= (r'/a)^{-q^2/T} . \tag{19-25}$$

Therefore, the number density of molecules larger than r is

$$\int_r^\infty dr' \, 2\pi r' \, f(r') \quad . \tag{19-26}$$

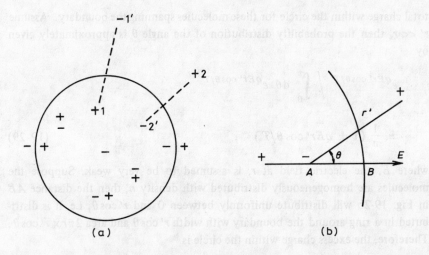

Fig. 19-2 Small molecules can be looked upon as dipoles, but big molecules like 11', 22' are more or less like single free charges.

The number of large molecules with one end within the circle is approximately

$$\pi r^2 \int_r^\infty dr' \, 2\pi r' \, (r'/a)^{-q^2/T}$$

$$\sim r^{4-q^2/T} . \qquad (19\text{-}27)$$

If this number tends to zero as $r \to \infty$, there is no screening; otherwise there will be. Therefore when

$$T < T_0 = q^2/4 , \qquad (19\text{-}28)$$

no screening occurs, i.e. the positive and negative charges combine in pairs. The above analysis needs some correction. In (19-25) we used $q^2 \ln r/a$ as ϕ. But there may be molecules between the dipoles, so this ϕ must be corrected. But when the temperature is not close to T_0 this correction is not important.

Let us now calculate the dielectric constant ϵ for $T < T_0$. Suppose each molecule is much smaller than r. Figure 19-2 shows a molecule with the negative charge inside and the positive charge outside of the circle. We shall calculate the total charge within the circle for these molecules spanning the boundary. Assume $r' \ll r$, then the probability distribution of the angle θ is approximately given by

$$e^{qEr'\cos\theta/T} \Big/ \int_0^\pi d\theta \, e^{qEr'\cos\theta/T}$$

$$\simeq \frac{1}{\pi}(1 + qEr'\cos\theta/T) \quad , \tag{19-29}$$

where E, the electric field at r, is assumed to be very weak. Suppose the molecules are homogeneously distributed with density n, then the distance AB in Fig. 19-2b will distribute uniformly between 0 and $r'\cos\theta$, i.e. it is distributed in a ring around the boundary with width $r'\cos\theta$ and area $2\pi r \times r'\cos\theta$. Therefore, the excess charge within the circle is

$$n(-q) \times 2\pi r \times r'\cos\theta \quad . \tag{19-30}$$

If $\cos\theta$ is negative, it indicates a positive charge in the circle and a negative charge outside. Averaging over r' and θ and using (19-29), we obtain the charge within the circle

$$-2\pi r E n \langle r'^2 \rangle \frac{q^2}{2T} + q = q/\epsilon \quad . \tag{19-31}$$

From Gauss' law, we know that $2\pi rE = 2\pi q/\epsilon$, so

$$\epsilon = 1 + \pi n \langle r'^2 \rangle q^2/T \quad . \tag{19-32}$$

The average value of r'^2 can be calculated from (19-25)

$$\langle r'^2 \rangle = \frac{\int_0^\infty f(r')r'^3 \, dr'}{\int_0^\infty f(r')r' \, dr'}$$

$$= a^2 \left[\frac{1-(2T/q^2)}{1-(4T/q^2)}\right] = a^2 \left[\frac{1-(T/2T_0)}{1-(T/T_0)}\right] . \tag{19-33}$$

This formula cannot be used when T is close to T_0.

If the temperature is high, then we can use the method of Secs. 19.1 and 19.2. The reader can prove that when r is very large

$$\phi(\mathbf{r}) \sim \frac{q}{\sqrt{r/b}} e^{-r/b}$$

$$1/b^2 \equiv 4\pi n q^2/T \ . \tag{19-34}$$

This is similar to (19-5), showing the property of a plasma gas. These are very simple results and require no complicated analysis.

If we want to study the properties for T close to T_0, then calculations of a slightly more complicated nature would be required. The reader may refer to the literature.[e]

Problems

1. The statistical model of an atom (the Fermi-Thomas model). A heavy atom has many electrons, whose density can be derived from (19-15) − (19-17). The nucleus resides at the origin. The potential $\phi(r)$ and the electron density $n(r)$ both have spherical symmetry.
   (a) Express $n(r)$ in terms of μ, and $\phi(r)$. [$n(r) \propto (\mu - e\phi)^{3/2}$].
   (b) From (19-15) obtain the differential equation for ϕ.
   (c) Impose the condition

$$Z = \int d^3 r \, n(r) \ , \tag{19-35}$$

to derive μ, where Ze is the charge of the nucleus. The above question is not easily solved analytically; a numerical procedure is necessary.

2. Model of the metal surface.

Consider the nuclei in a metal as a collection of homogeneously distributed positive charges. For $x < 0$ the density is N/V and for $x > 0$ the density is zero. Use the method of the preceding problem to solve for $n(x)$ i.e. the density distribution of the electrons. As $x \to -\infty$, $n(x) \to N/V$. Near $x = 0$, $n(x)$ is relatively complicated, because as $x \to \infty$, $n(x) \to 0$.

---

[e] Young (1979). This is clearer than the paper of Kosterlitz and Thouless.

3. Charge q is situated outside an insulator (see Fig. 19-3). Obtain the force on the charge. The insulator has a dielectric constant ϵ.
Hint: Suppose there is an image charge q' in the insulator

$$q' = -q \frac{\epsilon - 1}{\epsilon + 1} . \qquad (19\text{-}36)$$

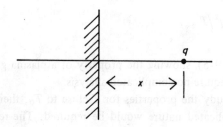

Fig. 19-3

4. From the result of the preceding problem, obtain the potential energy $U(x)$ of the electron at x.

(a) Write down the wave equation for this potential, assuming that the electron cannot enter the insulator.

(b) Solve for the energy and the wavefunction of the electron.

(c) Substitute the ϵ value of liquid He^4. Calculate the zero-point energy of the electron and its mean position.

5. Review Problem 7.11. This problem shows that when $T < q^2/2$, Z does not exist. Note that $q^2/2$ is $2T_0$ (see Eq. (19-28)), not T_0. Nevertheless, if we require that the distance between the particles is not less than a, then Z is still defined at low temperatures (see Eq. (19-23)).

6. The content of the preceding problem is that when $T < 2T_0$ the positive and negative charges form a bound state.

(a) Let the charge q be situated at the origin and $-q$ at r. Let $T < 2T_0$. Prove that the probability for r to be finite is 1, i.e. these two charges are a bound state.

(b) Prove that for $T_0 < T < 2T_0$ the dielectric constant ϵ does not exist, i.e. although the system is bound, $\langle r^2 \rangle \to \infty$. This is very obvious from (19-33). Therefore, between T_0 and $2T_0$, a pair of positive and negative charges are separated by quite a large distance. So other charges can come in between. So the conclusion that r is finite is untenable. The analysis for this range of temperatures is very complicated.

7. In three dimensions the Coulomb gas does not have a critical temperature like T_0. At low temperatures, the positive and negative charges combine together

while at high temperatures they dissociate into a plasma state. It is a continuous change from the combined molecules to the plasma gas.

(a) Let the interaction between q_i and q_j be

$$U(\mathbf{r}) = q_i q_j/r , \qquad r > a ,$$
$$= \infty , \qquad r \leqslant a ,$$
$$q_i = \pm e . \tag{19-37}$$

Let the average density of the particles be n and $na^3 \ll 1$. Under what conditions can these particles be considered as a plasma gas? Try to perform a rough analysis.

(b) Let the positive charges be the nuclei, the negative charges the electrons. Taking quantum mechanics into account, express the degree of ionisation (i.e. how many atoms are ionised) in terms of the temperature and pressure.

(c) From the above results, discuss the ionic solution and the conclusion of Sec. 19.2.

PART V
DYNAMICS

The basic assumption of statistical mechanics (Chapter 5) is a very effective tool for analysis. The assumption says nothing directly about the motion of the molecules, the effect of which serves only to establish the validity of this assumption. But we have emphasised throughout that this assumption takes on different forms under different situations, according to the nature of the motion. In the end, molecular motion determines everything. This part focuses on the motion, but with a view to further the understanding of the equilibrium states. The Boltzmann equation (Chapter 20) is most important. We use it to discuss some phenomena of small fluctuations, such as sound waves and heat conduction. Chapter 21 discusses the diffusion equation, emphasising metastable states. The series of examples in these two chapters, through phenomena displaying slow changes, enable us to understand the meaning of equilibrium. Chapter 22 discusses methods for obtaining the equilibrium properties directly from the motion of the molecules. These methods rely on the use of computers, and are a great achievement of modern physics, serving not only as computational tools but also as aids which permit a deeper understanding of the basic meaning of statistical mechanics.

Chapter 20

THE EQUATION OF MOTION FOR GASES

The equilibrium properties of a thermodynamical system can be explained quite satisfactorily by the basic assumption of statistical mechanics. However, nonequilibrium properties are beyond this simple tool. The theory of nonequilibrium phenomena is but in its infancy. The dynamical equation of motion for gases, i.e. the Boltzmann equation, is one of the few useful tools in this respect, as demonstrated by the examples in this chapter. This equation makes a clear separation between the effects of irregular collisions and the effects of regular motion. Regular motion is slowly varying, while collisions are rapid. The Boltzmann equation averages out the fast, irregular collisions. Therefore, the scales of time and length of interest in this equation are much larger than the collision time and the molecular size. This chapter uses some examples to illustrate the meaning and application of the equation of motion. The examples include the plasma oscillation, zero sound, scattering by impurities and the propagation of sound waves. Finally we mention the H-theorem.

20.1. Flow and Collision

The equation of motion of gases, called the Boltzmann equation, is the basic tool for discussing the motion of an ideal gas. Many-body motion is extremely complicated and the ideal gas is no exception. The equation of Boltzmann and his hypothesis concerning entropy are unparalleled achievements in the history of physics.

This equation can be written down by intuition. But it is extremely difficult to derive it from Newton's laws or the other laws of mechanics. The following is a simple argument leading to this equation.

THE EQUATION OF MOTION FOR GASES 347

Let $f(\mathbf{r}, \mathbf{p}, t)$ be the distribution function at time t of a collection of particles. This is a distribution in the six-dimensional $(\mathbf{r}, \mathbf{p})$ space. Let the number of particles be a conserved quantity, then the equation of probability conservation is

$$\frac{\partial f}{\partial t} + \frac{\partial}{\partial \mathbf{r}} \cdot (\mathbf{v} f) + \frac{\partial}{\partial \mathbf{p}} \cdot (m \mathbf{a} f) = 0 \ , \qquad (20\text{-}1)$$

where the velocity $\mathbf{v}$ is $\mathbf{p}/m$, and $\mathbf{a}$ is the acceleration, so that $(\mathbf{v}, m\mathbf{a})$ is the velocity in $(\mathbf{r}, \mathbf{p})$ space. Equation (20-1) is the generalisation of the conservation equation of particles in three dimensions:

$$\frac{\partial \rho}{\partial t} + \nabla \cdot \mathbf{v}\rho = 0 , \qquad (20\text{-}2)$$

where ρ is the density of the particles. Equations (20-1) and (20-2) assume that particles do not jump from one place to another, but only flow continuously. If we take collisions into account, then the momenta of the particles can suffer abrupt changes, so (20-1) will not be correct. (However, the positions of the particles still cannot change abruptly.) To take account of these abrupt changes, we have to add a correction term to (20-1):

$$\frac{\partial f}{\partial t} + \mathbf{v} \cdot \frac{\partial f}{\partial \mathbf{r}} + \mathbf{F} \cdot \frac{\partial f}{\partial \mathbf{p}} = \left(\frac{\partial f}{\partial t}\right)_C . \qquad (20\text{-}3)$$

Here $\mathbf{F} = m\mathbf{a}$ is the external force on every particle. The correction term on the right is the difference between the number of particles scattered into $(\mathbf{r}, \mathbf{p})$ and that scattered out of $(\mathbf{r}, \mathbf{p})$:

$$\left(\frac{\partial f}{\partial t}\right)_C = \int (f'' f''' - f f') \, R \, d^3 p' \, d^3 p'' \, d^3 p''' ,$$

$$R \propto \delta(\mathbf{p} + \mathbf{p}' - \mathbf{p}'' - \mathbf{p}''') \, \delta(\epsilon + \epsilon' - \epsilon'' - \epsilon''') , \qquad (20\text{-}4)$$

where R is the rate of the two-particle reaction $\mathbf{p} + \mathbf{p}' \to \mathbf{p}'' + \mathbf{p}'''$, containing δ functions for the conservation of energy and momentum, and the functions $f, f', \ldots$ etc. stand for

$$f \equiv f(\mathbf{r}, \mathbf{p}, t) ,$$

$$f' \equiv f(\mathbf{r}, \mathbf{p}', t) ,$$

$$f'' \equiv f(\mathbf{r}, \mathbf{p}'', t) ,$$

$$f''' \equiv f(\mathbf{r}, \mathbf{p}''', t) . \tag{20-5}$$

In the discussion of detailed balance in Chapter 3, we pointed out that the value of R for the forward and reverse directions of the reaction are the same. For fermions at high densities, the terms $f''' f'' - f f'$ must be changed to

$$f'' f''' (1-f)(1-f') - ff'(1-f'')(1-f''') , \tag{20-6}$$

(see Sec. 3.2.) Each f is the number of particles in each state in phase space; in $d^3p\, d^3r$ there are $d^3p\, d^3r/h^3$ states, i.e.

$$f d^3p\, d^3r/h^3$$

is the number of particles in $d^3p\, d^3r$. If $f \ll 1$, then quantum effects can be ignored, and the factor $1/h^3$ can be absorbed into the definition of f.

In equilibrium, collisions do not change f, hence (20-4) and (20-6) are zero. Then we can get the equilibrium distribution of the particles. (See Chapter 3.) Notice that if f is a function of the energy ϵ of the particles alone, $f(\mathbf{r}, \mathbf{p}) = f(\epsilon)$,

$$\epsilon \equiv p^2/2m + u(\mathbf{r}) , \qquad (\mathbf{F} = -\partial u/\partial \mathbf{r}) , \tag{20-7}$$

i.e. $f(\mathbf{r}, \mathbf{p}) = f(\epsilon(\mathbf{r}, \mathbf{p}))$, then the left-hand-side of (20-3) is zero. For the right-hand-side to vanish, $f(\epsilon)$ must take some special form, e.g. $e^{-\alpha - \beta \epsilon}$ to make (20-4) vanish, or $1/(e^{\alpha + \beta \epsilon} + 1)$ to make (20-6) vanish.

Notice that $(\partial f/\partial t)_c$ is not related to equilibrium; the Boltzmann equation can be used to calculate any physical quantities, including the equilibrium ones. However, equilibrium quantities should not depend on $\mathbf{r}$. If $\mathbf{F}$ is purely an external force, then the equilibrium properties will be those of the ideal gas. Nevertheless we may regard f as a function of $\mathbf{F}$ in order to incorporate part of the interaction.

THE EQUATION OF MOTION FOR GASES 349

The left-hand-side of (20-3) describes slow and gradual motion, and the right-hand-side the effects of fast collisions. In a volume element d^3r and a time interval dt, $(\partial f/\partial t)_C\, dt\, d^3p\, d^3r$ must be a large number. Every collision can be regarded as happening instantaneously and at a point. We first examine some cases with $(\partial f/\partial t)_C = 0$, and then consider some cases with $(\partial f/\partial t)_C \neq 0$.

20.2. Case of No Collisions – Plasma Oscillations

We start from the simplest example, assuming $\mathbf{F} = 0$ and no collisions, then

$$\frac{\partial f}{\partial t} + \mathbf{v} \cdot \frac{\partial f}{\partial \mathbf{r}} = 0 \quad . \tag{20-8}$$

The solution of this equation is

$$f(\mathbf{r}, \mathbf{p}, t) = \phi(\mathbf{r} - \mathbf{v}t, \mathbf{p}) \quad , \tag{20-9}$$

where $\phi(\mathbf{r}, \mathbf{p})$ is the distribution at $t = 0$. If one particle is at $\mathbf{r}$ at time t, then at time $t = 0$, its position is $\mathbf{r} - \mathbf{v}t$.

Example 1:

A gas at temperature T is concentrated at the origin at time $t = 0$. Calculate its subsequent density distribution $n(\mathbf{r}, t)$.

$$n(\mathbf{r}, t) = \int d^3p \, f(\mathbf{r}, \mathbf{p}, t)$$

$$= \int d^3p \, \phi(\mathbf{r} - \mathbf{v}t, \mathbf{p})$$

$$\phi(\mathbf{r}, \mathbf{p}) = N(2\pi mT)^{-3/2} e^{-p^2/2mT} \delta(\mathbf{r}) \quad . \tag{20-10}$$

Therefore,

$$n(\mathbf{r}, t) = N\left(\frac{m}{t}\right)^3 (2\pi mT)^{-3/2} e^{-mr^2/2Tt^2} \quad . \tag{20-11}$$

The gas disperses because the velocities of the molecules are different.

Example 2: Plasma oscillations (planar case).

Now we consider a slightly more complicated example, that is, electrons in two dimensions, (see Sec. 19.4) which are assumed to form a dilute gas. Let

$n'(\mathbf{r}, t)$ be the density distribution of the electrons minus the average N/L^2 where L^2 is the area of the plane. Then,

$$\mathbf{F}(\mathbf{r}, t) = e \nabla \phi \quad,$$

$$\frac{\partial f}{\partial t} + \mathbf{v} \cdot \nabla f + \mathbf{F} \cdot \frac{\partial f}{\partial \mathbf{p}} = 0$$

$$\phi(\mathbf{r}, t) = - \int d^2 r' \frac{e}{|\mathbf{r} - \mathbf{r}'|} n'(\mathbf{r}', t) \quad. \tag{20-12}$$

We now make the following assumptions:

$$f = f_0 + f' e^{-i\omega t + i\mathbf{k} \cdot \mathbf{r}} \quad,$$

$$f_0 \equiv n(2\pi mT)^{-1} e^{-p^2/2mT} \quad,$$

$$f' \ll f_0 \quad. \tag{20-13}$$

The other quantities ϕ, n' are also plane waves with small amplitudes. From (20-12) and (20-13) we get

$$-i(\omega - \mathbf{k} \cdot \mathbf{v}) f' + i \frac{2\pi e^2 n'}{k} \mathbf{k} \cdot \mathbf{v} f_0 / T = 0 \quad. \tag{20-14}$$

Notice that

$$\int d^2 r \frac{1}{r} e^{-i\mathbf{k} \cdot \mathbf{r}} = \frac{2\pi}{k} \quad,$$

$$\frac{\partial f_0}{\partial \mathbf{p}} = - \mathbf{v} f_0 / T \quad. \tag{20-15}$$

From (20-14) we get

$$\int d^2 p \, f' = n' = n' \frac{2\pi e^2}{k} \int d^2 p \frac{\mathbf{k} \cdot \mathbf{v} f_0 / T}{\omega - \mathbf{k} \cdot \mathbf{v}} \tag{20-16}$$

Therefore, the condition of oscillation is

$$1 = \frac{2\pi e^2}{k} \int d^2p \, \frac{\mathbf{k} \cdot \mathbf{v} f_0 / T}{\omega - \mathbf{k} \cdot \mathbf{v}} . \qquad (20\text{-}17)$$

When k/ω is very small, we can expand in terms of $\mathbf{k} \cdot \mathbf{v}/\omega$ on the right and get

$$1 = \frac{2\pi e^2}{k} \int d^2p \, \frac{(\mathbf{k} \cdot \mathbf{v})^2}{\omega^2} f_0/T + \ldots$$

$$\simeq \frac{2\pi e^2 n}{m} \frac{k}{\omega^2} , \qquad (20\text{-}18)$$

$$\omega^2 = \frac{2\pi e^2 n}{m} k . \qquad (20\text{-}19)$$

Therefore, the oscillation frequency is proportional to $\sqrt{k}$, and with imaginary terms $O(k^{3/2})$ representing the rate of attenuation. When k is very small, k/ω is indeed small, and the expansion in terms of k/ω is feasible.

The result of (20-19) applies not only to a planar electron gas, but also to electron crystals on a plane. The above has not considered collisions but even if there are collisions, (20-19) is still true. Equation (20-19) can be directly derived from the equations of fluid mechanics. (See problems 1, 2, 3.) The characteristic of (20-19) is this: it comes from the electrostatic interaction between the electrons which produces a very strong restoring force similar to the elastic forces in solids.

Notice that the **F** in (20-12) is an averaged force. Because of the motion of the molecules, the forces acting on the molecules will change abruptly, but **F** in (20-12) does not include these changes. It only includes the slowly varying $n'(\mathbf{r}, t)$, in which the scale of t is much larger than the collision time. Of course the purpose of $(\partial f/\partial t)_c$ is to include the effect of collisions, but it does not take into account the density correlation of the particles.

20.3. Zero Sound

Zero sound in a Fermi gas is similar to the plasma oscillation discussed in the preceding example. However, zero sound comes from short-range interactions, and occurs only for $T/\epsilon_F \ll 1$. Among neutral fermion systems easily produced in the laboratory, only low temperature liquid ^{3}He satisfies this condition.

352 STATISTICAL MECHANICS

(At standard atmospheric pressure, the Fermi momentum is $p_F \sim 5\,\text{K}$. Interatomic interactions are strong. Nevertheless the properties still resemble those of free particles. These results were first obtained by Landau.) Now proceeding as in (20-12), we can write (20-12) and (20-13) as

$$\mathbf{F} = -\nabla\phi \;,$$

$$\frac{\partial f}{\partial t} + \mathbf{v}\cdot\nabla f + \mathbf{F}\cdot\frac{\partial f}{\partial \mathbf{p}} = 0 \;,$$

$$\phi = \int d^3r'\, u(\mathbf{r}-\mathbf{r}')\, n'(\mathbf{r}',t) \;,$$

$$f = f_0 + f'\, e^{-i\omega t + i\mathbf{k}\cdot\mathbf{r}} \;,$$

$$f_0 = 1/(e^{\epsilon_p/T} + 1) \;,$$

$$\epsilon_p \equiv p^2/2m - \epsilon_F \;. \tag{20-20}$$

where $u(\mathbf{r})$ is the interaction energy between a pair of atoms at separation $\mathbf{r}$. Now assume that as $T \to 0$, $f_0 = \theta(-\epsilon_p)$, then (20-14) and (20-15) become

$$(\omega - \mathbf{k}\cdot\mathbf{v})f' - n' u_k\, \mathbf{k}\cdot\mathbf{v}\,\delta(\epsilon_p) = 0 \;,$$

$$\int d^3r\, u(r)\, e^{-i\mathbf{k}\cdot\mathbf{r}} = u_k \;, \tag{20-21}$$

$$\frac{\partial f_0}{\partial \mathbf{p}} = -\mathbf{v}\,\delta(\epsilon_p) \;. \tag{20-22}$$

Using the same steps as before, (20-16) and (20-17) become

$$n' = 2\int \frac{d^3p}{(2\pi)^3} f' \;,$$

$$1 = u_k\, 2\int \frac{d^3p}{(2\pi)^3}\, \frac{\mathbf{k}\cdot\mathbf{v}}{\omega - \mathbf{k}\cdot\mathbf{v}}\,\delta(\epsilon_p) \;,$$

$$= u_k\, g(0)\int \frac{d\Omega}{4\pi}\, \frac{k v_F \cos\theta}{\omega - k v_F \cos\theta} \;,$$

$$d\Omega \equiv 2\pi \sin\theta\, d\theta \;. \tag{20-23}$$

where $g(0) = mp_F/\pi^2$ is the density of states at the top of the Fermi sea (see Chapters 3 and 4). The integration in (20-23) is simple, and gives

$$1 = u_k \, g(0) \left[\frac{1}{2} \ln \frac{s+1}{s-1} - 1 \right] ,$$

$$\omega = v_F \, s \, k \ . \tag{20-24}$$

The oscillation frequency can be obtained from this equation. When the interaction is not strong, i.e. $u_k \, g(0)$ is small, s is close to 1.

$$\frac{1}{u_k \, g(0)} \simeq \frac{1}{2} \ln \frac{2}{s-1} ,$$

$$s \simeq 1 + 2 \, e^{-2/u_k g(0)} . \tag{20-25}$$

The wave velocity $s \, v_F$ is very close to v_F. This oscillation is the "zero sound".[a]

What is the difference between this "zero sound" and ordinary sound waves? Sound waves in solids are transmitted via the interaction between the atoms, i.e. they are elastic waves, similar to this zero sound. The phonons in crystals are ordinary sound waves (see Chapter 6), but can also be called zero sound,

$$c^2 = \frac{\partial p}{\partial n} . \tag{20-26}$$

If we use the formula for ordinary sound to calculate the velocity of sound c for a Fermi gas, we get (see the zero-point pressure formula in Sec. 4.1).

$$c = \frac{v_F}{\sqrt{3}} . \tag{20-27}$$

This result is different from (20-24) and (20-25). A careful analysis will show that this ordinary sound wave is the sound wave for the electrons outside the Fermi surface and the holes inside. In other words, at low temperatures, a Fermi gas can be regarded as a dilute gas of electrons and holes. This gas has a sound velocity $v_F/\sqrt{3}$. This sound wave is very similar to the "second sound" discussed in Chapter 6. The second sound is the sound wave of the phonon gas in a crystal. Its velocity is

$$c_2 = c_0/\sqrt{3} , \tag{20-28}$$

[a] For details, see Lifshitz and Pitaevskii (1980).

where c_0 is the velocity of the phonons. This results from the same effects as (20-27), i.e. propagation by collisions.

So we need not worry about the terminology "second" or "ordinary" (usually "ordinary" is called the "first"). It is better to understand the physical mechanism responsible for oscillation. These terms themselves are of limited scientific value.

Zero sound is caused by the interaction $u(\mathbf{r})$. Even though there is no collision, this sound wave still propagates. Then how do the sound waves in (20-27) and (20-28) propagate by collisions? By this we mean that in the absence of collisions propagation is not possible. The reader will understand this after reading Sec. 20.5.

20.4. Collisions and Diffusion

Now let us look at the effect of collisions. This is a complicated problem, so for the time being we do not discuss the collisions between phonons and particles but examine first of all a relatively simple problem: the collision of the particles with the fixed impurities.

Suppose there are many impurity atoms fixed in a metal. The equation of motion of the free electrons is (20-3), and the collision term is

$$\left(\frac{\partial f}{\partial t}\right)_C = \frac{1}{4\pi}\int d\Omega' \, \gamma(f' - f) \; ,$$

$$f' \equiv f(\mathbf{r}, \mathbf{p}', t) \; ,$$

$$p' = p \; , \tag{20-29}$$

where $\mathbf{p}$ and $\mathbf{p}'$ differ only in direction, $d\Omega'$ is the integration over the direction of $\mathbf{p}'$, and the rate of scattering γ is proportional to the density of the impurities. We assume that γ is not a function of the scattering angle but is a constant. Notice that

$$f(1 - f') - f'(1 - f) = f - f' \; . \tag{20-30}$$

Thus (20-29) is much simpler than (20-4).

Assume that f is not too much different from the equilibrium value of f_0 and that the difference is a plane wave of long wavelength

$$f \equiv f_0 + f_1 = [1 + \psi(\mathbf{p}, t) e^{i\mathbf{k}\cdot\mathbf{r}}] f_0 \; ,$$

$$\psi \ll 1 \; . \tag{20-31}$$

Using (20-3) and (20-29) we get

$$\frac{\partial \psi}{\partial t} = -(K_1 + K)\psi ,$$

$$K_1 \equiv i\mathbf{k}\cdot\mathbf{v} , \qquad \mathbf{v} = \mathbf{p}/m , \qquad (20\text{-}32)$$

$$-K\psi \equiv \frac{\gamma}{4\pi}\int d\Omega' (\psi' - \psi) ,$$

$$\psi' \equiv \psi(\mathbf{p}', t) . \qquad (20\text{-}33)$$

Because we assume that energy is conserved in collisions, only the direction of $\mathbf{p}$ is a variable while the magnitude of $\mathbf{p}$ is a constant, so ψ can be regarded as a function of the direction of $\mathbf{p}$ which is denoted by $\hat{p}$. Now we use the techniques of linear spaces to solve (20-32). The linear space in question is the set of all functions of $\hat{p}$. The inner product of two functions ϕ and ψ is defined as

$$(\psi, \phi) \equiv \frac{1}{4\pi}\int d\Omega \, \psi(\hat{p}) \, \phi(\hat{p})$$

$$= (\phi, \psi) . \qquad (20\text{-}34)$$

We now want to obtain the eigenvalues γ_α and eigenvectors ψ_α of $K_1 + K$.

$$(K_1 + K)\psi_\alpha = \gamma_\alpha \psi_\alpha . \qquad (20\text{-}35)$$

The solution of (20-32) is then

$$\psi = \sum_\alpha a_\alpha(t)\psi_\alpha ,$$

$$a_\alpha(t) = a_\alpha(0) e^{-\gamma_\alpha t} . \qquad (20\text{-}36)$$

Because we assume k to be very small, K_1 may be regarded as a perturbation. We first obtain the eigenvalues of K. Obviously from (20-33),

$$\psi_{00} = 1 , \qquad \gamma_{00} = 0 , \qquad \text{i.e.} \quad K\psi_{00} = 0 , \qquad (20\text{-}37)$$

is a solution and any $\psi_{0\alpha}$ whose integral is zero is an eigenvector of K with eigenvalue γ.

$$K\psi_{0\alpha} = \gamma\psi_{0\alpha} \quad,$$

$$(\psi_{00}, \psi_{0\alpha}) = \int \frac{d\Omega}{4\pi} \psi_{0\alpha} = 0 \quad,$$

$$\psi_{0\alpha} = \sqrt{4\pi}\, Y_{lm} \quad, \qquad \alpha = (l, m) \quad, \qquad l \neq 0 \quad. \tag{20-38}$$

That is to say, K has only two eigenvalues, 0 and γ. The eigenvector of 0 is 1 while the eigenvalue γ has infinitely many eigenvectors.

Now expand in powers of K_1, i.e. in powers of k. The first order correction to the eigenvalues is zero because K_1 is proportional to $\mathbf{v} = \mathbf{p}/m$, and $\psi_{0\alpha}^2$ has $\pm\mathbf{p}$ symmetry. So

$$\gamma_{1\alpha} = (\psi_{0\alpha}, K_1 \psi_{0\alpha}) = 0 \quad. \tag{20-39}$$

Next we look at the second order terms. From elementary quantum mechanics the second order correction to γ_0 can be immediately written down as

$$\gamma_{20} = \sum_{\alpha \neq 0} \frac{(\psi_{00}, K_1 \psi_{0\alpha})^2}{0 - \gamma_{\alpha 0}} \quad. \tag{20-40}$$

The only nonzero term from this sum comes from $\psi_{0\alpha} = \sqrt{4\pi}\, Y_{10} = \sqrt{3}\cos\theta$ where θ is the angle between $\mathbf{k}$ and $\mathbf{p}$,

$$(\psi_{00}, K_1 \sqrt{3}\cos\theta) = \frac{i\sqrt{3}}{4\pi} \int d\Omega\, kv \cos^2\theta$$

$$= \frac{i}{\sqrt{3}} kv \quad. \tag{20-41}$$

Substituting into (20-40) we get

$$\gamma_0 = \gamma_{20} = -\frac{k^2 v^2}{3\gamma} \equiv -Dk^2 \quad,$$

$$\gamma_\alpha = \gamma + O(k^2) \quad, \qquad \alpha \neq 0 \quad. \tag{20-42}$$

THE EQUATION OF MOTION FOR GASES 357

If k is very small, γ_α ($\alpha \neq 0$) are nearly all equal to γ. We can then ignore the $O(k^2)$ terms; however, the $O(k^2)$ term is the whole of γ_0 and cannot be neglected here. Therefore (20-36) is

$$\psi \simeq a_0(0) e^{-Dk^2 t} + \sum_\alpha a_\alpha(0) e^{-\gamma t} \quad ,$$

$$D = \frac{v^2}{3\gamma} \equiv \frac{v^2 \tau}{3} \quad . \tag{20-43}$$

The factor $1/\gamma$ is the mean free time τ, so D is proportional to τ. The last term in (20-43) vanishes very rapidly, on a time scale $\tau = 1/\gamma$. But for small k the first term on the right vanishes very slowly. This naturally is the result of the conservation of particle number: when $k = 0$ this term represents the change of total particle number, which is not affected by collisions. From (20-31) and (20-43), if $t \gg \tau$, then f_1 will satisfy the diffusion equation

$$\frac{\partial}{\partial t} f_1 = D \nabla^2 f_1 \quad , \tag{20-44}$$

where f_1 represents a collection of particles with the same energy. Because we assume that the impurities are fixed, the energy is not changed during the collision. Therefore the energy of each particle is an invariant.

The above analysis exhibits the relation between diffusion and collisions. For the sake of simplicity, we have not considered the problem of the oscillation of the impurities, the vibration of the crystals and the collision between electrons. Now that the steps of this analysis have been clearly explained, we can use it to look at slightly more complicated cases.

20.5. Collisions and Sound Waves, Viscosity and Heat Conduction

Now we use the equation of motion (20-3) and (20-4) to analyse the relation between collisions and sound waves. This is done only for an ideal gas, considering only collisions but not mutual interaction.

We write f as

$$f = f_0 + f_1 = [1 + \psi(\mathbf{p}, t) e^{i\mathbf{k}\cdot\mathbf{r}}] f_0(\mathbf{p}) \quad . \tag{20-45}$$

For f_0 see (20-13). Assume $\psi \ll 1$. Substituting (20-45) into (20-3) and (20-4), we get

$$\frac{\partial \psi}{\partial t} = -(K_1 + K)\psi \quad,$$

$$K_1 \equiv i\mathbf{k} \cdot \mathbf{v} \quad, \tag{20-46}$$

$$K\psi \equiv \int d^3p' \, d^3p'' \, d^3p''' \, f_0' \, R(\psi + \psi' - \psi'' - \psi''') \quad, \tag{20-47}$$

$$f_0' \equiv f_0(\mathbf{p}') \quad,$$

$$\psi' \equiv \psi(\mathbf{p}', t) \quad,$$

$$\psi'' \equiv \psi(\mathbf{p}'', t) \quad, \quad \text{etc.} \tag{20-48}$$

Now we use the techniques of the last section to solve (20-46). The set of all functions of $\mathbf{p}$ is the function space. The inner product of two vectors ψ and ϕ is defined as

$$(\phi, \psi) \equiv \int d^3p \, f_0(\mathbf{p}) \, \phi(\mathbf{p}) \, \psi(\mathbf{p}) \quad,$$

$$= (\psi, \phi) \quad. \tag{20-49}$$

If we can find the eigenvalues γ_α and eigenvectors ψ_α of $K_1 + K$, i.e.

$$(K_1 + K)\psi_\alpha = \gamma_\alpha \psi_\alpha \quad, \tag{20-50}$$

then the solution of (20-46) is

$$\psi = \sum_\alpha a_\alpha(t) \psi_\alpha \quad,$$

$$a_\alpha(t) = a_\alpha(0) \, e^{-\gamma_\alpha t} \quad. \tag{20-51}$$

Therefore, we only have to obtain ψ_α and γ_α. The operator K_1 is simple, but K is now rather complicated. The magnitude of the eigenvalue of K can be estimated from $(\psi, K\psi)$:

$$(\psi, K\psi) = \frac{1}{4} \int d^3p \, d^3p' \, d^3p'' \, d^3p''' \, f_0 \, f_0' \, R(\psi + \psi' - \psi'' - \psi''')^2 \, . \tag{20-52}$$

This formula is obtained from (20-49) and (20-47); notice the conservation of energy and momentum (see δ-function in R of (20-4)). From (20-52) it can be seen that the eigenvalues of K can never be negative. (Prove $(\psi, K\phi) = (\phi, K\psi)$.) The factor R is the rate of collisions and hence the eigenvalues of K are comparable to $1/\tau$, where τ is the mean free time.

We now consider the long wavelength motion, i.e. when k is very small. We expand γ_α and ψ_α with respect to K_1

$$\gamma_\alpha = \gamma_{0\alpha} + \gamma_{1\alpha} + \cdots ,$$

$$\psi_\alpha = \psi_{0\alpha} + \psi_{1\alpha} + \cdots ,$$

$$K\psi_{0\alpha} = \gamma_{0\alpha} \psi_{1\alpha} , \tag{20-53}$$

$$\gamma_{1\alpha} = O(k) ,$$

$$\gamma_{2\alpha} = O(k^2) , \quad \text{etc.} \tag{20-54}$$

If each $\gamma_{0\alpha}$ were comparable to $1/\tau$, there would be no long time oscillation. The main characteristic of K is that, because of the conservation of number, energy and momentum (see the δ-function in (20-4)), it has five eigenvectors with zero eigenvalue,

$$\psi_{00} = 1/\sqrt{A_0} ,$$

$$\psi_{0i} = p_i/\sqrt{A_i} , \quad i = 1, 2, 3$$

$$\psi_{04} = (\epsilon - \bar{\epsilon})/\sqrt{A_4} , \tag{20-55}$$

$$K\psi_{0\alpha} = 0 ,$$

$$(\psi_{0\alpha}, \psi_{0\beta}) = \delta_{\alpha\beta} , \quad \alpha, \beta = 0, 1, 2, 3, 4 \, . \tag{20-56}$$

The above $\sqrt{A_\alpha}$ and $\bar{\epsilon}$ are constants which satisfy (20-56):

$$A_0 = n \quad ,$$

$$A_{1,2,3} = n \langle p_i^2 \rangle = nmT \quad ,$$

$$A_4 = n \langle (\epsilon - \bar{\epsilon})^2 \rangle = \frac{3}{2} n T^2 \quad ,$$

$$\bar{\epsilon} \equiv \langle \epsilon \rangle = \frac{3}{2} T \quad . \tag{20-57}$$

The average $\langle \ldots \rangle$ is with respect to f_0. These five $\gamma_{0\alpha} = 0$, $\alpha = 0, 1, \ldots, 4$, are the lowest eigenvalues. Now we examine $\gamma_{1\alpha}$, $\alpha = 0, 1, \ldots, 4$, which are the eigenvalues of the 5×5 matrix

$$K_{1\alpha\beta} \equiv (\psi_{0\alpha}, K_1 \psi_{0\beta}) \quad . \tag{20-58}$$

Let **k** be along the third direction, then

$$K_1 = i\mathbf{k} \cdot \mathbf{v} = ikp_3/m \quad . \tag{20-59}$$

Obviously $K_{1\alpha\beta} = 0$, for $\alpha, \beta = 1, 2$ because $\mathbf{p}_1, \mathbf{p}_2$ and $\mathbf{p}_3$ are mutually perpendicular. So γ_1 in these two directions is zero. The other three $\psi_{0\alpha}$, $\alpha = 0, 3, 4$ form a 3×3 matrix

$$K_1 = ik \begin{pmatrix} 0 & u_0 & 0 \\ u_0 & 0 & u_4 \\ 0 & u_4 & 0 \end{pmatrix} \quad ,$$

$$u_0 = \frac{1}{m} (A_3/A_0)^{\frac{1}{2}} = (T/m)^{\frac{1}{2}} \quad ,$$

$$u_4 = \frac{2}{3} (A_4/A_3)^{\frac{1}{2}} = (2T/3m)^{\frac{1}{2}} \quad . \tag{20-60}$$

The eigenvalues of K_1 are

$$0, \pm ikc,$$

$$c \equiv \sqrt{u_0^2 + u_4^2} = (5T/3m)^{1/2} . \quad (20\text{-}61)$$

Therefore, except for $\gamma_{1\pm} = \pm ikc$, the other three $\gamma_{1\alpha}$ are zero. The next order term $\gamma_{2\alpha}$ is a bit more involved:

$$\gamma_{2\alpha} = \sum_{\beta > 4} \frac{K_{1\alpha\beta}^2}{-\gamma_{0\beta}} . \quad (20\text{-}62)$$

This calculation requires knowledge of the other eigenvalues and eigenvectors of K. Therefore, we need the details of R. Because $\gamma_{0\beta} \sim 1/\tau$, $K_1 \sim ikv$, thus

$$\gamma_{2\alpha} \sim k^2 v^2 \tau \sim k^2 \tau T/m . \quad (20\text{-}63)$$

We do not attempt to calculate (20-62) here, but simply use (20-63) as an approximation. The results are as follows:

$$\gamma_\pm = \pm ick + D_3 k^2 ,$$

$$\gamma_{1,2} = D_1 k^2 ,$$

$$\gamma_4 = D_4 k^2 ,$$

$$\gamma_\beta \sim 1/\tau , \quad \beta > 4 ,$$

$$D_{1,3,4} \sim v^2 \tau \sim \tau T/m . \quad (20\text{-}64)$$

Thus $\psi_\pm$ make periodic motion and are the sound waves in the $\pm \mathbf{k}$ directions:

$$f_1 \propto e^{-\gamma_\pm t} e^{i\mathbf{k} \cdot \mathbf{r}} = e^{i\mathbf{k} \cdot \mathbf{r} \pm ickt - D_3 k^2 t} ,$$

then

$$\frac{\partial^2}{\partial t^2} f_1 = c^2 \nabla^2 f - D_3 \frac{\partial}{\partial t} \nabla^2 f_1 , \quad (20\text{-}65)$$

so $D_3 k^2$ is the attenuation. Because $D_3 \propto \tau$, the sound wave will attenuate rapidly as τ is increased. If there are no collisions then $\tau \to \infty$ and there is no sound wave. That is to say, sound wave is propagated by collisions. This result is quite obvious. If there are no collisions, the particles will go in straight paths and there is no wave motion. The functions $\psi_\pm$ are linear combinations of ψ_{00}, ψ_{03} and ψ_{04}, and so describe a combination of the oscillations of particle density, energy and momentum (along **k**).

The function $\psi_{1,2}$ expresses the change of the momentum of the particles (perpendicular to **k**) and is a kind of "shear" or "twist". The coefficient D_1 is proportional to the viscosity. If

$$f_1 \propto \psi_{1,2}\, e^{i\mathbf{k}\cdot\mathbf{r}-\gamma_{1,2}t} \sim e^{i\mathbf{k}\cdot\mathbf{r}-D_1 k^2 t},$$

then

$$\frac{\partial}{\partial t} f_1 = D_1 \nabla^2 f_1 . \tag{20-66}$$

The function ψ_4 is the change of energy, together with a small density variation. If

$$f_1 \propto \psi_4\, e^{i\mathbf{k}\cdot\mathbf{r}-\gamma_4 t} \sim e^{i\mathbf{k}\cdot\mathbf{r}-D_4 k^2 t},$$

then

$$\frac{\partial}{\partial t} f_1 = D_4 \nabla^2 f_1 , \tag{20-67}$$

and this is the diffusion or conduction of heat. The above differential equations (20-65) − (20-67) can be regarded as another expression for (20-64). Equation (20-64) summarises the properties of the sound wave, viscosity and conduction of an ideal gas. The different coefficients can, in principle, be calculated from (20-62).

The above analysis shows that Eqs. (20-65) − (20-67) have a time scale much larger than τ and a length scale much larger than the mean free path l:

$$kv \ll \frac{1}{\tau} ,$$

$$\frac{1}{k} \gg \tau v = l . \tag{20-68}$$

For further analysis one may consult the literature.[b]

[b] Foster (1975) is a good place to start.

Now let us compare the results of this section with those of the preceding two. The analysis of this section regarded collisions as the main part. In (20-46) K_1 was treated as a perturbation of K. Although the final result was established mainly in the space $K = 0$, the expansion in this section would not be possible without K. This is an expansion in terms of the small quantity $kv/(1/\tau) = kl$. In Secs. 20.2 and 20.3 we did not consider collisions but rather the force arising from the variation of particle density.

This section discussed the ideal gas but the same analysis can be applied to a gas of dense fermions, with similar conclusions. There is however a main difference. For fermions at low temperatures, the number of holes and electrons outside the Fermi surface is proportional to T, and hence not constant. Therefore, at low temperatures, there is still little chance for particles to collide. Even if they collide, they are influenced by the exclusion principle: a particle must enter an empty hole. So the probability for two particles to meet is proportional to T and their energy is about T. If there is a collision, their energy after collision must be more or less evenly distributed, because states much lower that T are forbidden. So the states after the collision are limited. Therefore,

$$\frac{1}{\tau} \propto T^2 . \tag{20-69}$$

As $T \to 0$, there is no way to propagate a sound wave by collisions. Only zero sound can propagate, as it does not require collisions for propagation.

20.6. H-theorem

We mention here Boltzmann's H-theorem. This theorem says that according to Eq. (20-3) and if $(\partial f/\partial t)_C \neq 0$, then

$$H \equiv \int d^3p \, d^3r \, f \ln\left(\frac{1}{f}\right) , \tag{20-70}$$

must increase with time. The proof of this theorem is very simple. Indeed we have proved most of it in the last section, where we have pointed out that eigenvalues of the operator K are all positive (see Eq. (20-52)). This already proves that $f_1 \equiv f - f_0$ will tend to zero and f will tend to f_0. H will attain its maximum at $f = f_0$. (This too is very easy to prove.) Although this proof is limited to very small f_1, the main points are revealed here.

The reader may point out that H is just the entropy (apart from a constant). We would like to remark, however, that if the collisions are frequent, H is indeed the entropy; otherwise it is not.

Notice that even if there are no collisions, f can still tend to f_0. An initial irregular distribution will become a uniform distribution because the velocities of the particles are different. For example at $t = 0$, if

$$f(\mathbf{r}, \mathbf{p}, 0) = f_0(\mathbf{p})(1 + ae^{i\mathbf{k}\cdot\mathbf{r}}) \quad ,$$

then

$$f(\mathbf{r}, \mathbf{p}, t) = f_0(\mathbf{p})(1 + ae^{i\mathbf{k}\cdot\mathbf{r}} e^{-ikvt}) \quad , \tag{20-71}$$

i.e. when t is very large, the integral of the last term over any interval of $\mathbf{p}$ is zero because its frequency is infinite. So it is not wrong to say that f tends to f_0. But we cannot say that $H(f)$ tends to $H(f_0)$, i.e. the entropy does not increase. According to our view, to identify H as the entropy we need to have collisions to change continually the momenta of the particles. If there are no collisions, H will not be the entropy because in momentum space, i.e. the $3N$-dimensional space $(\mathbf{p}_1, \mathbf{p}_2, \ldots, \mathbf{p}_N)$, the trajectory is just a point. Only the collisions will disperse the trajectory. If there are no collisions each $\mathbf{p}_i$ is an invariant and H naturally is not the entropy. Notice that the kind of collisions discussed in Sec. 20.4 is not enough to allow H to be identified as the entropy because such collisions cannot change the energy of the particles.

Of course, if we substitute (20-71) into (20-70) and calculate dH/dt, we get zero (if there are no collisions). This is because the last term of (20-71) is oscillating so rapidly that dH/dt vanishes.

In Chapter 23 we shall discuss the "echo phenomena" where the problem of collisions and entropy shall be further considered. These phenomena roughly show that even if the particles are dispersed, the entropy may not increase. In Chapter 24 we shall carefully discuss how to calculate entropy from the trajectories of the motion. These are all very important problems.

Problems

1. Plasma oscillations are discussed in books on plasma physics and solid state physics. Suppose the positive charges are fixed while the electrons vibrate with a small amplitude.

(a) From fluid mechanics derive the following equations:

$$\frac{\partial n'}{\partial t} + n \nabla \cdot \mathbf{v} = 0 ,$$

$$m \frac{\partial \mathbf{v}}{\partial t} = -e\mathbf{E} ,$$

$$\nabla \cdot \mathbf{E} = -4\pi e n' . \qquad (20\text{-}72)$$

where n is the average density, n' the change of electron density and $\mathbf{v}$ is the flow velocity. Discuss the limits of validity of (20-72), taking into consideration the effects of collisions.

(b) Solve (20-72) to get the oscillation frequency

$$\omega_p = \left(\frac{4\pi e^2 n}{m} \right)^{1/2} . \qquad (20\text{-}73)$$

2. The above problem is appropriate for three dimensions. Change (20-72) so that it is suitable for planar conditions and then derive (20-19). Then use the methods of Sec. 20.2 to rederive the results for three dimensions.

3. Add a pressure gradient ∇p to the second equation of (20-72).

(a) Prove that

$$\frac{\partial^2 n'}{\partial t^2} + \omega_p^2 n' - \frac{1}{m} \frac{\partial p}{\partial n} \nabla^2 n' = 0 . \qquad (20\text{-}74)$$

Then solve it.

(b) Apply (20-17) to three dimensions and expand it terms of k, keeping the first two terms. Compare the results with (20-74).

4. Prove that the dielectric constant of a plasma gas is

$$\epsilon = 1/(1 - \omega_p^2/\omega^2) . \qquad (20\text{-}75)$$

The definition of ϵ is: if the distribution of an external charge is $q e^{i\mathbf{k} \cdot \mathbf{r} - i\omega t}$, then it will produce an electric potential $(4\pi q/k^2 \epsilon) e^{i\mathbf{k} \cdot \mathbf{r} - i\omega t}$. This assumes q is small.

5. A charge moves over a metal surface with a velocity parallel to the surface, and at a distance x from it.

Discuss the response of the electrons inside the metal. This problem can be considered together with Problem 3 of the last chapter. Prove that there is an oscillation of frequency $\sqrt{2}\,\omega_p$ on the metal surface. (For details see Rudnick (1967).)

6. The reader can understand the true meaning of collisions by carefully performing the calculations in Secs. 20.4 and 20.5 and comparing with Sec. 20.3. Also generalise the calculation of Sec. 20.5 to fermions at low temperature.

7. Prove that c^2 in (20-61) is the adiabatic compressibility coefficient $\frac{1}{m}\left(\frac{\partial p}{\partial n}\right)_s$ of an ideal gas.

Chapter 21
THE DIFFUSION EQUATION

Changes over long time scales are the accumulation of changes over shorter time scales. The cumulative result is usually independent of some of the details of the motion over short time scales. Therefore, we can simplify these details to avoid complicated calculations without affecting the outcome of the motion over long time scales. The diffusion equation and its various generalisations are such simplifications. Although crude and limited in applicability, these equations are very useful if we require only rough answers. In this chapter, emphasis is not on the theoretical basis, but rather on some simple applications, with many examples related to the metastable state. The problem of barrier penetration is the main point of analysis of the metastable state.

These equations themselves are interesting mathematical problems, to which mathematicians have already made many contributions. From the view point of physics, how crude are these equations? How can we improve on them? These questions are still unanswered. We still cannot start from the more basic principles (Newton's laws, quantum mechanics) to understand these equations. In some special cases, we can have a better understanding, e.g. in the last chapter we used the Boltzmann equation to introduce diffusion. But generally speaking, the understanding of collective motion is still in its infancy. These equations are just convenient to use; we know how to use them but do not fully understand why they can be so useful.

21.1. Simple Examples

In Chapter 12 we started from the equations of random motion to derive the diffusion equation (see Eqs. (12-55) to (12-68)). The diffusion equation can be regarded as a consequence of the central limit theorem. This section discusses some simple properties and examples of this equation. Section 21.4 will be devoted to a review of its relation with random motion.

The diffusion equation can be understood in terms of conservation of particle number:

$$\frac{\partial \rho}{\partial t} = - \nabla \cdot \mathbf{J} ,$$

$$\mathbf{J} = r \mathbf{F} \rho - D \nabla \rho , \qquad (21\text{-}1)$$

where ρ is the number density, $\mathbf{J}$ is the flux, $\mathbf{F}$ is the external force on each particle and D is the diffusion coefficient. The flow has two terms — one produced by diffusion proportional but antiparallel to the density gradient, while the other is proportional to the external force, with proportionality constant r. The parameters D and r are regarded as known, either measurable or calculable from other theories. They are temperature-dependent functions and reflect the conditions of the environment. The parameter r is a kind of coefficient of friction. If we ignore D, then

$$\mathbf{J} = r \mathbf{F} \rho , \qquad (21\text{-}2)$$

i.e. the flux is proportional to $\mathbf{F}$ where $r\mathbf{F}$ is the flow velocity. If the particles are spheres with radius a immersed in a viscous fluid, then

$$r = \frac{1}{6 \pi a \eta} , \qquad (21\text{-}3)$$

where η is the viscosity of fluid. The parameters r and D are closely related. In equilibrium $\mathbf{J} = 0$. Let $\mathbf{F} = -\nabla U$, where U is the potential of the external force, then

$$-r(\nabla U) \rho - D \nabla \rho = 0 ,$$

$$\rho \propto e^{-(r/D)U} . \qquad (21\text{-}4)$$

In equilibrium ρ should be proportional to $e^{-U/T}$, so

$$\frac{r}{D} = \frac{1}{T} . \qquad (21\text{-}5)$$

This is a very important result and has been derived in Chapter 12 (see (12-66)). This is also the condition for detailed balance in Chapter 3. Now let us consider some more examples.

Example 1:

The effect of a certain electronic component is shown as $U(x)$ in Fig. 21-1. The electron experiences a force $-F$ between 0 and b, and at b it suffers a very large force to the right.

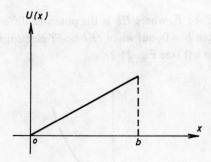

Fig. 21-1

$$U(x) = xF, \quad 0 < x < b,$$
$$= 0, \quad \text{otherwise}. \tag{21-6}$$

This is a simple "diode" model. Electrons flowing to the right experience a slight force, while those flowing to the left experience a large force.

In equilibrium there is no current. If we add an electric field E, what is the current?

Adding the electric field we change $U(x)$ to

$$U'(x) \equiv U(x) - exE. \tag{21-7}$$

From (21-1) and (21-5) we get

$$J = D\left(-\frac{\partial U'}{\partial x}\frac{\rho}{T} - \frac{\partial \rho}{\partial x}\right)$$

$$= -D e^{-U'/T}\frac{\partial}{\partial x}\left(\rho e^{U'/T}\right). \tag{21-8}$$

As the current becomes steady, J and ρ will not change. J is not a function of x while ρ is a constant outside $(0, b)$. Therefore integrating (21-8) we get

$$J = \rho D (1 - e^{-eEb/T}) \Big/ \int_0^b dx \, e^{U'/T} \quad ,$$

$$\simeq \frac{\rho D F}{T} e^{-bF/T} (e^{eEb/T} - 1) \quad . \tag{21-9}$$

The above assumes $eE \ll F$, where Eb is the potential difference across $(0, b)$. There is no current when $E = 0$, but when $eEb \gg T$ electrons can flow easily to the right but not to the left (see Fig. 21-2).

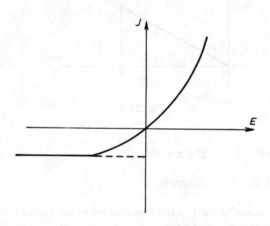

Fig. 21-2

Example 2: Barrier penetration.

The potential $U(x)$ as shown in Fig. 21-3 is a barrier. Suppose ρ is nonzero to the left of the barrier and $\rho = 0$ to the right. Moreover, assume that the barrier is sufficiently high, i.e. $U_c = U(c) \gg T$. Gradually, particles will penetrate the barrier to the right. What is the flow rate? Because the flow rate is small, the situation on the left is roughly steady. So J and $\rho(0)$ do not change. Using steps as in the last example we immediately get

$$J = \rho(0) D \Big/ \int_0^b dx \, e^{U(x)/T} \quad . \tag{21-10}$$

THE DIFFUSION EQUATION 371

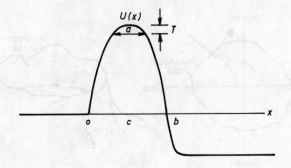

Fig. 21-3

Expanding $U(x)$ about $x = c$,

$$U(x) \simeq U_c - \frac{1}{2}\alpha(x-c)^2 \ . \tag{21-11}$$

Substituting into (21-10) we get

$$J \simeq \rho(0) D \frac{e^{-U_c/T}}{\sqrt{2\pi T/\alpha}} \ . \tag{21-12}$$

The height of the barrier U_c is seen to be a determining factor. Notice that the factor $e^{-bF/T}$ in (21-9) comes from the same reasoning. The parameter $T/\alpha \equiv a$ is the width of the barrier at $U_c - T$. The highest point is the most difficult place to penetrate while the other places are insignificant by comparison.

Example 3:

Figure 21-4 shows a two-dimensional barrier. A region is surrounded by a very high potential $U(\mathbf{r})$. At the north-east direction there is a place of lower altitude forming a mountain pass or "saddle point". The particle density is ρ inside the region. Calculate the outward flux.

To get out, the particle must overcome the barrier. From the last example, we know that the most important factor is the highest point on the way out. Obviously the mountain pass is the easiest way out, because the other routes require overcoming higher altitude. Place the x-axis through the mountain

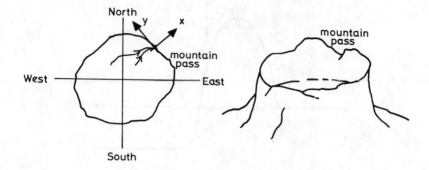

Fig. 21-4 A 2-dimensional barrier with a saddle point.

pass; then we can use (21-12). The flow rate of the particle out of the region is approximately given by

$$J \simeq \rho D \left[\sqrt{\alpha \beta}/(2\pi T) \right] e^{-U_c/T} \; ,$$

$$\alpha \equiv -(\partial^2 U/\partial x^2)_c \; ,$$

$$\beta \equiv (\partial^2 U/\partial y^2)_c \; . \tag{21-13}$$

The subscript c above denotes the value at the pass and ρ is the density inside the region.

21.2. The Metastable State

The above are examples of the metastable state. The particles at the left in Fig. 21-3 or those inside the region in Fig. 21-4 are in the metastable state. If the barrier is very high they can remain inside for a very long time. This time scale is of the order

$$\tau = \tau_0 \, e^{\Lambda/T} \; , \tag{21-14}$$

$\Lambda = U_c$ being the height of the barrier to be overcome. (We use the symbol Λ because it resembles the barrier.) The factor τ_0 is a time scale unrelated to Λ. We can treat it as the relevant time scale when the barrier is low. The exponential Λ/T is the most important factor. Equation (21-14) can be used to estimate the life-time of a metastable state.

Example 4: How long can an egg stand on its end?

A standing egg is a metastable state. Now let us estimate the barrier height Λ of this metastable state. An egg stands because its centre of mass is on a vertical line passing through the surface of contact of the egg with the table; the surface is created by the weight of the egg itself. For the egg to fall, we need to raise the centre of mass of the egg by a bit. Let a be the radius of the area of contact, h the height of the centre-of-mass and θ the angle between the axis of the egg and the perpendicular. When θ is very small, the rise of the centre-of-mass is

$$\Delta h \simeq a\theta + h(\cos\theta - 1) ,$$
$$\simeq a\theta - \frac{1}{2}h\theta^2 . \qquad (21\text{-}15)$$

This formula assumes that the area of contact is a hard surface. The first term $a\theta$ is the rise of the centre of the surface of contact and the second term $-\frac{1}{2}h\theta^2$ is the change of height of the centre-of-mass with respect to the centre of the surface of contact. This assumption is not too accurate. A correct analysis must take into account the elasticity of the egg shell, the smoothness of the shell surface and the flowing of the fluid inside the egg, etc.

From (21-15) we get the potential energy of the egg (Fig. 21-5).

$$U(\theta) \simeq mg\Delta h . \qquad (21\text{-}16)$$

Its maximum occurs at $\theta = a/h$, i.e.

$$\Lambda = mga^2/2h . \qquad (21\text{-}17)$$

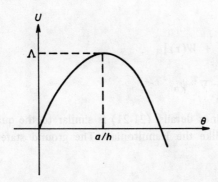

Fig. 21-5 Potential energy of an egg as a function of the angle θ between the perpendicular and axis of the egg.

An egg weighs about 30 g. Assume that $a \sim 0.03$ cm, $h \sim 1.5$ cm, then $\Lambda \sim 10$ erg. At room temperatures,

$$\Lambda/T \sim 10^{15} \quad . \tag{21-18}$$

According to (21-14), no matter how small τ_0 is, $e^{\Lambda/T}$ makes τ very large. That is to say, the motion of molecules in equilibrium cannot by itself make the egg fall. For the egg to topple, a certain molecule must have a kinetic energy of at least 10 erg. Therefore, to analyse the time that the egg can stand, we must include other factors such as rotting of the egg, wind, etc.

Generally speaking, the use of (21-14) is limited to situations in which Λ/T is neither too large nor too small. If Λ/T is very large it indicates that we must also consider other more complicated processes.

21.3. Transformation to the Wave Equation

The above examples focus on the flow **J**. Let us now examine the time derivative of (21-1). The most convenient method is to transform (21-1) into forms similar to the wave equation in quantum mechanics and then use our knowledge of quantum mechanics. The term $\nabla \cdot \mathbf{J}$ in (21-1) contains the $\nabla \rho$ term but the wave equation does not have this term. The first derivative is also not a symmetric operator. Now let us devise a method to eliminate it. Let

$$\rho(\mathbf{r}) \equiv \psi_0(\mathbf{r}) \, \psi(\mathbf{r}) \quad , \tag{21-19}$$

$$\psi_0^2(\mathbf{r}) \equiv e^{-U(r)/T} \quad . \tag{21-20}$$

Substituting into (21-1) and rearranging we get

$$\frac{\partial \psi}{\partial t} = -K\psi \quad ,$$

$$K \equiv -D[\nabla^2 + W(\mathbf{r})] \quad ,$$

$$W(\mathbf{r}) \equiv -\frac{1}{\psi_0} \nabla^2 \psi_0 \quad . \tag{21-21}$$

Other than some minor details, (21-21) is similar to the quantum mechanical equation and K is like the Hamiltonian. The ground state of K is ψ_0 with energy zero:

$$K\psi_0 = 0 \tag{21-22}$$

We can now use techniques developed in quantum mechanics.

Example 5:

Let us solve the diffusion equation for $U(x) = \alpha^2 x^2$. From (21-20) we get

$$\psi_0(x) = \exp(-\tfrac{1}{2}\alpha^2 x^2/T) \quad . \tag{21-23}$$

This is, of course, the ground state wave function of the simple harmonic oscillator. The other eigenfunctions of K are then the other eigenfunctions of the simple harmonic oscillator. Needless to say, the solution of (21-1) is

$$\rho(x,t) = \psi_0(x)\,\psi(x,t) \quad ,$$

$$\psi(x,t) = \sum_{n=1}^{\infty} A_n \psi_n(x)\,e^{-r_n t} + A_0\,\psi_0(x) \quad ,$$

$$r_n = 2nD\alpha^2/T \quad ,$$

$$\psi_n(x) = \sqrt{2/T}\,H_n(\alpha x)\,e^{-\tfrac{1}{2}\alpha^2 x^2/T} \quad . \tag{21-24}$$

The coefficients A_n are constants, depending on the density at $t = 0$.

Example 6:

Let us consider a hard-wall box

$$\begin{aligned} U(x) &= 0 \quad, & 0 < x < L \quad, \\ &= \infty \quad, & \text{otherwise} \quad . \end{aligned} \tag{21-25}$$

This example is given to remind the reader of the importance of the boundary conditions. The boundary condition of the diffusion equation is that at the walls

$$\mathbf{J}_\perp = 0 \quad , \tag{21-26}$$

where $\perp$ means the direction perpendicular to the wall. This is different from the condition that $\psi = 0$. Therefore, the solution of this problem is

$$\rho(x,t) = \frac{N}{L} + \sum_{n=1}^{\infty} A_n \psi_n(x)\,e^{-r_n t} \quad ,$$

where

$$\psi_n(x) = \cos\frac{n\pi}{L}x \quad , \qquad r_n = D\frac{n^2}{L^2} \quad . \tag{21-27}$$

and N is the total number of particles.

21.4. Derivation of the Diffusion Equation

The above-mentioned $\rho(\mathbf{r})$ is the distribution of particles in $\mathbf{r}$ space. However $\mathbf{r}$ need not denote the position but can be any set of variables and $\rho(\mathbf{r})/N$ is the probability distribution in the variable $\mathbf{r}$. The meaning of probability comes from the statistics of the information of N particles, i.e. the sample set is $(\mathbf{r}_1, \mathbf{r}_2, \ldots, \mathbf{r}_N)$. The number N need not be the particle number; it can be the number of times the experiment is repeated, where $\mathbf{r}_i$ is the result of the i-th outcome.

We now review the results of Chapter 12. We shall consider the displacement due to the random distribution of velocity $v(t)$ and derive the distribution of the displacement.

Starting from t', the displacement in Δt is

$$\Delta x = \int_0^{\Delta t} dt'' \, v(t' + t'') \quad . \tag{21-28}$$

If the correlation time of v is very short, much shorter than Δt, then the central limit theorem shows that the distribution $p(\Delta x)$ of Δx is normal, with

$$\langle \Delta x \rangle = \langle v \rangle \Delta t \, ,$$

$$\sigma^2 \equiv \langle (\Delta x)^2 \rangle_c = \Delta t \int_{-\infty}^{\infty} dt'' \, \langle v(t'' + t') \, v(t') \rangle_c \quad . \tag{21-29}$$

The distribution of $x(t)$ can be deduced from that of $x(t')$,

$$x(t) = x(t') + \Delta x \, , \qquad t \equiv t' + \Delta t \, ,$$

$$\rho(x, t) = \int d(\Delta x) \, p(\Delta x) \, \rho(x', t') \, , \tag{21-30}$$

$$p(\Delta x) = \frac{1}{\sqrt{2\pi\sigma^2}} \exp\left\{ -\frac{1}{2\sigma^2} (\Delta x - \langle v \rangle \Delta t)^2 \right\} \quad . \tag{21-31}$$

By direct differentiation, we can prove that $p(\Delta x)$ satisfies the following equation:

$$\frac{\partial p}{\partial \Delta t} = \langle v \rangle \frac{\partial p}{\partial (\Delta x)} + \frac{\sigma^2}{2 \Delta t} \frac{\partial^2 p}{\partial (\Delta x)^2} \quad . \tag{21-32}$$

Because $\Delta t = t - t'$, $\Delta x = x - x'$, from (21-30) we get

$$\frac{\partial \rho}{\partial t} = \int dx' \frac{\partial p}{\partial (\Delta t)} \rho(x', t')$$

$$= \int dx' \left[\langle v \rangle \frac{\partial p}{\partial (\Delta x)} + \frac{\sigma^2}{2\Delta t} \frac{\partial^2 p}{\partial (\Delta x)^2} \right] \rho(x', t')$$

$$= \frac{\partial}{\partial x} \int dx' \langle v \rangle p(\Delta x) \rho(x', t')$$

$$+ \frac{\partial^2}{\partial x^2} \int dx' \frac{\sigma^2}{2\Delta t} p(\Delta x) \rho(x', t') \quad . \quad (21\text{-}33)$$

Now assume that Δt is not too large, then x and x' are nearly the same, and $\langle v \rangle$ does not change with x', and (21-32) becomes

$$\frac{\partial \rho}{\partial t} = \frac{\partial}{\partial x} \langle v \rangle \rho + \frac{1}{2} \frac{\partial^2}{\partial x^2} \frac{\sigma^2}{\Delta t} \rho \quad . \quad (21\text{-}34)$$

The above steps are more or less the same as those in Chapter 12 ((12-55) to (12-68)). Here we consider $\langle v \rangle \neq 0$, and let $\langle v \rangle$ and $\sigma^2/(\Delta t)$ be slowly varying functions of x and t. If there is more than one variable, it is not difficult to generalise the above derivation. Let x_α, $\alpha = 1, 2, \ldots$, be the variables, then

$$\frac{\partial \rho}{\partial t} = - \sum_\alpha \frac{\partial}{\partial x_\alpha} J_\alpha \quad ,$$

$$J_\alpha = - \frac{\langle \Delta x_\alpha \rangle}{\Delta t} \rho - \frac{1}{2} \sum_\beta \frac{\partial}{\partial x_\beta} \left(\frac{\langle \Delta x_\alpha \Delta x_\beta \rangle_c}{\Delta t} \rho \right) \quad . \quad (21\text{-}35)$$

The relation of (21-1) and (21-34), (21-35) is now clear. These results come from the central limit theorem. If this theorem is not appropriate, then they are likewise invalid. Equation (21-35) is the so-called Fokker-Planck equation. This is the diffusion equation (21-1), written in a slightly more detailed form and pointing out how to obtain D and $r\mathbf{F}$ from the details of the random motion.

The x_α in (21-35) are continuous variables. If the variables are discrete, then the form of the diffusion equation will be changed. We give an example to illustrate this.

Let $s_i = \pm 1$, $i = 1, 2, \ldots, N$. The Ising variable s_i is the simplest discrete variable. The change of s_i is discrete. Let ρ_+ and ρ_- be the population or probability of the states $+1$ and -1 respectively. Then,

$$\frac{d\rho_+}{dt} = -r_{-+} \rho_+ + r_{+-} \rho_- ,$$

$$\frac{d\rho_-}{dt} = r_{-+} \rho_+ - r_{+-} \rho_- . \tag{21-36}$$

where $dt\, r_{-+}$ is the probability in time dt for s_i to change from $+1$ to -1 and $dt\, r_{+-}$ is the probability from -1 to $+1$, r_{+-} and r_{-+} are the reaction rates. In equilibrium $\rho_\pm$ is unchanged, so

$$\frac{r_{-+}}{r_{+-}} = \frac{\rho_-}{\rho_+} . \tag{21-37}$$

If the variable takes on many values, then the generalisation of (21-36) is

$$\frac{d\rho_\alpha}{dt} = -\sum_\beta (r_{\beta\alpha} \rho_\alpha - r_{\alpha\beta} \rho_\beta) , \tag{21-38}$$

where α, β label the various states.

If there is an external force so that the energy of state α is ϵ_α, then in equilibrium

$$\rho_\alpha \propto e^{-\epsilon_\alpha/T} . \tag{21-39}$$

Therefore, the different $r_{\alpha\beta}$ must satisfy

$$\frac{r_{\alpha\beta}}{r_{\beta\alpha}} = e^{-(\epsilon_\alpha - \epsilon_\beta)/T} . \tag{21-40}$$

This is the condition of detailed balance as discussed in Chapter 2.

Equations like (21-1), (21-35) and (21-35) are crude models of the molecular motion. They regard the cause of motion as an irregular disturbance or a sudden change at any moment. These disturbances or changes must satisfy some conditions, the most important ones being independence and detailed balance. These models are crude, but nevertheless valuable because they contain some of the main characteristics of molecular motion, while a more refined analysis will be too complicated.

Example 7: The elimination of high energy states.

Let s_i have three values $+1, 0, -1$. The energies are $-\epsilon_+, 0, -\epsilon_-$, respectively (see Fig. 21-6). Let $\epsilon_\pm \gg T$. Suppose s_i cannot change from $+1$ to -1 directly, i.e. $r_{+-} = r_{-+} = 0$.

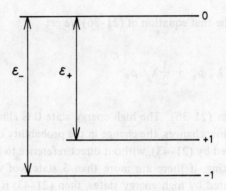

Fig. 21-6 Energy levels for a hypothetical system.

We first define the various reaction rates, which must satisfy the condition for detailed balance (21-40). We use the simplest model: $r_{\alpha\beta} = 1$ if $\epsilon_\alpha < \epsilon_\beta$ and $r_{\alpha\beta} = \exp[-(\epsilon_\alpha - \epsilon_\beta)/T]$ if $\epsilon_\alpha > \epsilon_\beta$, i.e. the reaction rate for a decrease of energy is 1, and $e^{-\epsilon/T}$ for an increase, where ϵ is the energy needed for the reaction. Then

$$\frac{d\rho_\pm}{dt} = -\lambda_\pm \rho_\pm + \rho_0 ,$$

$$\frac{d\rho_0}{dt} = -2\rho_0 + \lambda_+ \rho_+ + \lambda_- \rho_- ,$$

$$\lambda_\pm \equiv e^{-\epsilon_\pm/T} . \qquad (21\text{-}41)$$

The state 0 has a high energy while ± 1 are states with low energy. The time will be short for a particle to stay in a high energy level, but long in a low energy one. We can solve the equation for ρ_0 assuming $\rho_\pm$ are unchanged.

$$\rho_0(t) = \int_{-\infty}^{t} dt' \, e^{-2(t-t')} [\lambda_+ \rho_+(t') + \lambda_- \rho_-(t')]$$

$$\simeq \frac{1}{2}\lambda_+ \rho_+(t) + \frac{1}{2}\lambda_- \rho_-(t) \quad . \tag{21-42}$$

Substituting into the first equation of (21-36) we get

$$\frac{d\rho_\pm}{dt} = -\frac{1}{2}\lambda_\pm \rho_\pm + \frac{1}{2}\lambda_\mp \rho_\mp \quad . \tag{21-43}$$

This is the same as (21-36). The high energy state 0 is eliminated. That is to say, in discussing slow changes, the change in the probability of the two states + and − is determined by (21-43), without direct reference to the 0 state.

Generally speaking, if there are more than 3 states, of which two are low-lying states, separated by high energy states, then (21-43) is also true, although $\lambda_\pm$ will be more complicated.

Notice that $\lambda_\pm = \exp(-\epsilon_\pm/T)$ are, like (21-12), caused by having to penetrate a barrier. Any reaction involving barrier penetration will have this factor in the reaction rates.

21.5. Two-State Cluster Model

Amorphous substances like glass and rock or amorphous magnetic substances mentioned in Chapter 18 have very complicated structure. Nevertheless, in some special cases the low temperature properties of these complicated structures can be explained by the two-state cluster model. This model assumes that most of the body is frozen and only a small part, e.g. some special molecules or spins dispersed throughout the body, can move freely. Each movable element is a small cluster composed of several molecules or spins. Each cluster has two low-lying states which are most important at low temperatures. Assume that there is a barrier between the two states, so that the rate of change is slow.

Each cluster is characterised by ϵ, the energy difference between the two states, and Λ, the barrier height. Figure 21-7 plots the energy versus coordinates in the configuration space of the small clusters. To change from the + state to the − state, the particle must overcome a barrier of height Λ.

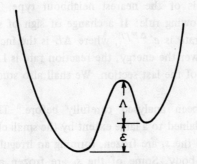

Fig. 21-7 A plot of energy versus coordinates for a cluster of molecules.

There are many clusters all with different Λ and ϵ. Suppose the number of clusters is sufficiently large, then we can calculate the distribution of Λ and ϵ and discuss the probabilities $\rho_\pm$ of the two states using equations like (21-38). Assume

$$\frac{d\rho_+}{dt} = -\frac{1}{\tau}\rho_+ + \frac{1}{\tau}e^{-\epsilon/T}\rho_- \; ,$$

$$\frac{d\rho_-}{dt} = \frac{1}{\tau}\rho_+ - \frac{1}{\tau}e^{-\epsilon/T}\rho_- \; , \qquad (21\text{-}44)$$

$$\frac{1}{\tau} \equiv \frac{1}{\tau_0}e^{-\Lambda/T} \; , \qquad (21\text{-}45)$$

where $1/\tau_0$ is a constant, essentially the reaction rate at high temperatures. From the solution of this equation and the distributions of ϵ and Λ we can deduce the various properties of this model.

The reader may be impatient to know what these clusters are. How are Λ and ϵ distributed? We now use a simple model to clarify these points.

Let H be the energy of an Ising magnetic model.

$$H = -\sum_{i,j} J_{ij} s_i s_j \; , \qquad (21\text{-}46)$$

$s_i = \pm 1$, $i = 1, 2, \ldots, N$. The interaction J_{ij} can be positive or negative, large or small, but does not change with time. The J_{ij} form a normal distribution and the interaction is of the nearest neighbour type. The motion of s_i is according to the following rule: If a change of sign of s_i increases the total energy, the reaction rate is $e^{-\Delta E/T}$, where ΔE is the increase of energy. If a change of sign will lower the energy, the reaction rate is 1. This is also the rule used in the example of the last section. We shall also study it carefully in the next chapter.

This model has been analysed carefully before.[a] The low temperature properties can be explained to a large extent by the small cluster model. At low temperatures most of the s_i are frozen, forming an irregular magnetic body or amorphous magnetic body. Some of the s_i are frozen at $+1$, others at -1. To change s_i requires an energy $2|h_i|$, where h_i is the magnetic field produced by the neighbours of s_i:

$$h_i = \sum_j J_{ij} s_j \quad . \tag{21-47}$$

At low temperatures, spins s_i with small h_i will move, while those with large h_i are frozen in. Even if some h_i is not small, the simultaneous change of sign of s_i together with one neighbouring spin may not increase the energy very much, it may even decrease the energy. If s_1 and s_2 change together, the increase of energy is

$$\epsilon = 2|h_1| + 2|h_2| - 4|J_{12}| \quad . \tag{21-48}$$

This can be obtained from (21-47). When the two change together, J_{12} does not come in. If J_{12} is very strong, even though h_1 and h_2 are very large, ϵ can be very small or negative. Nonetheless, the motion of the element is one at a time and not two at the same time. Therefore, the energy must first increase by $2|h_1|$ or $2|h_2|$ and then decrease, as in Fig. 21-7. Let $|h_1| < |h_2|$, $\epsilon < 0$, then it is easier to first flip s_1 and then s_2. The barrier is

$$\Lambda = 2|h_1| - \epsilon \quad , \tag{21-49}$$

where ϵ is defined by (21-48). This is a two-state cluster. The analysis of a multi-state cluster follows similarly from that of the two-state cluster. The distribution of Λ and ϵ is calculated by statistics.

This example shows the origin of small clusters, giving the reader a sense of "reality" for the abstract model. Let us return to (21-44) and see the conclusion of this cluster model.

---

[a] For details see Dasgupta, Ma and Hu (1979).

THE DIFFUSION EQUATION

The solution of (21-44) is very simple. Let $\rho_+ + \rho_- = 1$, then

$$\frac{d\rho_+}{dt} = -\frac{1}{\tau}\rho_+(1 + e^{-\epsilon/T}) + \frac{1}{\tau}e^{-\epsilon/T} ,$$

$$\rho_+(t) = f + e^{-rt}(\rho_+(0) - f) , \qquad (21\text{-}50)$$

$$f \equiv \frac{1}{(e^{\epsilon/T} + 1)} ,$$

$$r \equiv \frac{1}{\tau_0} e^{-\Lambda/T}(1 + e^{-\epsilon/T}) . \qquad (21\text{-}51)$$

Here $\rho_+(0)$ is the probability of being in the + state at $t = 0$, and its value depends on the structure of the material. Now let us estimate the change of energy as an application of this solution.

The total energy is

$$E(t) = \int d\epsilon \, d\Lambda \, N(\epsilon, \Lambda) \, \rho'_+(0) \, \epsilon \, e^{-rt} + \text{constant} ,$$

$$\rho'_+(0) \equiv \rho_+(0) - f . \qquad (21\text{-}52)$$

Here $N(\epsilon, \Lambda)$ are the distributions of ϵ and Λ for the various clusters, r is defined by (21-51) and rt can be written as

$$rt = e^{-(\Lambda - x)/T} ,$$

$$x \equiv T \ln(t/\tau') ,$$

$$\tau' \equiv \tau_0(1 + e^{-\epsilon/T})^{-1} \simeq \tau_0 . \qquad (21\text{-}53)$$

The times τ' and τ_0 are nearly equal if $\epsilon > 0$. Obviously the integration with respect to Λ gives a function depending on x. Since τ' appears in x as $\ln \tau'$ we can replace it by τ. If $\epsilon < 0$ and $|\epsilon|/T$ is very large, then r will not be small.

We can neglect e^{-rt}. Now we ignore the situation $\epsilon < 0$, then from (21-52) we get

$$E(t) = \int_0^\infty d\Lambda \, g(\Lambda) \, e^{-rt} + \text{constant} \,,$$

$$g(\Lambda) = \int_0^\infty d\epsilon \, N(\epsilon, \Lambda) \, \rho'_+(0) \, \epsilon \,. \qquad (21\text{-}54)$$

The function e^{-rt} (see Fig. 21-8) is nearly 1 when $\Lambda \gg x$, and nearly 0 when $\Lambda \ll x$:

$$e^{-rt} = \exp(-e^{-(\Lambda-x)/T}) \longrightarrow 1 \,, \qquad (\Lambda - x)/T \gg 0 \,,$$
$$\longrightarrow 0 \,, \qquad (\Lambda - x)/T \ll 0 \,.$$
$$(21\text{-}55)$$

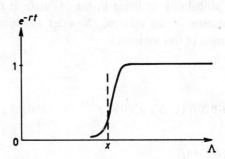

Fig. 21-8 Plot of e^{-rt} as a function of barrier height Λ.

The width of the change from 0 to 1 is about T. If T is very small, i.e. when Λ changes to $\Lambda + T$, $g(\Lambda)$ does not change much, then

$$e^{-rt} \simeq \theta(\Lambda - x) \,. \qquad (21\text{-}56)$$

That is to say, if the Λ of a small cluster is greater than $x = T \ln t/\tau_0$, the small cluster has no time to change but is frozen in. If $\Lambda < x$, it has sufficient time to change, and be freed from the frozen state. Notice that time appears as $\ln t$. Substitute (21-56) into (21-54) and we get

$$E(t) \simeq \int_x^\infty d\Lambda \, g(\Lambda) + \text{constant} \,. \qquad (21\text{-}57)$$

If $g(\Lambda)$ is a well-behaved function, $E(t)$ will be a slowly decreasing function of x (see Fig. 21-9). That is to say, E decreases slowly with $T \ln t$ and this is a very slow change. This is true not only for the energy but also for other quantities. For example, the magnetic moment, once established by the magnetic field, does not vanish immediately if the magnetic field disappears; rather, it decreases slowly as $T \ln t/\tau_0$. This is familiar to those studying the magnetic properties of ancient rocks. The recent study of the amorphous magnetic state also depends heavily on such analysis. This small cluster model is extremely crude, with many assumptions. The overlapping of the clusters and their interaction have not been considered. If the cluster is very large, it is not enough to study only the two-state model. But the above conclusion on the logarithmic change $T \ln t/\tau_0$ is simple, and not sensitive to the details of the model. Indeed this is the analogue of the result $r = (1/\tau_0) e^{-\Lambda/T}$. If we have a distribution of barrier heights Λ, we shall have a logarithmic rate of change. So it does not matter whether we study the two-state, three-state clusters or even the no-cluster situation. This collective model mainly tells us the possible origin of the distribution of Λ.

Similar quantum mechanical models have appeared in the theory of glass.[b]

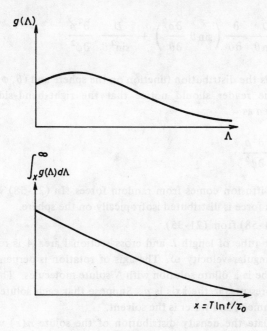

Fig. 21-9

[b] See Anderson, Halperin and Varma (1975).

Problems

1. Suppose we have a mirror with diameter much smaller than the mean free path of the molecules. It is suspended by a thread in a dilute gas at temperature T and pressure p. Prove that the torque acting on the mirror is $-\zeta \omega$, where

$$\zeta = \frac{2 m \bar{v} p I}{\sigma T}$$

where $\bar{v} = \sqrt{8T/\pi m}$, I is the moment of inertia of the mirror, σ is the mass per unit area of the mirror, m is the mass of the gas molecule and ω is the angular velocity.

2. A ball of smooth surface moves with constant velocity in a dilute gas of temperature T. The radius of the ball is bigger than the radius of the molecules but smaller than the mean free path of the molecules. Assuming that the collisions between the ball and the molecules are elastic, calculate the retarding force on the ball.

3. Prove that the diffusion equation on the surface of a sphere takes the form

$$\frac{\partial \rho}{\partial t} = \frac{D}{\sin \theta} \frac{\partial}{\partial \theta} \left(\sin \theta \frac{\partial \rho}{\partial \theta} \right) + \frac{D}{\sin^2 \theta} \frac{\partial^2 \rho}{\partial \phi^2} \quad , \quad (21\text{-}58)$$

where $\rho(\theta, \phi)$ is the distribution function on the sphere and (θ, ϕ) are the polar coordinates. The reader should notice that the right-hand-side of (21-58) cannot be written as

$$\frac{\partial^2 \rho}{\partial \theta^2} + \frac{\partial^2 \rho}{\partial \phi^2}$$

The origin of diffusion comes from random forces. In (21-58) we assume the direction of this force is distributed isotropically on the sphere.

4. Derive (21-58) from (21-35).

5. A circular tube of length L and cross-sectional area A is revolving about one end with angular velocity ω. The axis of rotation is perpendicular to the tube. In the tube is a dilute solution with N solute molecules. The temperature is T, and the pressure at the axis is p. Suppose that each solute molecule has mass m and volume v, and water is the solvent.

   (a) Calculate the density distribution of the solute $n(r)$ where r is the distance of the molecules from the axis. Notice especially the effect of the value of m/v.

   (b) Calculate the entropy and other thermodynamic quantities.

(c) If at $t = 0$ all the solute molecules are concentrated at the end of the axis. Analyse the subsequent change of the density distribution $n(r, t)$ and estimate the time taken to reach equilibrium.

6. Certain large molecules can be regarded as small rods of length l and radius a, with $l \gg a$. These molecules dissolve in a solvent and the viscosity of the solvent is η.

At the time $t = 0$, all the rods are pointing in the same direction. Calculate the subsequent distribution of directions. (Hint: use Eq. (21-58).)

7. A rock has many small magnetic moments m_i, $i = 1, 2, 3, \ldots, N$, $m_i^2 = m^2 = $ constant. The energy is

$$H = -\sum_i (h_i \cos \theta_i + A_i \cos^2 \theta_i) \ . \tag{21-59}$$

Each A_i and h_i are not the same but distributed as

$$p(h_i) = (2\pi b^2)^{-1/2} e^{-h_i^2/2b^2} ,$$

$$p(A_i) = \frac{\theta(A_i)}{a} e^{-A_i/a} . \tag{21-60}$$

Assume $a \gg b$. The parameters A_i and h_i describe the environment of each magnetic moment and θ_i is the angle made by m_i with a fixed direction. Every magnetic moment is continually influenced by disturbances in its environment, performing irregular motion.

Suppose at $t = 0$ all the $\theta_i = 0$. Calculate the change of the total magnetic moment assuming $T \ll a$. The reader should first seek a clear understanding of the meaning of the quantities in the model and then simplify the model. This is a metastable model. A_i causes m_i to prefer two directions, i.e. $\theta_i = 0$ or π.

Chapter 22
NUMERICAL SIMULATION

The invention of the computer has had a profound influence on statistical mechanics. Not only it enables us to solve problems requiring complicated calculation, but it has also made an important contribution to the understanding of the basic concepts. This chapter introduces two computational methods. One is the numerical solution of molecular motion, i.e. solving the equation of motion and calculating the positions and velocities of all the molecules. The second is "Monte Carlo simulation". This replaces the equation of motion by random sequences, and is used to calculate equilibrium properties. This chapter is mainly devoted to the second method.

The calculation of the molecular motion is indeed an numerical experiment: to calculate the trajectory of motion explicitly and then make all kinds of observations. The Monte Carlo simulation also calculates a trajectory, but uses a model of random motion rather than Newton's laws. It can also be regarded as a numerical experiment. What are the advantages of these numerical experiments? First, these calculations do not employ approximations. Once the model is set up, the calculations are fixed. All the results can be regarded as the experimental outcome of the model. (Of course, the experimental steps must be correct, the duration of the observation time and the size of the model must also be considered.) This is more powerful than analytical treatments, not relying on assumptions, approximations, or the discarding of "small" terms. Therefore, the result of the calculations can be compared with real experiments to determine the validity of the model. We can also compare the result with analytical solutions to justify the validity of various approximations or assumptions. Secondly, these calculations give the details of motion of each molecule at each instant. This information is not available in real experiments. Of course, numerical experiments are easier to manage without having to worry about the problems of the purity of the sample, or heat loss at low temperature, etc.

Because of these advantages, numerical experiments have become widely used. These kinds of calculations, being "experimental", can only be fully appreciated if one has taken part in them. The design of the various programmes rely heavily on experience, and are seldom recorded in journals and indeed are difficult to explain. This chapter discusses only some problems of principle, to give the reader some mental preparation for this topic.

These calculations, besides being useful tools, provide a deeper understanding of statistical mechancs, making the concepts clearer. This aspect will occupy much of this chapter.

22.1. Numerical Solution of Molecular Motion

The principle of this method is very simple: we just delegate the job of solving the equation of motion of the molecules (Newton's laws) to the computer.

Before the invention of the computer, to solve the equation of motion of more than 3 particles was an inconceivable job, not to mention the collective motion of many particles. Therefore, some scholars of statistical mechanics were tempted to abandom the notion of molecular motion and the physical meaning of the trajectory of motion, because it seemed impossible to solve for these quantities. This attitude tends to bias statistical mechanics towards abstraction and pure mathematics. Even today traditional statistical mechanics still utilises the abstract concept of probability to define equilibrium states, and avoids the discussion of trajectories.

Soon after the advent of the computer, those interested in molecular motion almost immediately used it to solve the equation of motion and to see whether equilibrium would appear. Fermi and Ulam solved such oscillation equations but failed to find equilibrium. This brought attention to the problem. Gradually more people used more powerful machines to solve the equation of motion in order to understand equilibrium at a more basic level. The difficulty encountered by Fermi and Ulam was that the oscillation elements were too few. With more elements, equilibrium states would emerge. Nowadays, using the computer to solve the collective equation of motion is a branch of statistical mechanics in itself. Each calculation involves thousands of particles and millions of collisions.

This kind of calculation is a type of experiment, calculating the trajectory of the system in a certain period by the computer. The positions and velocities of each molecule are calculated. Any information about the system in this period can be obtained from the trajectory. Each equilibrium quantity can be obtained by averaging. Temperature comes from averaging the kinetic energy of each particle. This numerical experiment is often easier to manage than real experiments and we can make observation on any scale, and discuss both equilibrium states and non-equilibrium ones.

Of course, this calculation is still quite different from real experiments.

(a) A collection of thousands of particles is still much smaller than a macroscopic object. The time for millions of collisions is still too short compared to macroscopic times.

(b) We cannot consider quantum mechanics. Even today, the computer still cannot solve many-body quantum mechanical wave equations.

These two weaknesses seem insurmountable. But because of the independence of the various parts of a body, (a) is not too serious. To analyse the equilibrium properties we need to consider models much larger than the correlation length ξ and observation times much longer than the correlation time τ. If ξ is small enough we need not consider very many particles, and if τ is short enough we do not need a long observation time. The magnitudes of ξ and τ depend on particular situations. Experience tells us that a model of a thousand particles is sufficient for most cases and the time is sufficient also. In some special cases, e.g. in the neighbourhood of the critical point, ξ and τ are very large and we need larger models, which usually go beyond the capacity of modern computers. Point (b) is more serious. Many important problems cannot avoid quantum mechanics, e.g. electrons in the metal, superfluidity, etc. This greatly limits the use of computers. Nevertheless, these computer calculations still have a wide range of applicability. Many phenomena such as melting, crystallisation and some collective properties of heavy atoms or molecules are not essentially quantum mechanical in nature.

From the viewpoint of the basic concepts, the success of this kind of calculation has had a profound significance since it reinforces the fundamental role of Newton's law. Every thermodynamic phenomenon and all concepts in statistical mechanics have their root in mechanics. Equilibrium is a special situation of motion, and must obey the laws of mechanics. The trajectory of motion carries all the information about the motion, and any concept or any quantity can be expressed by this information and calculated from it. Some thermodynamic concepts have no obvious meaning in mechanics. These are quantities related to entropy. The relation of entropy with the trajectory is a very important topic. In Chapter 25 we shall give a detailed discussion and point out a method to determine entropy from the trajectory.

Solving the equation of motion is a complicated computation. There are many problems that do not require the motion to be represented realistically.

Therefore, we can use a simpler model of motion to replace Newton's laws. Monte Carlo simulation is such a simplified model of motion.

22.2. Random Sequences

Monte Carlo simulation uses a random sequence rather than Newton's laws to determine motion. The use of the random sequences is to produce random forces.

From Chapter 10 to Chapter 12 we have discussed random sequences. This is a sequence of numbers satisfying certain conditions (see Sec. 11.3). The computers today are usually equipped with programmes that generate random number sequences, the most common being a sequence of numbers between 0 and 1. The random sequences mentioned in the following will be of this type unless otherwise specified.

The programmes that generate a random sequence are like a fast die and are designed very much from experience. At present there is no standard programme. Notice that the programme is a fixed rule like Newton's laws. Why are the numbers so generated random? This problem is very similar to our basic problem: Why can Newton's laws produce random molecular motion? The answer is still unknown.

The random sequence is a defining tool for probability. If there is no such rapidly generated number sequence, then the concept of probability will not be a useful computational tool. Today many scientific researchers rely heavily on the random sequence. This was unexpected thirty or forty years ago.

Having the random number program we can choose "at random". For example to choose between 1, 2 and 3 we can first generate a random number x. If x is smaller than $\frac{1}{3}$, we then take 1; we choose 2 if x falls between $\frac{1}{3}$ and $\frac{2}{3}$; 3 if x is greater than $\frac{2}{3}$. (We have assumed that the probability distribution of x is uniform between 0 and 1.) If we require probability p_1 to choose 1, p_2 to choose 2 and p_3 to choose 3, how do we proceed? The answer is: if $x < p_1$ take 1; $p_1 < x < p_1 + p_2$, take 2 and if $p_1 + p_2 < x$, take 3.

Notice that this random choice is meaningless for a few trials. Probability is meaningful only after repeating many times.

22.3. Monte Carlo Simulation

We can regard the rules of Monte Carlo simulation as a model of random motion. The steps of calculation depend upon the model. Now we use the Ising model as an example to illustrate the principles.

Let the total energy be

$$H(s) = -\frac{1}{2} \sum_{i,j} J_{ij} s_i s_j \, , \tag{22-1}$$

where $s_i = \pm 1$, $i = 1, 2, \ldots, N$. The simulation proceeds as follows:
(a) First randomly take a set of values of s_i.
(b) Take one s_i. Calculate h_i and ΔE,

$$h_i \equiv \sum_j J_{ij} s_j \, ,$$

$$\Delta E \equiv 2 s_i h_i \, . \tag{22-2}$$

(c) If $\Delta E \leq 0$, then change s_i to $-s_i$. If $\Delta E > 0$, then use the probability $e^{-\Delta E/T}$ to change s_i to $-s_i$ (i.e. take a random number x, and if $x < e^{-\Delta E/T}$ change s_i; otherwise not).
(d) Repeat step (b).

This simulation process is quite simple. Step (a) determines the initial state and any configuration is possible. Step (b) picks an element to flip. Step (c) determines whether to flip or not. Here, ΔE is the energy change due to the flip, and h_i is the magnetic field on s_i created by the neighbours. If the energy decreases, then change s_i to $-s_i$. If the energy increases, the change to $-s_i$ occurs with a probability $e^{-\Delta E/T}$. This method satisfies the condition of detailed balance, i.e. the probability of $s_i \to -s_i$ (under the same h_i) to that of $-s_i \to s_i$ is $e^{-2s_i h_i/T}$. Therefore

$$\frac{P(s_i | h_i)}{P(-s_i | h_i)} = e^{2s_i h_i/T} \tag{22-3}$$

$P(s_i | h_i)$ is the conditional probability under a fixed h_i. Every time we perform processes (b) and (c), time increases by one unit. After repeating many times, the probability in (c) has a sufficiently large sample for its justification. Repeated calculation will produce a sequence of configurations, i.e. the trajectory.

Let $s(t) \equiv [(s_1(t), s_2(t), \ldots, s_N(t)]$ be the configuration at time t, then the average value of any $\hat{A}(s)$ can be obtained as

$$\langle \hat{A} \rangle = \frac{1}{\mathcal{J}} \sum_{t=t_0}^{t_0 + \mathcal{J}} A(s(t)) \, , \tag{22-4}$$

where $\mathcal{J}$ is some large number. We discard the configurations before t_0. The reason is that some configurations in the beginning are too far away from the typical equilibrium states. The configuration of $t=0$ was chosen at random and it requires some time to "reach equilibrium". This problem will be discussed below. Now let us look at another model of motion.

Let $W_i dt$ be the probability for s_i to change in a time interval dt.

$$W_i = \frac{e^{-s_i h_i/T}}{\cosh(h_i/T)} . \tag{22-5}$$

This is also a model of motion. Notice that

$$\frac{W_i(s_i)}{W_i(-s_i)} = e^{-2s_i h_i/T} , \tag{22-6}$$

i.e. this model also satisfies the condition of detailed balance. Using this model for simulation is very simple. Let

$$\Omega \equiv \sum_i W_i . \tag{22-7}$$

Starting from t and after an interval t' the probability of having no change is $e^{-\Omega t'}$. So the probability of having no change up to t', and then s' changing in dt' is

$$e^{-\Omega t'} W_i dt' . \tag{22-8}$$

According to (22-5), (22-7) and (22-8), the steps of simulation are:
 (a) From the configuration at time t, calculate W_i and Ω.
 (b) Take two random numbers x and y and let

$$t' = (-\ln x)/\Omega , \tag{22-9}$$

and let the new time be $t + t'$.
 (c) Divide the interval $(0, 1)$ into N sections, each of length W_i/Ω, $i = 1, 2, \ldots, N$. If y falls in the j-th section, change s_j to $-s_j$. Now return to (a).

The steps of simulation here are more complicated than the above one. In the above example, in each step we pick an element and then determine whether to flip it. This example determines the interval of rest at each step first, and then which element to flip. The calculation of the average value is somewhat different.

$$\langle \hat{A}(s) \rangle = \frac{1}{\mathcal{J}} \sum_{k=k_0}^{k_0+n} t'_k A(s(t_k)) \quad,$$

$$\mathcal{J} = \sum_{k=k_0}^{k_0+n} t'_k \quad, \tag{22-10}$$

where t'_k is the interval of rest before the k-th change and $s(t_k)$ is the configuration at that moment. In time $\mathcal{J}$, there are n changes.

22.4. Conceptual Problems to be Noted

The steps of simulation in the last section have very obvious meaning. Each element is acted upon by random but uncorrelated forces. Each element moves due to these forces and is also influenced by the other elements. Therefore, each element is not independent. The influence is through neighbouring elements whose interaction is not zero. So, elements far apart are uncorrelated. Therefore, these steps are a very reasonable model of motion.

These steps do not mention the basic assumption. The only link with thermodynamics is through the probability of change $e^{-\Delta E/T}$. Temperature is thus introduced. In Chapter 3 the discussion of detailed balance was limited to the ideal gas. Here we have interaction between the elements, so the situation is relatively complicated and requires further discussion.

The change of s_i is due to a fixed h_i, i.e. the probability of change of s_i is determined by the environment of s_i. The condition of detailed balance is that under the same environment the probabilities of the forward and reverse directions of $s_i \to -s_i$ are equal. Therefore we get (22-3). Equation (22-3) can be slightly generalised. If the neighbours s_j of s_i are fixed (neighbours in the sense $J_{ij} \neq 0$), h_i is fixed:

$$\frac{P(s_i, s'_i, s''_i, \dots)}{P(-s_i, s'_i, s''_i, \dots)} = e^{2s_i h_i/T} \quad. \tag{22-11}$$

Here $s'_i, s''_i, \dots$ include all spins having interaction with s_i and possibly other elements as well. From (22-11) we can get (22-3). $P(s_i, s'_i, \dots)$ is a joint probability.

If $s_i, s_i', s_i'', \ldots$ include all the elements, then the conclusion of (22-11) is very obvious, i.e.

$$P(s_1, s_2, \ldots, s_N) \propto e^{-H(s)/T} \qquad (22\text{-}12)$$

This is usually regarded as a proof of equilibrium. Nevertheless there are faults in this approach. We have repeatedly emphasised that the definition of probability requires a sufficiently large sample to do the statistics. The sample here is the configurations of the trajectory, i.e. those simulated configurations. The above probabilities must be defined from these configurations. If N is not a small number, then the definition of $P(s_1, \ldots, s_N)$ requires many configurations i.e. $\sim 2^N$ configurations. If $N = 100$, even if the trajectory includes millions of configurations, that would still be far from 2^{100}. Equation (22-12) is only a mathematical conclusion, i.e. a conclusion from an infinitely long trajectory. But "infinitely long" has no physical meaning. Equation (22-11) is suitable for small numbers of elements like $(s_1, s_2, \ldots, s_{10})$ or (s_1, s_2, s_3). Notice that $(s_1, \ldots, s_N)$ represents one point in configuration space and $(s_1, \ldots, s_{10})$ represents a region in this space. This region is determined by the values of $s_1, \ldots, s_{10}$ with $2^{(N-10)}$ points in this region, which is therefore quite large. Likewise (s_1, s_2, s_3) represents a region of $2^{(N-3)}$ points. The probability of such a large region is meaningful because the trajectory remains in it for a sufficiently long time to define probability. These have been discussed at the end of Chapter 10. The reader can review it. The equilibrium under the basic assumption does not involve the single configuration probability in (22-12). It involves the volume of large regions and equilibrium properties are determined by large regions. In the chapters on the basic assumption and probability, we have said much about these. In carrying out the numerical simulation, the result of this abstract discussion becomes crystal clear. The reader will benefit from doing some of these calculations.

Can we use simulation to calculate the thermodynamic potential? The partition function can be written as an average value

$$e^{-F/T} = Z = \sum_s e^{-H/T} \quad ,$$

$$\langle e^{H/T} \rangle = \frac{1}{Z} \sum_s e^{-H/T} e^{H/T} = \frac{2^N}{Z} \quad ,$$

hence

$$Z = 2^N / \langle e^{H/T} \rangle \quad . \qquad (22\text{-}13)$$

Averaging $e^{H/T}$ with respect to time, we can obtain Z and F. But this is impractical unless N is very small. The reason is this: H is a large number of order N, and $e^{H/T}$ is a superlarge number. Its change is also superlarge. Unless we use an infinitely long time, (22-13) is useless. This "infinitely long" time indicates a time longer than the superlarge number $e^{H/T}$. If we want to calculate $F = E - TS$ we must directly calculate $E = \langle H \rangle$ and S. The calculation of entropy from the trajectory is the main theme of Chapter 25.

In Chapter 6 we mentioned that in the region sampled by the motion nearly all the configurations have the same macroscopic properties. If $A(s)$ is a macroscopic variable, then in the region of motion, the majority of the s have nearly the same $A(s)$:

$$A(s) = \langle A(s) \rangle \left(1 + O\left(\frac{1}{\sqrt{N}}\right) \right). \qquad (22\text{-}14)$$

Therefore, it seems that we need only one configuration s and then we know $A(s)$. Why do we need to do the simulation?

In fact, it is not easy to take a point from the region of motion. The region of motion is very large, of volume

$$\Gamma = e^S. \qquad (22\text{-}15)$$

If $S \sim 0.5 N \ln 2$ then $\Gamma \sim 2^{0.5N}$. But this Γ is much smaller than the whole configuration space

$$\frac{\Gamma}{2^N} \sim 2^{-0.5N}. \qquad (22\text{-}16)$$

If $N = 100$, this ratio then is extremely small. It is, therefore, hopeless to choose one point out of the configuration space and hope that it comes close to the region of motion. (Because of this, using (22-12) as the probability it is still not possible to calculate the average value by random integration.) The only way to enter the region of motion is to let the system evolve, gradually bringing the configurations into the region of motion.

In Sec. 22.1 we mentioned that in calculations we typically take N to be 100 or 1 000, but not 10^{20}. Is it too much to hope that thermodynamic properties could be deduced from this relatively small N? The situation is actually not so bad. Macroscopic bodies has the property of independence of parts. We only need to have a model much larger than the correlation length, which can then be regarded as part of the large body. Of course, the boundary becomes an important problem. When N is not large, then $N^{-1/3}$ and $N^{-1/2}$ will not be very small numbers. The usual practice is to use cyclic boundary conditions, i.e. models with no boundary. This method is quite successful. Because N is not a macroscopic number, many conclusions (e.g. that of (22-14)) must be handled with care. At present, simulation is in the stage of accumulating experience and does not have a very solid theoretical basis yet.

Simulation is a solution of the equation of motion so its application is not limited to equilibrium states. (The Monte Carlo simulation on a larger time scale can be regarded as a very realistic model of motion.) It is also ideal for dealing with metastable states. The metastable state is an equilibrium state on short time scales. For longer time scales, changes appear. In addition to the metastable state, for those situations when the basic assumption of statistical mechanics is not too useful, simulation is an invaluable and the only dependable tool for analysis. Many non-equilibrium phenomena, e.g. crystallisation and melting, etc. can be analysed by simulation and many interesting results have been obtained.[a]

Problems

1. Review Problems 11.5 and 11.7.

2. There are many technical problems in numerical simulation. The main one is that the model is too small or the time of simulation too short. However, many interesting problems involve long correlation time and length. Therefore in reading the literature, or doing the simulation ourselves, we must be careful. Usually a wrong interpretation of the result will lead to improper conclusions. Use the Ising model as an example to discuss the problem of time for $h = 0$ and $h \neq 0$ at low temperature.

3. Discuss how to modify the various calculational rules in Chapter 7 for not too large N and V.

4. Discuss the boundary problem of the models.

5. Section 22.3 discusses the motion under a constant temperature. Devise a simulation for the constant energy and the restricted energy methods. (Refer to Problem 3.)

6. Devise a random simulation of the motion of a gas. Notice that we can ignore the kinetic energy of the molecules. Each step moves a molecule by a small distance. Then calculate the difference in energy. Assume a constant temperature situation.

---

[a] For bibliography we can start from Binder (1979), Berne (1977) and Abraham (1979).

7. How do we simulate constant pressure? Under a constant pressure, the volume becomes a variable of motion. If we use the cyclic boundary it is not easy to change the volume.

There is a method of writing the positions of the molecules as br_i, $i = 1, 2, \ldots, N$. Then we take $(b, r_1, r_2, \ldots, r_N)$ as its phase space. Use

$$H(br_1, \ldots, br_N) + p\, b^3 L^3$$

to do the simulation. Here L^3 is a fixed volume, p is the pressure, and r_i are limited within the volume L^3. Devise the programme of simulation and discuss the usefulness of this method.

8. Each method of Chapter 7 has its own steps for random simulation. Devise these programmes. Remember the discussion of Problem 3.

9. The low temperature random simulation gives configurations of low energy. If the temperature is very low, then zero-point energy (the ground state) can appear. So, as we lower the temperature and do the simulation at the same time, we can simulate the zero-point and obtain the minimum value of the total energy function $H(s)$. Therefore, random simulation can be used as a tool for finding the minimum value, and is not limited to statistical mechanical problems. $H(s)$ can be any function and s any variable. For example, s can represent a tour around the world and $H(s)$ the cost of this tour. The tour is required to pass through certain places. To determine the lowest cost we can use the method of random simulation. Devise a step to change s and then we do the simulation. Starting from high temperatures, we then lower the temperature gradually. (This method of obtaining the minimum value is very useful in industry and was initiated by physicist S. Kirkpatrick.) This method has many difficulties depending on the problem. The reader can discuss the problem of metastable states i.e. the result may be a metastable state but not a ground state. Random simulation has many applications; the reader can devise some himself.

PART VI
THEORETICAL BASES

The preceding parts gave a general introduction to the concepts and applications of statistical mechanics. Now we return and analyse some basic concepts in greater depth.

Chapter 23 discusses the three laws of thermodynamics, aiming at a further understanding of their meaning. Chapter 24 gives some examples of the "echo phenomena", clearing up some misconceptions about entropy. These phenomena are also interesting by themselves. Chapter 25 introduces a method to calculate the entropy directly from the trajectory, thus linking all the basic concepts of statistical mechanics with molecular motion. Chapter 26 discusses some problems of principles awaiting solution.

Chapter 23

LAWS OF THERMODYNAMICS

This chapter discusses again the basic concepts of thermodynamics first introduced in Chapter 2, and further explains the role of the time scale. It summarises the relation between the basic assumption of statistical mechanics and thermodynamics, including the third law. We first discuss why the entropy is unchanged in an adiabatic process and then address the problem of the increase of entropy. That the entropy must be a maximum is part of the basic assumption. Lastly we discuss the third law and the role of reversible processes. The usual violations of the third law are due to the occurrence of irreversible processes.

23.1. Adiabatic Processes

The first law of thermodynamics is, first of all, a re-statement of the law of conservation of energy. Secondly, it divides the energy transferred into two kinds — heat and work. In Chapter 2 we have discussed this. The distinction between heat and work is defined through adiabatic processes. Now we consider the adiabatic process.

Let the total energy function of a body be $H(s, L)$, where s is the configuration of the body and L is an invariant quantity. For example, consider a gas in a container. The influence of the container walls on the gas molecules can be represented by a potential energy U. Then

$$H = \text{kinetic energy} + \text{interaction between molecules} + \sum_{i=1}^{N} U(\mathbf{r}_i) \ .$$

(23-1)

The shape of $U(\mathbf{r})$ is depicted in Fig. 23-1, and is similar in the y, z directions. Near the walls, U increases, representing the repulsion of the wall on the molecules. When x is near L,

$$U(\mathbf{r}) = v(L-x) \quad , \tag{23-2}$$

i.e. the potential energy is a function of the distance between the wall and the molecule. Here L is regarded as an invariant quantity or a parameter.

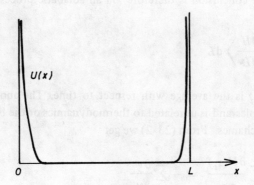

Fig. 23-1

If the environment of this gas is adiabatic, then the total energy is (23-1) without exchanging energy with the outside and U is a fixed function. Under this situation $H(s, L)$ is a conserved quantity:

$$\frac{\partial}{\partial t} H(s(t), L) = 0 \quad . \tag{23-3}$$

Now we change L, moving one wall so that L is changed uniformly to $L + \Delta L$. Assume that this takes time $\mathfrak{I}$, then after the movement the change of energy is

$$\Delta E = \int_0^{\mathfrak{I}} dt \, \frac{dH}{dt} = \int_0^{\mathfrak{I}} dt \, \frac{\partial H}{\partial L} \frac{dL}{dt}$$

$$= \frac{\Delta L}{\mathfrak{I}} \int_0^{\mathfrak{I}} dt \, \frac{\partial H}{\partial L} = \Delta L \left\langle \frac{\partial H}{\partial L} \right\rangle \quad , \tag{23-4}$$

$$\left\langle \frac{\partial H}{\partial L} \right\rangle \equiv \frac{1}{\mathcal{J}} \int_0^{\mathcal{J}} dt \, \frac{\partial H}{\partial L} \qquad (23\text{-}5)$$

Notice that $dL/dt = \Delta L / \mathcal{J}$ and we have used (23-3), i.e. if L is unchanged, then the energy will not change. The change of energy is caused by the change of L alone. The averages in (23-4) and (23-5) are over the duration of the process. If ΔL is very small, then the L in $H(s, L)$ of (23-5) can be regarded as unchanged. The conclusion is therefore: in an adiabatic process, the change of energy is

$$dE = \left\langle \frac{\partial H}{\partial L} \right\rangle dL \, , \qquad (23\text{-}6)$$

where $\langle \partial H/\partial L \rangle$ is the average with respect to time. The above is purely the result of mechanics and is unrelated to thermodynamics or the basic assumption of statistical mechanics. From (23-2) we get

$$\frac{\partial H}{\partial L} = \sum_i \frac{\partial U(\mathbf{r}_i)}{\partial L} = -\sum_i \frac{\partial v}{\partial x_i} \, ,$$

$$\left\langle \frac{\partial H}{\partial L} \right\rangle dL = -p \, dV \, ,$$

$$p \equiv \frac{1}{A} \left\langle -\sum_i \frac{\partial v}{\partial x_i} \right\rangle \, , \qquad dV \equiv A \, dL \qquad (23\text{-}7)$$

where A is the cross-section of the container in the y, z directions and $-\partial v/\partial x_i$ is the force of the wall acting on the i-th particle. Equations (23-6) and (23-7) describe of course one of the adiabatic processes as discussed in Chapter 2.

If the time for the process $\mathcal{J}$ is much longer than the correlation time of the molecular motion, then $\langle \partial H/\partial L \rangle$ is an equilibrium property (assuming that the body is in equilibrium before and after the process). However $\mathcal{J}$ cannot be too long, otherwise the process is not adiabatic. Hence on the one hand $\mathcal{J}$ must be longer than the correlation time, but on the other hand $\mathcal{J}$ must be short in a macroscopic sense.

Do not confuse the adiabatic process here with the adiabatic process in quantum mechanics. Let us explain. In quantum mechanics the total energy operator H may depend on a parameter λ. Hence each $E_n(\lambda)$ is a function of λ, where n

is the quantum number of the stationary state. If we change λ slowly we have the following results:

(a) Each energy level keeps the same form and it is only necessary to substitute $\lambda(t)$, i.e. each energy is $E_n(\lambda(t))$.

(b) If the initial state is n, then it will stay in the state n. This is the theorem of the adiabatic process in quantum mechanics and the condition is that the time $\mathcal{T}$ over which λ changes must be very long

$$\mathcal{T} \gg \hbar/\Delta E \quad , \tag{23-8}$$

where ΔE is the energy difference between two neighbouring E_n. This theorem is valid for any isolated body, including the adiabatic system discussed here.

From (a) and (b), we can get (23-6). But the condition (23-8) of this theorem is not valid for a macroscopic system. If we apply (23-8) to a macroscopic system, ΔE will be very small and $\mathcal{T}$ must be larger than $e^{10^{20}}$ (i.e. the time required for the trajectory to pass through each configuration in the region of motion). In other words

$$\Delta E \sim E/\Gamma(E) \quad ,$$

$$\mathcal{T} \gg \hbar \Gamma(E)/E \sim e^N \quad . \tag{23-9}$$

This is, of course, an unrealistic condition. Therefore, this theorem in quantum mechanics cannot shed any light on the adiabatic process in thermodynamics. The condition of an adiabatic process requires $\mathcal{T}$ to exceed the correlation time of fluctuation, and this is a small time unrelated to (23-9).

We can now define work as the change caused by changes on a macroscopic time scale and call heat the energy change caused by microscopic motion.

23.2. Adiabatic Process and Entropy

The above analysis does not involve the concept of entropy. Starting from the basic assumption, we now look at the relation between the change in entropy and adiabatic processes. We use the method of "restricted energy" to calculate the entropy (see Eq. (7-27)):

$$S(E, L) = \ln \sum_s \theta(E - H(s, L)) \quad . \tag{23-10}$$

This method is used purely for the sake of convenience. Differentiating (23-10), we get

$$dS = \left(\frac{\partial S}{\partial E}\right)_L dE + \left(\frac{\partial S}{\partial L}\right)_E dL \quad ,$$

where

$$\left(\frac{\partial S}{\partial L}\right)_E = -\frac{\sum_s \delta(E-H)\frac{\partial H}{\partial L}}{\sum_s \theta(E-H)}$$

$$= -\frac{\sum_s \delta(E-H)\frac{\partial H}{\partial L}}{\sum_s \delta(E-H)} \cdot \frac{\sum_s \delta(E-H)}{\sum_s \theta(E-H)} \quad . \quad (23\text{-}11)$$

The last factor in the last line is just $(\partial S/\partial E)_L$. The other factor is the average value of $\partial H/\partial L$, i.e., the quantity

$$\left\langle \frac{\partial H}{\partial L} \right\rangle \qquad (23\text{-}12)$$

calculated by the method of constant energy. Therefore,

$$dS = \left(\frac{\partial S}{\partial E}\right)_L \left[dE - \left\langle \frac{\partial H}{\partial L} \right\rangle dL \right] \quad . \qquad (23\text{-}13)$$

The quantity $(\partial H/\partial L)$ is a macroscopic variable, so the average value of (23-12) should be the same as the time average (23-5). This is the result from the discussion of the previous chapters. If the basic assumption is established, this result follows. Now from (23-6) we can see that if dL and dE are caused by an adiabatic process, then $dS = 0$. This is a most important result directly derived from the basic assumption and the result (23-6) of mechanics. The basic assumption does not mention the process and (23-6) does not mention the entropy. In Chapter 2 the concept of entropy was derived from adiabatic process. As seen from the above result, thermodynamics and the basic assumption of statistical mechanics agree perfectly.

Through (23-13) we obtain

$$\left\langle \frac{\partial H}{\partial L} \right\rangle = \left(\frac{\partial E}{\partial L} \right)_S . \qquad (23\text{-}14)$$

Hence force is the isentropic derivative of energy with respect to displacement. Equation (23-13) can be written in a more familiar form

$$dE = T\, dS + \left\langle \frac{\partial H}{\partial L} \right\rangle dL ,$$

where

$$\frac{1}{T} \equiv \left(\frac{\partial S}{\partial E} \right)_L . \qquad (23\text{-}15)$$

23.3. The Second Law of Thermodynamics

The second law of thermodynamics goes beyond equilibrium and points out the direction entropy changes; changes are of course concerned with non-equilibrium states.

First, we must note that the definition of entropy requires some observation time $\mathcal{J}$. The entropy is defined by the region of motion of the trajectory in phase space during this period of time. There is no instantaneous meaning for entropy. Therefore, to speak of entropy at a certain time we must use a time scale larger than $\mathcal{J}$. To take an example, let the gas in Fig. 23-2 be separated into two parts by a wall with a small hole. Initially the right half is empty and the gas slowly leaks into the right. The observation time $\mathcal{J}$ must satisfy

$$\mathcal{J} \gg \tau , \qquad (23\text{-}16)$$

where τ is the mean free time of the molecules. If we want to discuss the rate of change of entropy dS/dt, this dt must be greater than $\mathcal{J}$, i.e. the rate of leakage must be small:

$$\left| \frac{1}{N_1} \frac{dN_1}{dt} \right| \ll \frac{1}{\mathcal{J}} , \qquad (23\text{-}17)$$

where N_1 is the number of molecules in the left. Thus in time $\mathcal{J}$, N_1 can be regarded as unchanged and the entropy can be calculated. If the leakage is too rapid, entropy cannot be defined in this process. Of course we can discuss the entropy before or after leakage and avoid discussing entropy during the process. We have sufficient time to define entropy before and after the process.

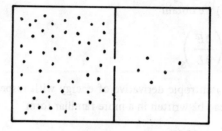

Fig. 23-2

According to the second law, under adiabatic conditions, the entropy of a system cannot decrease. From the basic assumption, the explanation of this law is quite obvious. The basic assumption says: entropy is $\ln \Gamma$, where Γ is the total number of configurations in the region of motion, including all configurations permitted by conservation laws. That is to say, entropy has already been assumed to be maximum because all possibilities are included.

Within a short time, it may be true that the region of motion is restricted because time is not long enough for the system to go through all configurations; but once time is lengthened, the region will grow. The leaking gas is a good example. Because the hole is small, the molecules are restricted to only one part in a short time. When the time is long, molecules can go to the other part; the region is large and so is the entropy. When the numbers are equal on the two sides, the hole will no longer matter. (See the discussion on the various esembles in Chapter 6.) So the fact that the entropy can only increase is an ingredient of the basic assumption. The previous section pointed out that entropy is unchanged before and after an adiabatic process. Hence to perform work on a system or to allow the system to do external work cannot decrease the entropy. Of course, the adiabatic process discussed above is a slow process. If the motion is very fast, much faster than the correlation time of the molecules, will this change decrease the entropy? This problem is beyond the basic assumption. At least no man or machine can achieve such rapid motion, but to prove its infeasibility is also difficult. There are proofs in the literature that this cannot be done. These proofs assume that the system and the machine are in equilibrium, and use the basic assumption on the system plus the machine. This naturally gives the result that entropy cannot decrease. Let us consider an example. Suppose a pair of flywheels are fitted to the hole in Fig. 23-2 (see Fig. 23-3). There is a hole in each flywheel. The flywheels are revolving rapidly. Suppose initially the right chamber is empty. Those rapidly moving molecules in the left can

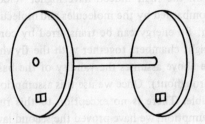

(a) Flywheel with a hole

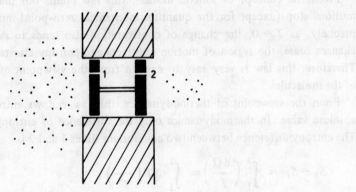

(b) The flywheels fitted to the hole in Fig. 23-2

Fig. 23-3

pass through the hole easily. If the velocity v is large enough, i.e.

$$v > a\omega ,$$

where a is the width of the small hole, and ω is the angular velocity of the flywheels, the molecules after passing through hole 1 can pass through hole 2. If $v < a\omega$, then the particle after passing through hole 1 will not pass through hole 2. Of course there is still a chance to go to the right after many collisions.

At first sight, this pair of flywheels allow fast moving particles to go through while keeping slow moving particles to the left. Hence the temperature on the right is higher than that on the left, violating the second law. In fact this is not so.

Initially the molecules on the right indeed have higher velocities. But the flywheels are continually bombarded by the molecules and molecules accumulate in the space between them. So energy can be transferred by conduction; after a long time, the left and right chambers together with the flywheels will be at the same temperature. The above assumes the validity of the basic assumption (temperature is the same throughout). Once we use this assumption, no violation of the second law is possible. There is no exception to this rule. Therefore, once we prove the basic assumption, we have proved the second law.

23.4. The Third Law of Thermodynamics

The content of this law is: the entropy tends to zero as the temperature T tends to zero.[a]

From the concept of kinetic motion, this law points out that as $T \to 0$, motions stop (except for the quantum mechanical zero-point motion). More precisely, as $T \to 0$, the change of configuration also tends to stop. Because changes cease, the region of motion contracts and entropy also tends to zero. Therefore, this law is very easy to explain from the viewpoint of the motion of the molecules.

From the viewpoint of thermodynamics, this law endows entropy with an absolute value. In thermodynamics only the difference of entropy is defined. The entropy difference between two equilibrium states 1 and 2 is

$$S_2 - S_1 = \int_1^2 \left(\frac{dQ}{T}\right) = \int_1^2 dT \left(\frac{C}{T}\right),$$

$$C = \frac{dQ}{dT} = \text{heat capacity}, \qquad (23\text{-}18)$$

where dQ is the heat added and the integration is along the reversible process from 1 to 2. Hence entropy is defined up to an integration constant. The choice of this constant was discussed in Chapter 2 to Chapter 5.

Because we are now quite familiar with statistical mechanics, it is not difficult to determine this constant. The following is a standard example illustrating the content of the third law.

If atoms A and B can combine to form molecules AB, at very high temperatures a gas AB will dissociate into a mixture of gases A and B. The entropy is

$$S_{AB} = S_A + S_B + N \ln 2. \qquad (23\text{-}19)$$

[a] Proposed by Nernst in 1905.

The term $N\ln 2$ is the mixing entropy, N being the number of A and also that of B. The difference of the entropy at high temperature and that at $T = 0$ can be determined by (23-18):

$$S_A(T) = \int_0^T dT' \frac{C_A(T')}{T'} + S_A(0) \; ,$$

$$S_B(T) = \int_0^T dT' \frac{C_B(T')}{T'} + S_B(0) \; ,$$

$$S_{AB}(T) = \int_0^T dT' \frac{C_{AB}(T')}{T'} + S_{AB}(0) \; . \tag{23-20}$$

In these equations C_A, C_B and C_{AB} are the heat capacities of A, B and AB respectively. During the change of phase, we have to add the latent heat terms. These three heat capacities and latent heats can be determined experimentally. Equation (23-19) is the result of thermodynamics (see Chapter 2). Once we have chosen the constant $S_A(0)$ and $S_B(0)$, then $S_{AB}(0)$ is fixed. For the same reason, for any possible compound e.g. A_2B_3, AB_4, etc. their entropy constants can be determined by $S_A(0)$ and $S_B(0)$. Obviously if we fix the entropy constants of all the elements, then the entropy constants of all the compounds are fixed. Since every element is composed of electrons and the nucleus, and the nucleus is composed of neutrons and protons, if we fix the entropy constants for neutrons, protons and electrons, then the entropy constant of all substances are fixed. Obviously from the definition of entropy there should be a basic constant independent of the structure of the substances. The third law puts all $S(0)$ to be zero. This choice is identical to the choice made in Sec. 3.3. We have indeed no other choice. When the third law was proposed (1905), statistical mechanics had not yet come into being.

According to the above conclusion the entropy of any body, from $T = 0$ to high temperature, can be obtained by measuring the heat capacity, i.e.

$$S(T) = \int_0^T dT' \frac{C}{T'} \; , \tag{23-21}$$

together with the latent heat contributions in phase changes. At high temperatures, the calculation of entropy is simple, i.e. the entropy of the ideal gas plus some corrections. Now we can use experimental results to see whether (23-21) agrees with the theory at high temperatures. In fact, even before the advent of statistical mechanics, people had used the third law and (23-21) to derive the entropy of an ideal gas,[b] identical to that determined later in statistical mechanics. They discovered that the Planck's constant appears among the constants.

Now let us return to the discussion on (23-19) and (23-20). Put the entropy at $T = 0$ to be zero. Then the three integrals of (23-20) must agree with (23-19) when T is very large. This is a very stringent conclusion. Many experiments can be performed to test this property. Note that $A + B$ and the combination AB are different phases of the same system. Equation (23-20) indicates that there exist relations between the heat capacities of these different phases. To generalise, if a molecule exhibits different structures, there is a relation between the heat capacities of the various structures.

The solid phase of many elements exhibits different structures. Within ordinary observation time these structures are equilibrium states. (If time is infinitely long, only one structure is stable while the other are metastable.) Their heat capacities can be measured. The most common example is tin. When metallic tin is cooled below $T_0 = 292$ K, it is in the metastable "white state". The stable structure is a semiconductor called the "grey state". At T_0 the grey and white states can coexist, just like the coexistence of water and vapour. The grey state absorbs heat to change to the white state. Below T_0 the white state is metastable, but changes very little within the usual observation time. Hence it suits our definition of equilibrium. At T_0, the entropy of the white state is

$$S_1(T_0) = \int_0^{T_0} dT \frac{C_1(T)}{T}, \qquad (23\text{-}22)$$

where C_1 is the heat capacity of the white state. The entropy S_1 can also be obtained by integrating the heat capacity C_2 of the grey state:

$$S_1(T_0) = \int_0^{T_0} dT \frac{C_2(T)}{T} + \frac{L}{T_0}, \qquad (23\text{-}23)$$

where L is the latent heat to change from grey to white. Equation (23-22) and (23-23) must be equal, and this is the relation between C_1 and C_2.

---

[b]Sackur (1911) and Tetrode (1912).

In the above discussion we have not been careful about the limit $T \to 0$. It is no trivial task to produce low temperatures in the laboratory. How low must T be in order to say that entropy becomes zero? This depends on the energy scale of the internal motion of the body. When the temperature is lower than all the energy scales of motion, it can be regarded as zero. In solids, the motion with the lowest energy is the change of the nuclear spins. If the temperature is higher than this energy, the nuclear spins can still have entropy. If the motion of the nuclear spins is unrelated to the other motions or the structure of the body, then it can be ignored. For example in the case of tin, the nuclear spins are unrelated to the white or grey states. If the temperature is not too low, each nucleus has an entropy of $\ln(2I+1)$ where I is the nuclear spin. Although this is not zero, because of the constancy of the the number of nuclei, it will not influence the above result. (If $N \ln(2I+1)$ is added to both (23-22) and (23-23), the relation between C_1 and C_2 is unchanged.) If the temperature is so low as to approach the interaction energy of the nuclei, C_1 and C_2 will be influenced by the nuclear motion. The meaning $T \to 0$ indicates a temperature lower than this interaction energy. At this time all the motions stop.

Notice that when we say entropy tends to zero, we mean

$$\lim_{T \to 0} \lim_{N \to \infty} \frac{S(T)}{N} \longrightarrow 0 \quad , \tag{23-24}$$

i.e. the entropy per molecule tends to zero. The energy scale therefore refers to the energy scale of each molecule.

23.5. The Amorphous State and the Third Law

About thirty years ago some chemists did experiments to determine the heat capacities of many relatively complex solids such as ice or solids made up of molecules (not monatomic solids). Later there were experiments to determine the heat capacity of glass. The results of all these experiments obviously violate the third law of thermodynamics. Let $S(T)$ be known for the high temperatures gaseous state (the statistical mechanical result of a gas). Experiments shows that

$$S(0) \equiv S(T) - \int_0^T dT \, \frac{C(T)}{T} \quad , \tag{23-25}$$

is not zero (in fact always greater than zero). For these experiments see references.[c]

[c] Wilks (1961).

Actually these experiments do not violate the third law because the definition of entropy in (23-18) involves integration over reversible processes. But during these experiments some elements of irreversibility are introduced. From this viewpoint the experiments do not fulfill the conditions of the third law, and the third law itself is not violated. From the viewpoint of kinetic motion, as $T \to 0$ motions stop and entropy also tends to zero. The third law cannot be violated. The explanation of these experiments are as follows.

At low temperatures, ice and glass are amorphous substances (see Chapter 17), having many time scales. Some variables of motion change rapidly and some slowly. These time scales are very sensitive to the temperature. As temperature is lowered a little bit, some rapidly changing variables are frozen. When temperature rises, some variables are unfrozen. Cooling from high temperatures can produce different frozen structures. Although different structures look the same on a large scale, on a small scale they are different. Therefore, freezing after unfreezing will not return to the original frozen structure. The process is irreversible.

As temperature increases, the movable variables will increase the amplitude of change. In addition, those variables which previously could not change during the observation time will now start to move. It is not hard to see that

$$dS > \frac{dQ}{T} . \qquad (23\text{-}26)$$

If there is no time limit, $dS = dQ/T$. According to (23-26), $S(0)$ defined in (23-25) must be positive.

Notice that if the ground state of a system is degenerate and the logarithm of the degeneracy is proportional to N, then $S(0)/N$ is not zero. This is unrelated to the amorphous states as discussed above. The simplest example is a system of N spins I without interaction. Their entropy is $N \ln(2I + 1)$, no matter how low the temperature becomes. In many models there are ground states with nonzero entropy. But substances in nature almost invariably interact.

The reader may now ask: if irreversible processes are unavoidable then entropy cannot be determined by heat capacities; so how should we define entropy? What is the dS in (23-26)? According to the basic assumption, entropy is determined by the volume of the region of motion in configuration space. If the time scale is not clear, can we define this region? These questions can be answered by analysing the trajectory of motion. See Chapter 25.

Problems

1. If $\langle \partial H/\partial L \rangle$ in (23-11) to (23-13) is the same as the time average in (23-5) and (23-6), then entropy is unchanged during an adiabatic process. The time during the process $\mathcal{J}$ must be much longer than the correlation time in order to establish the above conclusion.

In a gas, the correlation time can be taken to be the mean free time. But as pointed out by examples in Chapter 20, sound wave and diffusion can lengthen the correlation time considerably, e.g. long wavelength oscillations can persist for a long time.

Now consider the process in Sec. 23.1 (see Fig. 23-1). L is moved uniformly to $L + \Delta L$ in time $\mathcal{J}$.

(a) Write down the fluid equations for the motion in this process.

(b) These equations are differential equations, and the change L is the change of the boundary which can be regarded as a change of parameter. This problem is very similar to the adiabatic process in quantum mechanics. The differential equations are like the single particle wave equation.

(c) How small should $\Delta L/\mathcal{J}$ be so as to avoid exciting oscillations in the fluid?

(d) If the mean free distance of the molecules is larger than L, what will be the conclusion then?

2. Review the analysis of the hydrogen molecule gas in Sec. 8.2. We only consider the rotation of the molecules.

(a) Use the integration of the heat capacity to obtain the entropy when $T \to 0$. Prove

$$S(0) = N_1 \ln 3 \quad , \tag{23-27}$$

where N_1 is the number of molecules with spin 1.

(b) The ground state of the H_2 molecules is that with spin 0. So the entropy obtained by integrating the heat capacity may not be related to the ground state. This should be quite obvious. The reader can give several other examples.

3. A body is made up of N spins s_i, $i = 1, 2, \ldots, N$. The total energy is

$$H = -\sum_i s_i h_i + \sum_i (1-s_i^2)\Lambda_i \quad . \tag{23-28}$$

Each spin has three states. $s_i = +1, 0, -1$, and h_i and Λ_i are fixed, but are different at different sites. The distribution is

$$p(h) = (2\pi b^2)^{-\frac{1}{2}} \exp(-h^2/2b^2) \quad ,$$

$$p(\Lambda) = \frac{1}{a} e^{-\Lambda/a} . \qquad (23\text{-}29)$$

Assume $a \gg h$.

This model is nearly the same as Problem (21.6). Suppose the rules of motion are as follows:

If $\Lambda_i > T \ln t$, then s_i will not change.

If $\Lambda_i < T \ln t$, then s_i can change. Here t is a constant and can be regarded as the observation time.

(a) If $T \gg a$, calculate the total energy, heat capacity and entropy.

(b) Suppose this body is cooled gradually and at each period of time t, the temperature is lowered by a/n. Let $n = 100$ and $b = a/10$. Starting from $T = 2a$, calculate and sketch the relation of the various quantities with T. Notice that if s_i cannot move, then its entropy is zero.

(c) Use the heat capacity dE/dt in (b) to perform the integration to get the entropy, and make a comparison with the entropy directly derived in (b). Then discuss the third law of thermodynamics.

Chapter 24
ECHO PHENOMENA

Entropy comes from random motion. We have repeatedly emphasised the importance of motion. Irregular arrangement or random distribution will not be related to entropy if they do not change with time. We have given many examples showing that the region of motion and entropy must be determined by motion. In this chapter we shall discuss three dramatic experiments, i.e. "spin echo", "viscous liquid bottle" experiment and "plasma echo". The first is the experiment of nuclear spin resonance by Hahn.[a] He reversed the directions of the precessing spins, causing the dispersed spins to restore their directions. The viscous liquid bottle makes the dispersed colours of the viscous liquid return to their original positions. This is a very simple experiment for classroom demonstration. Hahn has used it to demonstrate his spin echo experiment. (Hahn claims that he is not the first to invent this demonstration, but he must at least be credited for popularising it.) After Hahn's experiment, many similar experiments were done including the "plasma echo".[b] Hahn's aim was not to discuss entropy; the same is true for the other experiments. These experiments are called the echo phenomena in the literature. The techniques and results of these experiments are extremely interesting and invaluable. Our aim is to use them to further discuss entropy. We shall study carefully the demonstration of the viscous liquid bottle, which is a simple classroom demonstration. Then we analyse the principles involved. After that we shall mention

---

[a] Hahn (1950).

[b] Malmberg *et al.* (1968), O'Neil and Gould (1968).

the spin echo and plasma echo experiments. These cannot be demonstrated in the classroom, and involve more complicated theory. What we present are simplified versions, involving thought experiments, so that the reader can grasp the essence of these phenomena.

24.1. Demonstration of the Viscous Liquid Bottle

This demonstration is extremely simple. The construction of the bottle is as follows: two concentric cylinders (made of plastic or glass) are installed so that the outer cylinder is fixed and the inner cylinder can be turned by a handle (see Fig. 24-1). In the space between the two cylinders is some viscous liquid, e.g., corn starch paste. The viscous liquid must be transparent and the more viscous the better. The distance between the two cylinders should be about 1 cm and the diameter of the outer cylinder about 10 cm. The dimensions are not crucial.

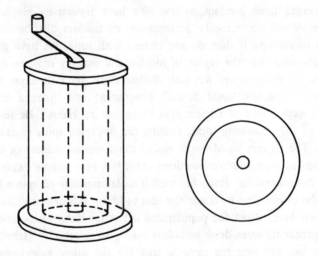

Fig. 24-1 Viscous liquid bottle demonstration.

Use a beaker and mix some dye with the viscous liquid (we can use food dye, ink or other dyes). The molecules of the dye are distributed uniformly in the viscous liquid. A small amount of dye, just enough to show the colour, is adequate.

The demonstration proceeds as follows:

(a) Use a fine glass tube to suck up some coloured viscous liquid and transfer it to the bottle, so that a coloured line is formed in the bottle.

(b) Turn the handle in one direction several times, so that the molecules near the inner wall are moved along while those near the outer wall do not move. Hence the coloured line will become dispersed. After one or two turns the colour disappears, and after ten or more turns the molecules of the dye are completely dispersed. At this time, the experiment demonstrates that the dye molecules are distributed randomly like the molecules of an ideal gas; however, unlike an ideal gas, they are not moving.

(c) Turn the handle in the reverse direction for the same number of turns. Surprise! The coloured line reappears. The molecules of the dye are restored to their original positions. It is unbelievable unless one sees the experiment himself.

What is surprising in this experiment? It is, of course, the restoration of the dye molecules from the dispersed state. The usual interpretation of entropy is the degree of randomness and the second law of thermodynamics expressed the inevitability of becoming more and more random. This demonstration seems to say that the second law is incorrect. Some people argue that when the dye molecules are restored, entropy flows through the handle of the bottle to the hand of the experimenter. This kind of explanation is a misconception about entropy and is untenable. When the hand pushes the handle, it does work no matter what the direction of motion is, and this is unrelated to the arrangement of the dye molecules.

In fact, there is no contradiction between this phenomenon and the second law.

24.2. Analysis of the Demonstration

We first present an incorrect analysis and then give the right one.

The incorrect analysis goes as follows. Before the dye is put into the bottle, the uncertainty in position of each dye molecule is the volume V_0 of the coloured line, and the probability of distribution is $1/V_0$. So it has an entropy

$$S_0 = \ln\left(\frac{V_0^N}{N!}\right) + N\sigma$$

$$= N \ln(V_0/N) + N + N\sigma \quad , \tag{24-1}$$

where N is the number of dye molecules, N/V_0 is the density of the dye molecules in the beaker, and σ is the entropy caused by the small amplitude vibration of the molecules.

418 STATISTICAL MECHANICS

After the handle is turned several times, the probability distribution of the position of the dye molecule is $1/V$ where V is the volume of the viscous liquid bottle. Hence the entropy is

$$S = \ln\left(\frac{V^N}{N!}\right) + N\sigma$$

$$= N\ln(V/N) + N + N\sigma \quad ,$$

i.e. $\quad S - S_0 = N\ln(V/V_0) \quad .$ (24-2)

This is the increase of entropy due to the dispersion. This analysis seems quite reasonable, but obviously there are faults because in turning the handle backwards the molecules of the dye are restored to V_0 and entropy decreases, violating the second law. (It is not correct to say that entropy flows to the hand, since this heat current would be easily felt.) The fault of the above analysis lies in misconceptions about entropy. In defining entropy, the region of motion and the probability distribution are determined by the amplitude of motion. The volumes V_0 and V are not related to the amplitude of motion of the dye molecules. Hence (24-1) and (24-2) are not correct.

We emphasise once more that entropy is unrelated to the uncertainty of observation or the knowledge or ignorance of the observer. If we use these concepts to define entropy we are naturally led to the wrong conclusion of (24-1) and (24-2). Now we look at the correct analysis.

This phenomenon of positional restoration is hardly surprising. Each dye molecule, except for a slight vibration, is stationary. This is quite unlike the molecules in a gas, which are moving incessantly. Within the observation time of several hours, diffusion does not occur. The time scale of this demonstration is the time of persistence of vision of our eyes, about 0.1 second. Under this time scale, entropy is that of small amplitude vibrations, unrelated to the distribution of the dye molecules in the liquid. Hence, turning the handle several times just heats up the whole system slightly. Because of friction, no matter in which direction the handle is turned the vibration of the viscous liquid and the dye molecules will be larger, and entropy increases slightly. This is unrelated to the distribution of molecules. Hence the entropy is $N\sigma$, independent of V_0 and V.

We can also look at the problem this way. Because the dye molecules are few (the dye is very dilute in the beaker), they are surrounded mostly by the molecules of the viscous liquid. Because the dye molecule does not move, its coordinates and surroundings are unchanged (except for a small amplitude of vibration). It is like a trapped impurity in a solid. Hence the entropy is related to the total number of dye molecules, but not to their position. This point has been discussed under frozen impurities.

Therefore, during our demonstration, the entropy of the dye molecules is the same in the dispersed state or the restored state (neglecting the heat produced by friction). Dispersed or restored, they are equilibrium states on this time scale. The positions of the dye molecules are invariant quantities. The dispersed state or the restored state have the same entropy. This does not contradict the second law. After a few days, the dye molecules will be spread out and cannot be returned to the original positions. Has entropy increased? The answer is no. The above conclusion is that entropy is unrelated to the distribution of positions of the dye molecules. The time scale after a few days is still the scale of human vision, i.e., 0.1 second to several minutes.

If the liquid in the bottle is not highly viscous, then the dye molecules can move rapidly. Equations (24-1) and (24-2) will then be correct. The important time scale is

$$\tau \sim l^2/D \quad ,$$
$$l = (V/N)^{1/3} \quad . \tag{24-3}$$

where D is the diffusion coefficient of the dye molecules, and l is the distance between the dye molecules. To change the viscosity is to change D and the time scale correspondingly. The diffusion coefficient D of this demonstration viscous liquid is very small, so τ can be about ten days or more, far exceeding the observation time of the equilibrium states. If we use water instead of corn starch paste, τ will be smaller than the observation time and, (24-1) and (24-2) would then be appropriate.

Note that diffusion is related not only to the viscosity of the liquid but also to the size and shape of the dye molecules. For the above demonstration, we should use dyes with large molecules. The usual food dye is quite ideal. Of course, the most important factor is the viscosity, which must be large enough that the viscous fluid itself undergoes negligible diffusion.

24.3. Spin Echo Experiment

Here we mention briefly the principle of Hahn's experiment. The reader can find the details in the original paper. In a fixed magnetic field, a magnetic dipole **m** will precess. The equation of precession is

$$\frac{d\mathbf{m}}{dt} = \lambda \mathbf{h} \times \mathbf{m} , \qquad (24\text{-}4)$$

where λ is a constant and **h** is the magnetic field (Fig. 24-2). The angular velocity of precession is λh. The nuclear dipole moment of hydrogen in water molecules can be used in this experiment.

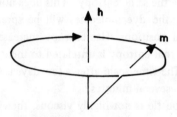

Fig. 24-2 Precession of a magnetic dipole **m** about the magnetic field **h**.

If we use a rotating coordinate frame with angular velocity λh, then **m** is stationary in this frame. Modern techniques enable us to do many experiments in this rotating frame. The steps of Hahn's experiment are as follows:

(a) Use a magnetic field **h** (along the z-direction) to align the magnetic dipoles of H nuclei in water along the **h** direction. (Fig. 24-3(a).) The magnetic field may be slightly inhomogeneous and the magnetic field at different parts of the body may be slightly different. Let h_0 be the average value.

(b) In a rotating frame with angular velocity λh_0 apply a magnetic field in the y-direction, so that **m** precesses to the x-direction. Then immediately switch off the field. This rapidly changing field flips **m** from the z- to the x-direction, turning through $90°$. (Fig. 24-3(b).)

(c) Because the magnetic field is slightly different at various places, in the rotating frame the spins precess slightly differently at various places. After a period t, the spin directions at different places are different, and they are randomly distributed with zero total magnetic moment. The arrows in Fig. 24-3(c) represent the directions of the different spins. On the planar diagram two spins and their precession velocities are drawn, the faster in front and the slower behind.

(d) Use the method of (b) to apply a rapid field in the y-direction, but lasting twice as long, so that all the spins precess by $180°$. This is like turning the whole set of the arrows upside down. Figure 24-3(d) shows the way that

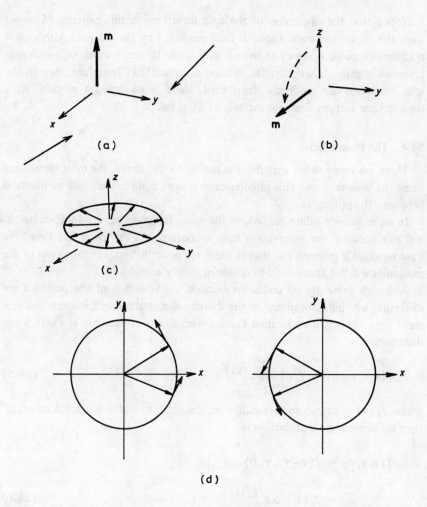

Fig. 24-3

the two arrows in (c) are turned. Notice that after turning, the precession direction remains the same, but now the slower is in front while the faster is behind. That is to say, all the relative angles of precession change sign, and after a period t, the spins are restored to their original positions and the total magnetic moment reappears, which is directly measurable. This is the spin echo.

The last step is like the reverse turning of the handle in the viscous liquid bottle experiment. The other steps are similar too.

Notice that the dispersion of the spin directions in this experiment comes from the inhomogeneous magnetic field produced by the magnet. Although it is inhomogeneous, it does not change with time. Hence, even though each spin precesses slightly differently, the motion is prescribed. Therefore, though the spin directions are randomly distributed, there is no random motion. As a consequence entropy does not increase or decrease.

24.4. The Plasma Echo

There are many other experiments similar to the above, the most interesting being the plasma echo. This phenomenon is more complicated, and we mention here only the principles.

In an extremely dilute gas, where the mean free time is very long, collisions will not occur if the observation time is shorter than the mean free time. We have repeatedly pointed out that if there are no collisions, the momenta of the molecules will not change and thus cannot supply entropy.

Although there are no collisions because the velocities of the particles are different, any inhomogeneity in the density distribution will smooth out in a short time. This is the so-called Landau damping. For example at $t = 0$, if the distribution is

$$f(\mathbf{r}, \mathbf{v}, t=0) = f_0 (1 + \alpha e^{i\mathbf{k} \cdot \mathbf{r}}) \quad , \tag{24-5}$$

where $f_0(\mathbf{v})$ is the uniform equilibrium distribution and α is a small constant, then the subsequent distribution is

$$f(\mathbf{r}, \mathbf{v}, t) = f(\mathbf{r} - \mathbf{v}t, \mathbf{v}, 0)$$

$$= f_0 (1 + \alpha e^{i\mathbf{k} \cdot (\mathbf{r} - \mathbf{v}t)}) \quad . \tag{24-6}$$

When t is very large, $e^{-i\mathbf{k} \cdot \mathbf{v}t}$ changes rapidly with $\mathbf{v}$ and if we integrate with respect to $\mathbf{v}$, it will be zero. The density of the particles can be obtained by integrating (24-6):

$$n(\mathbf{r}, t) = n_0 + \alpha e^{i\mathbf{k} \cdot \mathbf{r}} \int d^3v \, f_0(\mathbf{v}) \, e^{-i\mathbf{k} \cdot \mathbf{v}t}$$

$$= n_0 (1 + \alpha e^{i\mathbf{k} \cdot \mathbf{r}} \, e^{-\frac{1}{2} k^2 t^2 T/m}) \tag{24-7}$$

i.e. the inhomogeneous part will vanish after a time $1/k\bar{v}$ where $\bar{v} \sim \sqrt{T/m}$ is the average speed. This plasma echo experiment points out that even though the inhomogeneity vanishes, it can be made to appear again, only that the form of appearance is somewhat indirect. That is to say, even when t is very large $e^{-i\mathbf{k}\cdot\mathbf{v}t}$ still has its effect and cannot be neglected. This effect is measured by the changes in the plasma. The steps are as follows:

(a) At $t=0$, we add an inhomogeneous electric field $\propto e^{i\mathbf{k}\cdot\mathbf{r}}$ on the plasma, creating an inhomogeneous density distribution like (24-5). When the electric field is switched off, the distribution becomes (24-6). The inhomogeneity in the distribution vanishes after a time $1/k\bar{v}$.

(b) At time $t=t'$, add an inhomogeneous electric field $\propto e^{-i\mathbf{k}'\cdot\mathbf{r}}$, and then switch it off. Hence the distribution has one more term

$$f_0(\mathbf{v}) \, \alpha' \, e^{-i\mathbf{k}'\cdot(\mathbf{r}-\mathbf{v}(t-t'))} \quad . \tag{24-8}$$

After the electric field has been switched off for $1/k'\bar{v}$, the density is uniform again. But this is not the complete solution, being only the terms proportional to the external electric field, i.e. the first order term. There are higher order terms. Among the second order terms there is one proportional to $\alpha\alpha'$ (obtained by replacing $f_0(\mathbf{v})$ in (24-8) by (24-6)).

$$f_0(\mathbf{v}) \, \alpha \alpha' \, \exp\left[i(\mathbf{k}-\mathbf{k}')\cdot(\mathbf{r}-\mathbf{v}t) - i\mathbf{k}'\cdot\mathbf{v}t'\right] \quad . \tag{24-9}$$

Therefore if $\mathbf{k}'$ and $\mathbf{k}$ point in the same direction, then at a time

$$t = \frac{k'}{k'-k} t' \quad , \tag{24-10}$$

the exponential terms in (24-9) containing $\mathbf{v}$ cancel, and we get

$$f_0(\mathbf{v}) \, \alpha \alpha' \, e^{i(\mathbf{k}-\mathbf{k}')\cdot\mathbf{r}} \quad . \tag{24-11}$$

Its integral with respect to $\mathbf{v}$ is not zero. If k' and k are very close, then t in (24-10) can be much larger than $1/k\bar{v}$. At this time, (24-11) creates an inhomogeneous charge density which is measurable. This is a special echo phenomenon. That is to say, even when the inhomogeneous terms of (24-6) and (24-8) seem to disappear, their interference effect brings an inhomogeneous density distribution, confirming that they cannot be neglected.

Hence the disappearance of inhomogeneous density cannot be regarded as an increase of entropy and its reappearance cannot be explained as a decrease of entropy. On the time scale of this experiment there are no collisions between the particles. In momentum space the configuration is unchanged, i.e. there are no changes of the momenta of the particles, so this does not supply entropy. The second law is then not violated.

Problems

1. Try to analyse the viscous liquid bottle demonstration in configuration space. Let $(r_1, r_2, \ldots, r_N) \equiv R$ be the positions of the dye molecules.

   (a) Initially, what is the region of motion of R?

   (b) After the molecules are dispersed, what do the position, shape and size of the region of motion become?

   (c) What about these quantities when the dye is restored to the original position?

2. The plasma echo phenomenon is relatively complicated, but it can still be analysed in configuration space. Try this.

Chapter 25

ENTROPY CALCULATION FROM THE TRAJECTORY OF MOTION

If we know the details of motion of every molecule during the observation time, then any property of the system can be calculated. Many equilibrium properties, such as pressure, magnetisation, energy, etc., are averaged values during the observation time. Although the commonly seen macroscopic objects are too complicated, simple models with hundreds or thousands of particles or spins can be analysed on the computer (see Chapter 21). Now we want to discuss the following problem: Can entropy be calculated from information on the trajectory? The answer to this question should be yes. As the whole process of motion is known, all physical quantities should be calculable. But entropy is unlike quantities such as energy and pressure. It is not the averaged value over time of a dynamical variable. In mechanics and electromagnetism there is no such concept as entropy. Nevertheless, if the determination of entropy had to go beyond knowledge on the whole motion, then the concept of entropy would be outside the realm of science; that clearly is not the case. The problem is really one of finding a practical computational method. We shall devise a method of calculating the entropy step by step from knowledge about the whole motion. We introduce here a specially simple method called the "method of coincidence". Better methods must certainly exist, waiting to be discovered.

The reader may ask: Why must we discuss this problem of calculating entropy from the motion? We have two motives. One is a matter of principle. Starting from Chapter 5 we have discussed over and over again the relation of entropy with motion. Many examples were given to show that the definition of entropy is based on motion. But all these examples are very simple. Which quantities are invariant and which quantities are changing gradually are very clear. When we use the basic assumption, we only need to treat the changing quantities of motion as the variables and those unchanged quantities as constants. But this distinction may not always be so clear, especially for amorphous states. Until the problem is solved, we do not know which are variables and which are constants. With these problems remaining unsolved, the region of motion must be determined by the trajectory.

The second motive is practical. A method of calculating entropy from motion can be used to analyse various models, especially those exhibiting metastable states.

The Boltzmann formula for entropy is

$$S = \ln \Gamma(E) \quad ,$$

where S is the entropy, E is the total energy of the system and $\Gamma(E)$ is the volume of the region of motion Ω, which includes all configurations with energy E. The discussion of this chapter starts from this formula. Our basic viewpoint is this: $\Gamma(E)$ must be determined by the trajectory. The method of coincidence is simply a programme to calculate $\Gamma(E)$ from the trajectory.

We first explain the principles of the coincidence method and then use the Ising model as an example to illustrate the steps of the calculation.

The reader should note that the method and techniques of this chapter are in the stage of development and the conclusions are only preliminary. The reason for including this relatively new method in this book is that the method is very important but not yet commonly known.[a]

25.1. Number of Coincidences and the Size of the Region

The basic principle for calculating entropy from the number of coincidences is very simple, as explained in Fig. 25-1. A collection of dots is scattered randomly in a certain region. The region Ω has the shape of a ring. The area Γ of this region Ω is $\pi(a^2 - b^2)$ squares. In this area Γ, only a small number of the squares are occupied by the dots and the majority are empty. If we do not look at the picture and only know the number of dots in all the squares, this area is not so easy to obtain. But if we know that these dots are randomly distributed in Ω, then the area Γ of the region Ω can be estimated as follows.

Let n be the total number of dots, and Γ be the number of squares in Ω. Because the distribution in Ω is random, it is still possible for two or more dots

---

[a] This chapter comes from research reports, see Ma (1981a, b).

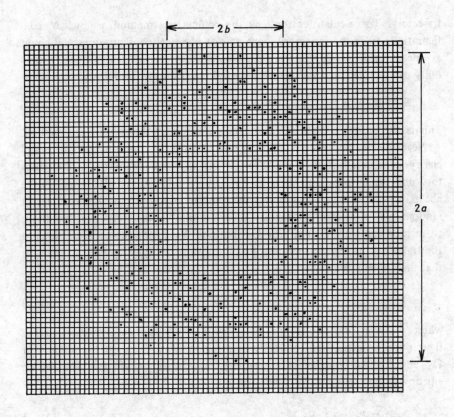

Fig. 25-1 A collection of dots randomly scattered in an annular region.

to occupy the same square, even though Γ is much larger than n. Take any pair of dots; the probability that these two are within the same square is $1/\Gamma$, which we call the coincidence probability $R = 1/\Gamma$. Take N_t pairs of black spots, then the "coincidence number" N_c is $N_t R$. The maximum value of N_t is $\frac{1}{2}n(n-1)$. Hence if the coincidence number is measured, then Γ can be calculated by

$$\Gamma = \frac{1}{R} = \frac{N_t}{N_c} \quad . \tag{25-1}$$

This method of estimating Γ is called the method of coincidence. This method is very simple. It need not consider the shape of the distribution region or other geometrical properties. This method can be used to calculate the entropy. Each square can be thought of as a configuration, and each dot a configuration in the trajectory. To calculate entropy by the coincidence method, we pick N_t configurations from the trajectory; each pair of identical configurations will count as one coincidence. If there are N_c coincidences then the coincidence probability can be calculated and the entropy follows from the formula

$$S = \ln(1/R) \quad . \tag{25-2}$$

This is the basic principle for calculating entropy by the coincidence method.

Nevertheless, for practical applications we have to solve some problems and clarify some details first. We first review the important properties of the trajectory and the entropy and then try to set up a calculational algorithm. In the above, we have assumed that the distribution of points in Ω is completely random, i.e. a uniform probability distribution. We now relax this condition to include cases of non-uniform distributions. We divide the configurations of the trajectory into groups and let $\mathcal{T}_\lambda$ be the time occupied by each group λ, so that the probability is

$$P_\lambda = \mathcal{T}_\lambda / \mathcal{T} \quad , \tag{25-3}$$

where $\mathcal{T}$ is the total observation time. Now assume that each group λ of configurations is distributed randomly in a region Ω_λ; let Γ_λ be the volume of Ω_λ and R_λ the coincidence probability in this region. We can now define entropy as the average value of $\ln(1/R)$

$$S = \sum_\lambda P_\lambda \ln(1/R_\lambda) \quad . \tag{25-4}$$

This is a generalisation of (25-2). The factor R_λ is the coincidence probability for two configurations in the λ group:

$$R_\lambda = P_\lambda / \Gamma_\lambda \quad , \tag{25-5}$$

where P_λ is the probability of any configuration in the λ group, and the probability of choosing one configuration coinciding with the one already chosen is $1/\Gamma_\lambda$. Let $N_{t\lambda}$ and $N_{c\lambda}$ be the relative number and total number of coincidence in the group λ, then

$$S = \sum_\lambda P_\lambda \ln(\Gamma_\lambda / P_\lambda) \quad ,$$

$$\Gamma_\lambda = N_{t\lambda} / N_{c\lambda} \quad . \tag{25-6}$$

The choice of grouping depends on the actual situation. For example, if there is exchange of energy with the environment, the total energy of a system is not fixed and we can use λ to label the different energies.

Two points should be noted. (i) The method of coincidence is used to calculate Γ_λ. The probability P_λ can be directly calculated in the numerical simulation from $\mathcal{J}_\lambda$ and (25-3), or by other methods. (ii) The grouping cannot be too small. In the limit of a single configuration for each group, (25-4) becomes

$$S = - \sum_s p_s \ln p_s \quad . \tag{25-7}$$

Here s denotes the various configurations. But to measure p_s we must measure the time $\mathcal{J}_s$ that the trajectory stays in the configuration s. This time is not measurable unless the total number of configurations is very small so as to make $\mathcal{J}_s$ relatively large. Usually it is said that statistical mechanics and information theory are the same. We want to clarify this here. In information theory, (25-7) is a basic formula where s denotes the symbols and p_s its probability of appearance in a message. The quantity S measures how much information is carried by these symbols. Each symbol is like a configuration in statistical mechanics. The appearance of the symbols in a message is like the trajectory. In information theory the number of symbols is very small and each symbol is used many times so that p_s can be accurately determined. The situation is quite different in statistical mechanics. The total number of configurations s is very large and the trajectory is short. In a trajectory only a very small fraction of possible configurations actually appear. Hence, each group λ must include many s in order to make P_λ a measurable quantity. Hence from the viewpoint of direct measurement, (25-7) is a practical definition in information theory but has no direct use in statistical mechanics, and is therefore not appropriate as a definition of entropy. If we use information theory on (25-7) to discuss statistical mechanics we would be missing the point.

25.2. Independence of Different Parts of a System

This coincidence method uses a relatively few number of points to determine the volume of a very large region, i.e. $n \ll \Gamma$. But this method obviously has its limitations. We have said that N_t cannot exceed $\frac{1}{2} n(n-1)$. Therefore

$$N_c \leqslant \frac{1}{2} n^2 / \Gamma \quad . \tag{25-8}$$

The coincidence number N_c must be quite large, i.e. $N_c \gg 1$ for Γ to be meaningful. Therefore n cannot be too small:

$$n \gtrsim \sqrt{2\Gamma} \quad . \tag{25-9}$$

Hence the trajectory must be very long, because the Γ to be determined is always very large. For a system consisting of N elements, Γ increases exponentially with N:

$$\Gamma = e^S \quad , \tag{25-10}$$

$$S \propto N \quad . \tag{25-11}$$

But the length of the trajectory n is proportional to N and the observation time $\mathcal{J}$:

$$n \sim N(\mathcal{J}/\tau) \quad , \tag{25-12}$$

i.e. the configurations of the system change n times in time $\mathcal{J}$. Here τ is the time for each element to change once. If N is very large and $\mathcal{J}$ of reasonable value, (25-9) will not be easily satisfied. This is indeed a problem, but not a serious one. The volume Γ increases exponentially with N because the various parts of the body are to a large extent mutually independent (see Chapter 12). The molecular motions far apart are uncorrelated. If $A(\mathbf{x}_1)$ is a variable of motion at $\mathbf{x}_1$ and $B(\mathbf{x}_2)$ that at $\mathbf{x}_2$ and $|\mathbf{x}_1 - \mathbf{x}_2| \gg \xi$, then

$$\langle A(\mathbf{x}_1) B(\mathbf{x}_2) \rangle - \langle A(\mathbf{x}_1) \rangle \langle B(\mathbf{x}_2) \rangle = 0 \quad . \tag{25-13}$$

Here $\langle \ldots \rangle$ denotes the averaged value within the observation time and ξ is the correlation length. If the system is much larger than ξ, then it is composed of many largely independent parts, each larger than ξ, with the total entropy being the sum of the entropies of the parts. The total number of parts is proportional to N. Hence $S \propto N$ and Γ increases exponentially with N. So, if ξ is small enough, this coincidence method can be used to calculate the entropy of the various parts of the system.

If a system has two independent parts A and B, then the coincidence probability R is $R_A R_B$ because the probability of coincidence in A and B simultaneously is the product of the separate probabilities. Therefore

$$S = \ln[1/(R_A R_B)]$$

$$= S_A + S_B \quad . \tag{25-14}$$

The product of independent coincidence probabilities and the additive property of entropy are parallel. If a system is divided mainly into independent parts, then

$$S = \sum_A S_A + \sum_{A,B} S_{AB} \ . \tag{25-15}$$

Here S_A is the entropy of part A. If A and B are neighbouring and not completely independent, we need S_{AB} as a correction, which can be obtained from S_A, S_B and S_{A+B} by

$$S_{AB} = S_{A+B} - S_A - S_B \ , \tag{25-16}$$

where S_{A+B} is the entropy for the combination of A and B. The independence of the various parts enables us to deduce the properties of the total system from those of the small parts (each larger than ξ). Because of this, the usual model of numerical simulation can be used to deduce many properties of the whole system, even though these models employ only hundreds or fewer elements of motion. Although these models are small, they are sufficient when they are larger than ξ. Using larger models merely shows clearly the independence property. Of course, the magnitude of ξ depends on the problem. If ξ is very large (as in critical phenomena) numerical simulation is not very efficient. The coincidence method discussed here is only suitable for small ξ.

25.3. Correlation Time

The coincidence method tests various uncorrelated configurations. Although the trajectory is not a sequence of completely random points, after a long time the configurations occurring will be independent and uncorrelated. If A is the variable of motion at an instant and $B(t)$ is that after time t, then for $t \gg \tau$,

$$\langle AB(t) \rangle - \langle A \rangle \langle B \rangle = 0 \ . \tag{25-17}$$

Here τ is the correlation time. If the observation time is much larger than τ we can choose configurations separated by τ for comparison. Equation (25-17) points out the independence of motion over a long time, while (25-13) expresses the independence over a long distance. This characteristic is a necessary condition for computation. We just assume that this condition is satisfied and do not look for its justification.

The exposition of the basic principles ends here. There are, of course, many problems of principle or of calculational techniques, which we hope to clarify through the actual calculation in the next section.

25.4. Process of Computation

Now we demonstrate the above method of coincidence by calculating the entropy of the Ising model, listing the computational steps. The first stage is the usual numerical simulation whose results are used as input for the second stage. Our main discussion is on the second stage. To be precise, we first clear up the steps of the first stage. (For the principle and details of simulation, see Chapter 20.)

A. Ising model of motion

The configuration of this model is determined by the values of N spins $s_1, \ldots, s_N$. Each spin can be $+1$ or -1, and they change according to the following rules. From the configuration at any time t we can calculate the rate of change of each spin (the probability of change per unit time)

$$W_i = e^{-s_i h_i/T} / \cosh(h_i/T) \quad , \tag{25-18}$$

where T is the temperature of the system and h_i is the force on s_i:

$$h_i = h + \sum_j J_{ij} s_j \quad , \tag{25-19}$$

in which h is the external magnetic field and the last term is the total force exerted by all the other spins on the i-th spin. The coefficient J_{ij}, the strength of interaction between i and j, is known and is constant over time. Now the rules of motion are completely defined.

The trajectory is simulated by the above probability and the random numbers generated by the computer. For details see Sec. 22.3.

The use of random numbers is convenient but not necessary. The principle of calculating entropy from the trajectory is suitable for any trajectory, the only condition being independence for long time-scale and long distances. This independence property is the characteristic of the equilibrium state.

B. Programme of calculation

This programme is to record a number of configurations in the trajectory and then pick them out pairwise for comparison to observe the number of coincidences.

The following describes a simple but not very efficient programme to illustrate the principle of the coincidence method. This programme does not record all the configurations in the trajectory but it stores the labels of the changing elements as a sequence $i_1, i_2, \ldots, i_n$. To compare two configurations we must consider the changes in between. For example to compare the configuration A before i_2 and B after i_{10}, we compute the number of elements which appear in the list $i_1, i_2, \ldots, i_{10}$ an odd number of times. An element is unchanged if it has changed an even number of times. The number of elements changing an odd number of times is called the "difference". If there is no difference, then A and B are the same configuration and we have a coincidence. We discuss the details below.

C. Energy and the effective volume of a configuration

This model is a constant temperature system with fluctuation in energy. So we have to classify the configurations by their energy to calculate $\Gamma(E)$ and $P(E)$, and then use (25-6) to get the entropy:

$$S = \sum_E P(E) \ln\left[\Gamma(E)/P(E)\right] . \qquad (25\text{-}20)$$

The probability of the E group is given by the simulation programme:

$$P(E) = \mathfrak{J}(E)/\mathfrak{J} , \qquad (25\text{-}21)$$

where $\mathfrak{J}(E)$ is the time that the energy E appears, and $\mathfrak{J}$ is the total time of the trajectory. So we have to record the energy also.

To increase the number of coincidences we can relax the rule of coincidence. For example, a pair of configuration with the same energy and a difference not exceeding m can be regarded as coincident. Here m is some small integer such as 0, 1 or 2. If $m = 0$, then it is same as before. Let V_s be the effective volume of the configuration s, i.e. the total number of configurations differing from s by not more than m. That is to say, if s' and s differ by not more than m and have the same energy, s' will be in this volume (if $m = 0$, then $V_s = 1$). To calculate the coincidence probability we have to make correction with V_s. V_s must be calculated in the simulation programme and recorded for use. No other quantities need to be stored.

D. Limitation of the programme

How large is the entropy that can be calculated from the programme? How many elements are analysed each time? The answer depends on the problem under investigation. But we can estimate roughly its general limitation. If the trajectory has n configurations in k groups, and with n/k configurations in each group having rougly the same energy, then the number of coincidences in each group is approximately

$$N_c(E) \sim \frac{1}{2}\left(\frac{n}{k}\right)^2 \frac{v}{\Gamma(E)} . \qquad (25\text{-}22)$$

In this formula v is a typical value of V_s. The larger N_c is, the more accurate $\Gamma(E)$ will be. If we require $N_c(E) > 25$, the limit on the entropy is

$$S \sim \ln\left(\frac{v}{k^2}\right) + \ln\left(\frac{n^2}{50}\right) . \qquad (25\text{-}23)$$

To increase this limit we can increase v or decrease k, but it all depends on the actual problem. If v is sufficiently large the first term on the right of (25-23) is not important and the total number of configurations n appearing determines (25-23). If $n \sim 10^6$, $S \sim 23$. Even if n is considerably larger, S will not be much larger. If n is 1 000, S will not be much smaller. If each element has an entropy of 0.5, then this programme can test $N \sim 40$ elements. So this programme can only be applied to very small systems, or to small independent parts of a system. Of course, this requirement limits the application of this programme, but not in a serious way. The range of application is still quite extensive. For example, it has been used to calculate the entropy of the amorphous magnetic state.[b]

The above programme can be improved in many places, the most important being the choice of the elements of motion and the grouping so that uncorrelated elements are in different groups. In this way, the method of coincidences is most effective.

25.5. Entropy of the Metastable State

Many metastable systems have time scales spanning an enormous range, e.g. in H_2 gas, the spins (1 or 0) of H_2 are usually unchanged by collisions, so their correlation time is very long. The correlation time of the velocities of the molecules, i.e. the mean free time, is on the other hand quite small. So if the observation time lies between these two time scales, equilibrium properties will be well defined. We simply disregard variables with change over the longer time scale, but consider those changing over the shorter one. (See Chapter 8.)

---

[b] Ma and Payne (1980).

In many systems there is no sharp demarcation between the long and the short time scales; rather there are a number of time scales, long and short. This troublesome distribution of time scales usually appears in amorphous systems. The configuration space is divided into regions separated by barriers of different heights. Hence, the region in which the trajectory is distributed depends on the length of the observation time and so does the entropy. We can make an estimate of how the entropy changes with this time scale. To penetrate a barrier of height Λ, we need a time of

$$t \sim \tau_0 \, e^{\Lambda/T} \quad , \tag{25-24}$$

where τ_0 is a short time scale, i.e. the correlation time without barrier. When $t \gg \tau_0$, barriers higher than

$$\Lambda_c \sim T \ln(t/\tau_0) \quad , \tag{25-25}$$

are impenetrable and those lower than Λ_c are easily overcome. So the value of the entropy changes logarithmically with the observation time. This change is very small, especially at low temperatures. Thus entropy is still unambiguously defined.

25.6. The Third Law of Thermodynamics

In thermodynamics, the entropy difference between two equilibrium states 1 and 2 is

$$S_2 - S_1 = \int_1^2 \frac{dT}{T} \frac{dQ}{dT} \quad , \tag{25-26}$$

where 1 and 2 must be linked by a reversible process and dQ/dT is the heat capacity determined experimentally along this process. Apart from an integration constant, (25-26) can be used as an experimental definition of entropy. This constant can be determined by the entropy of the high temperature gaseous state (by the theory of ideal gas) or from the fact that the entropy at $T = 0$ is zero (the third law of thermodynamics, see Chapter 23).

The heat capacity of most models can be calculated by numerical simulation. Using (25-26), we can calculate the entropy. Therefore entropy and the trajectory are connected. But this calculation is different from the above discussion. It requires a series of trajectories for the intermediate equilibrium states and not only those under observation, so it does not fulfill the goal of calculating the entropy from the observed trajectory corresponding to a given equilibrium state.

In (25-26), the process between 1 and 2 must be reversible. Otherwise (25-26) will not be defined. This condition is quite stringent. An example of an irreversible process is one in which the trajectory does not return to the original region when the temperature is first raised and then lowered to the original value. Sometimes irreversible phenomena are not too obvious, e.g. glass can have many different structures; after heating and recooling. Nevertheless these different structures exhibit the same macroscopic properties and are difficult to distinguish.

Of course, we can disregard the question of reversibility, and use (25-26) as a definition of the entropy. This definition is commonly used in discussing glasses.

We can determine the constant of integration by the high temperature entropy. If between the high temperature and $T = 0$ some irreversible processes occur, then the entropy obtained by integration may not be zero at $T = 0$, violating the third law. This is the so-called "frozen entropy". The entropy calculated from the trajectory cannot violate the third law because it directly measures the region of motion in phase space. When $T \to 0$, motion stops and this region will contract to a point. The fact that motion stops is the basic meaning of the third law.

Hence if some irreversible processes occur, then at some temperatures the entropy defined from the trajectory may be different from that defined by the specific heat (25-26). This difference is a measure of irreversibility. Notice that if an amount of heat dQ is added, the entropy (25-26) changes by dS':

$$dS' = \frac{dQ}{T}. \tag{25-27}$$

This is the definition of S in (25-26). But the change dS of the entropy defined from the trajectory may be larger,

$$dS \geq \frac{dQ}{T}, \tag{25-28}$$

because the process of heating may be irreversible.

Notice that in cooling, i.e. $dS < 0$, the decrease of entropy is larger than dQ/T, i.e. the reverse (25-28). As temperature is lowered, the time scale is changed because of the restriction of the trajectory. This does not violate the second law. If the entropy decreases upon cooling, but increases correspondingly upon heating, it is impossible to transform heat into work by this process.

Problems

1. The degree of randomness of a distribution of points on a plane or any continuous space can be estimated by the coincidence method. We see from Fig. 25-1 that Γ is proportional to the number of squares (see Eq. (25-1)) and inversely proportional to the size v of each square. Let $S(v) = \ln \Gamma(v)$

   (a) Plot $S(v)$ against $\ln(1/v)$. If the distribution is completely random, what should the graph be? (Answer: A straight line with gradient 1.)

   (b) Use this method to test the randomness of a random sequence generated by a computer.

   (c) Use this method to test the sequence.

$$X_m = \sin(2^m) \quad , \quad m = 1, 2, 3, \ldots \quad . \tag{25-29}$$

Note that if there are n points and M squares labelled by $i = 1, 2, \ldots, M$, with n_i points in square i, then N_t and N_c in (25-1) are

$$N_t = \frac{1}{2} n(n-1) \quad ,$$

$$N_c = \frac{1}{2} \sum_{i=1}^{M} n_i (n_i - 1) \quad . \tag{25-30}$$

Note that 2^m in (25-29) is really modulo 2π. It is intuitive that these remainders are distributed uniformly between 0 and 2π. So the x_m in (25-29) will not have a uniform distribution. Review Problems (11.3) and (11.4).

2. Purely from the viewpoint of Newtonian mechanics, phase space is continuous. Without quantum mechanics, entropy must be defined by drawing squares. So the value of the entropy is related to the size of the squares. But if the size of the squares is chosen suitably so that the entropy is related to the size of the square in the manner of Problem 1, the definition of entropy is unambiguous. Discuss this issue. Note that Newtonian mechanics must at least, within a certain scale, determine a sequence of independent trajectories so that a group of random points can be selected.

3. The details in Sec. 25.4 are for demonstration only and are not efficient. Computer specialists have many methods for comparing numbers. For example, given a sequence of numbers A_i, $i = 1, 2, \ldots, n$, there is the ordering programme to arrange these numbers in order of their magnitude. Identical numbers then appear in neighbouring positions. This type of programme requires a time proportional to $n \ln n$.

   (a) How does one represent the configuration of a system as a number?

   (b) How can one obtain the coincidence probability by making use of this programme?

   Discuss various details.

438   STATISTICAL MECHANICS

4. There is another method to calculate entropy from the trajectory. The derivation is as follows. Let s_i, $i = 1, 2, \ldots, N$ be the variables of motion and $P(s_1, s_2, \ldots, s_N)$ be the joint probability of these variables. Write the entropy as

$$S_N = - \sum_{s_1} \sum_{s_2} \ldots \sum_{s_N} P \ln P \; . \tag{25-31}$$

(a) Let

$$P(s_1, s_2, \ldots, s_N) = p(s_1 | s_2, \ldots, s_N) P(s_2, \ldots, s_N) \; , \tag{25-32}$$

where $p(s_1 | s_2 \ldots)$ is the conditional probability of s_1 with $(s_2, \ldots, s_N)$ fixed, and $P(s_2, \ldots, s_N)$ is the joint probability of $(s_2, \ldots, s_N)$.
Prove that

$$S_N = - \sum_{s_1} \sum_{s_2} \ldots \sum_{s_N} P(s_1, s_2, \ldots, s_N) \ln p(s_1 | s_2, \ldots, s_N) + S_{N-1} \; , \tag{25-33}$$

where S_{N-1} is the entropy of $(s_2, \ldots, s_N)$. If s_1 is related only to a small number of neighbouring elements $s_2, \ldots, s_m$, then

$$p(s_1 | s_2, \ldots, s_N) = p(s_1 | s_2, \ldots, s_m) \; , \qquad m \ll N \; . \tag{25-34}$$

(b) Let $p(s_1, s_2, \ldots, s_m)$ be the joint probability of a small number of m elements. Prove that

$$S = - \sum_{s_1} \ldots \sum_{s_m} p(s_1, s_2, \ldots, s_m) \ln p(s_1 | s_2, \ldots, s_m)$$

$$- \sum_{s_2} \ldots \sum_{s_m} p(s_2, s_3, \ldots, s'_m) \ln p(s_2 | s_3, \ldots, s'_m) + S_{N-2} \; . \tag{25-35}$$

(c) In the above, each step separates an element. Finally each term of S contains only conditional probabilities, related to a small number of neighbouring elements. If these conditional probabilities can be calculated from the trajectory then S can be calculated.
Apply the steps of (a) and (b) to the one-dimensional Ising model.

(d) Analyse the two-dimensional Ising model. If each conditional probability is the probability that the four neighbours are fixed, then this method is quite ideal. But we have to include more neighbours, and the conditional probability will include more elements further away, but within the correlation distance.

(e) Discuss the advantages and disadvantages of this method. This method has an extensive range of applications and originally it was not used in numerical simulation. The original derivation is somewhat different from that given above. See Kikuchi (1955), Alexandrowicz (1975) and Mierovitch (1977).

5. Apply the definition of entropy in this chapter to the example of the viscous liquid bottle. What is the trajectory of the dye molecules? How is the coincidence probability defined? Start from Problem 1 of the previous chapter.

6. The various parts of a system being more or less independent is a necessary condition for the definition of entropy. But these parts need not be divided according to position. We need only divide the N variables into m groups, $m = O(N)$, with each group being approximately independent.

(a) Discuss the ideal gas case. (Hint: group according to momentum.)
(b) Discuss the vibrations of a crystal.

Chapter 26
THE ORIGIN OF THE BASIC ASSUMPTION

The development of statistical mechanics is very strange. As mentioned in the preface, it is rather like an inverted pyramid. Starting from the basic assumption the development is upwards. Specialists of physics, chemistry and materials science, etc. apply it to various models, successfully explaining many phenomena. Mathematicians make it elegant and abstract and use rigorous proofs to derive some results. The development downwards is less remarkable. Physicists are certainly interested in the origin of the basic assumption, but are on the whole not enthusiastic in pursuing this problem because it is too difficult.

Physicists do not have the habit of avoiding difficult problems. Perhaps they are content with the success of the various applications and thus overlook the importance of understanding the origin. So the quest for the origin of the basic assumption, though certainly within the realm of physics, tends to fall into the hands of some great mathematicians. However, most mathematicians like elegant mathematics and they sometimes neglect some of the requirements of physics.

In fact the understanding of the basic assumption is not only a physics problem, but also an understanding of irregular phenomena in general. How do the regular rules of motion cause random results? This is a very common problem.

Matters of principle are difficult, but applications are not easy either. Although the applications are quite successful, most calculations involve approximations that may not be reliable. From the experience of the above chapters, whenever the interaction cannot be neglected, the problem becomes extremely difficult. This difficulty is linked to that encountered in the basic assumption.

In this chapter, we shall summarise our views on the basic assumption, i.e. our views on the bases of statistical mechanics. We then discuss the concepts of ergodicity, ensembles and mixing. Then we look into and criticise some conventional views. The instability of the trajectory is also introduced here.

The discussion in this chapter cannot be said to be an objective view of statistical mechanics, but is the subjective opinion of the author. These ideas, scattered in previous chapters, are summerised here.

26.1. The Basic Assumption

We now review the basic assumption as discussed in the previous chapters. We classify the variables of motion into two types: (1) the invariants, i.e. variables unchanged during the observation time, and (2) the rapidly changing variables. Then we define the region of motion as a set including all the configurations of the changing variable, i.e. the set of configurations satisfying the conditions of invariance. We have emphasised repeatedly that invariance means unchanging quantity during the observation time. What variables are invariant depends on the time scale and in some cases it is not obvious how to classify the variables. The strict definition of the region of motion depends on the detailed trajectory of motion.

Let Γ be the volume of the region of motion, i.e. the total number of configurations of the region. The basic assumption says that entropy is the logarithm of Γ:

$$S = \ln \Gamma \ . \tag{26-1}$$

This assumption must be supplemented by the following condition

$$S = O(N) \ , \quad \text{i.e.} \quad \Gamma = O(e^N) \ , \tag{26-2}$$

where N is the particle number or the total number of variables. From (26-1) and (26-2) the entire theory of thermodynamics can be deduced and all the equilibrium properties can be calculated. This is the conclusion of Chapter 2 and the content of Chapter 23. Notice that condition (26-2) is most important. Without it there would be no thermodynamics. After Chapter 4 we have come to regard this condition as that establishing the independence of the various variables of motion. Hence (26-2) is the hypothesis of independence. In discussing the central limit theorem (Chapter 12, and the end of Chapter 6), we pointed out that the time average of large value $(O(N))$ variables is the same as the average value in the region of motion. Secondly, the fluctuations of these variables are of $O(\sqrt{N})$ and normally distributed. This conclusion can be written as

$$P(R) \equiv \frac{\mathcal{J}(R)}{\mathcal{J}} = \frac{\Gamma(R)}{\Gamma} , \qquad (26\text{-}3)$$

where $\mathcal{J}$ is the total observation time and R is a subset whose volume $\Gamma(R)$ is comparable with Γ:

$$\Gamma(R) = O(e^N) . \qquad (26\text{-}4)$$

$\mathcal{J}(R)$ is the time that the trajectory stays in R. These also follow from (26-2). (See Chapter 12.) $P(R)$ can be interpreted as the probability of the configurations being in R. Notice that $\mathcal{J}$ is a finite time and cannot be regarded as a large number of $O(e^N)$. R must be a large subset. Otherwise (26-3) is meaningless. (See Sec. 12.9.) The fluctuation of a large variable is directly related to the differentials of the thermodynamical potentials. The term "large variable" here denotes the sum of the variables describing various parts of the system, e.g. the total magnetic moment is the sum of the various magnetic moments. Its average value may be zero. This type of variable also includes variables like

$$\int d^3 r \, \rho(r) \, e^{-i \mathbf{k} \cdot \mathbf{r}} .$$

It is the sum, over various parts of the system, of the product of the density with $e^{-i\mathbf{k}\cdot\mathbf{r}}$. To sum up, the above basic assumption is sufficient to analyse all experiments of thermodynamics and scattering.

Hence we see that (26-1) is a daring assumption and it links entropy with the microscopic motion through Γ. From the deductive viewpoint, (26-2) is the origin of this assumption. To understand statistical mechanics, we must first understand (26-2).

We now mention here ergodicity and ensembles, two most important concepts in the development of statistical mechanics.

26.2. Ergodicity and Ensembles

The above basic assumption is not too harsh a condition because we discuss only large variables and large regions. The reader may notice that this basic assumption may appear in different forms in the literature. When Boltzmann wrote down this assumption, his thinking went like this: the trajectory will pass through every configuration in the region of motion, i.e. it is ergodic, and hence the infinite time average is equal to the average in the region. This argument is wrong, because this infinite long time must be much longer than $O(e^N)$, while the usual observation time is $O(1)$. To require the time average of every variable to equal the average in the region is a harsh condition. We do not require this in this book, but we require that the time average of large variables equal the average over the region. This condition is not too harsh because the values of large variables are nearly the same everywhere in the region of motion and the trajectory need not pass through every configuration, i.e. it need not be truly ergodic.

The ergodicity of the trajectory is a major problem in mechanics.[a] Many mathematicians have devoted much effort in proving the ergodicity of some models. However these results are not too helpful to statistical mechanics.

The reader must have noticed that the definition of terms is not uniform in the literature. The term "ergodicity" may have different definitions in different papers. In some places the definition is as above. In other places it means that the trajectory "nearly" passes through every point. Others may say that whenever the basic assumption holds the motion is ergodic. This is a mess. The reader should be clear about the definition in any paper before delving into it.

The ergodic theory of Boltzmann met with immediate criticism and doubt was cast on his theory including the H-theorem. Boltzmann perhaps died of the resulting frustration.

The theory of Boltzmann, reformulated by Gibbs in terms of "ensembles",[b] forms the main trend of modern statistical mechanics.

The theory of ensemble says that the equilibrium states of the system form an ensemble, i.e. a system with infinitely many similar structures. Each configuration of the system is a point in phase space. These points distributed in phase space can be thought of as an ideal gas (each point is a "molecule"). We need at least $\sim e^N$ molecules to define the distribution.

Properties of the equilibrium states are average properties of this ensemble. This abstract way of thinking has many advantages, because we have a rather strong intuition about the flow and distribution of a gas. The ideas of Gibbs produce many useful formulas (equivalent to the calculational rules of Chapter 7) and applications. The ensemble theory has become the traditional basic concept.

Nevertheless the ensemble is an abstract artifact. In reality there is one system, not infinitely many. This was pointed out by Gibbs in his book, but is usually neglected nowadays. Ensemble is a mathematical concept and it cannot

---

[a] See Arnold (1968), Chap. 6 of Yang (1978) and other books on Ergodic Theory.
[b] Gibbs (1960).

solve the physical problem of Boltzmann's assumption. If the ergodic property proposed by Boltzmann is said to be unrealistic then ensemble theory is perhaps more so. The content of the ensemble theory is identical to the assumption of Boltzmann. It distributes the ensemble in the region of motion. The ensemble average is the average over the region of motion.

After Gibbs, many scholars have tried to axiomatise this theory of ensembles or at least give it a more definite meaning. Today the most commonly held idea is to define the ensemble as the knowledge of a system by the experimenter,[c] or "the degree of uncertainty". This degree of uncertainty is regarded as an *a priori* probability; that is to say, before the observation, the configurations of the system have a probability distribution. We have criticised this view in the above chapters (especially Chapter 24). In the next section we shall compare and discuss this view with that in our book. Here we first mention some basic concepts of the ensemble theory.

The ensemble is like a cloud in phase space and its density can be represented by a distribution function ρ. Now phase space is a $6N$ dimentional space of N particles all of which obey Newton's laws. Gibbs discussed the change of ρ and defined entropy as

$$S_G = - \int \rho \ln \rho \, d\Gamma \quad , \tag{26-5}$$

where ρ is normalised $\int \rho \, d\Gamma = 1$. The integral in (26-5) is over the whole of phase space.

The motion of each system in the ensemble obeys Newton's law, tracing out a trajectory. The trajectories of different systems are non-overlapping. (If there is intersection, that would mean different trajectories can come from the same initial conditon.) The reader has probably come across Liouville's theorem in mechanics, i.e. ρ is unchanged along the trajectory. Let us fix a region $R(0)$ in phase space and let every point in this region move according to the laws of mechanics. After a time t they form another region $R(t)$. According to Liouville's theorem the volumes of $R(0)$ and $R(t)$ are the same; only the shape changes. From this we can prove that the entropy defined by Gibbs in (26-5) is an invariant. Obviously (26-5) cannot be used to discuss phenomena of entropy increase. Many people tried to explain and patch up this unreasonable conclusion of constant entropy. Suppose $R(0)$ is a region like a lump of dough. After a long time, $R(t)$ will be stretched out like intertwining noodles made from the dough. If at $t = 0$, ρ is nonzero only in $R(0)$, then ρ is nonzero

[c] See Tolman (1962).

only in the noodles at time t. The expression in (26-5) is the volume of the dough or noodles, but it does not represent the randomness of these noodles or threads. We do not discuss the correction to (26-5) because it is built upon an incorrect theory. The equilibrium state is not an instantaneous concept and entropy needs a time scale to be defined. This point is emphasised from the beginning of this book. The ensemble theory attempts to use the instantaneous properties of infinitely many systems to represent equilibrium. This is an unreasonable approach. We will discuss it carefully in Sec. 26.5.

26.3. Mixing and Independence

Many scholars in the study of modern statistical mechanics devote themselves to the analysis of mixing. Mixing implies that the noodles or threads $R(t)$ will distribute uniformly in the region of motion. If $t \to \infty$, then

$$\lim_{t \to \infty} \frac{\Gamma(R(t) \cap R')}{\Gamma} = \frac{\Gamma(R(0))}{\Gamma} \cdot \frac{\Gamma(R')}{\Gamma} , \qquad (26\text{-}6)$$

where R' is any subset of the region of motion, Γ is the total volume of the region of motion, and $\Gamma(R')$ is the volume of R'. That is to say, the volume of $R(t)$ in R' is proportional to the volume R'.

The property of mixing is more reasonable than that of ergodicity. It requires that any region $R(0)$ be transformed to a patch of fine threads spreading uniformly over the region of motion. The analysis of mixing is a complicated mathematical problem and the author is no expert in this field. There are some recent books on this aspect.[d]

The theory of mixing is commonly accepted now. It is more advanced than the theory of Boltzmann or Gibbs, and it can be said to have descended from Gibbs' theory.

Now we rewrite the view adopted in this book in a form similar to that of mixing and then make a general discussion. The view of this book is: the basic assumption is an assumption of independence, i.e. an assumption on the correlation time and the correlation length.

In Sec. 26.1 we pointed out that the independence property (26-2) is the root of the basic assumption. The calculation of entropy from the trajectory discussed in the last chapter shows that the region of motion must be defined and understood from the trajectory, and the independence property is a necessary property for defining entropy from the trajectory. This independence property

---

[d] See Sinai (1979) and Krylov (1979).

is the independence beyond long time scales and long length scales. That is to say, if we have correlation time τ and correlation distance ξ and $|t_1 - t_2| \gg \tau$ or $|\mathbf{r}_1 - \mathbf{r}_2| \gg \xi$, then

$$\langle A(\mathbf{r}_1, t_1) B(\mathbf{r}_2, t_2) \rangle$$
$$\equiv \frac{1}{\mathfrak{J}} \int_0^{\mathfrak{J}} dt \, A(\mathbf{r}_1, t+t_1) B(\mathbf{r}_2, t+t_2) \longrightarrow \langle A(\mathbf{r}_1) \rangle \langle B(\mathbf{r}_2) \rangle \quad .$$
(26-7)

The average value is the time average over the observation time $\mathfrak{J}$. A small change of $\mathfrak{J}$ will not affect the average value. The variables A and B are local variables of motion near $\mathbf{r}_1$ and $\mathbf{r}_2$; "near" means distance smaller than ξ, and "local" means related to a small number of variables of motion.

If we can start from the principles of mechanics to derive (26-7), then the region of motion can be determined, (26-2) is established, and the whole basic assumption is built up. In the above, we have not supplied the mathematical steps to obtain (26-2) from (26-7). These steps are probably not easy, or may require other conditions. The main points here is to understand (26-7), i.e. how the independence property follows from mechanics.

What is the difference between mixing and independence? We look at a simple example here.

Consider two one-dimensional models: **(A)** a particle restricted in a rigid box between $0 < x < L$ and **(B)** a particle attached on a spring and executing simple harmonic motion with frequency ω. Let us look at the property of mixing in these two examples.

The black disc region in Fig. 26-1(a) becomes the black threaded region in Fig. 26-1(b) after a long time. The longer the time, the finer and denser the threaded region becomes. The total area of the black region is unchanged, but the black disc is stretched to threads. Obviously these black threads are distributed uniformly in the region

$$a < |p| < b \quad , \qquad 0 < x < L \quad , \tag{26-8}$$

where $\Delta p = b - a$ is the diameter of the black disc. This uniform distribution occurs when $t \gg \tau$,

$\tau =$ time interval to traverse length of box .

We can say that this model has the mixing property.

THE ORIGIN OF THE BASIC ASSUMPTION 447

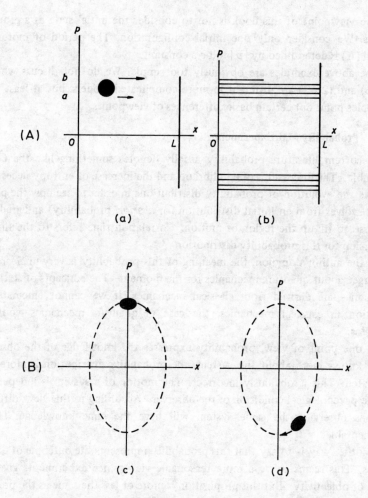

Fig. 26-1

Model (B) is completely different. The black disc in Fig. 26-1(c) goes in an elliptical path with slightly changing shape but never stretches into threads. No matter how long the time is, situations such as Fig. 26-1(b) cannot occur. (See Fig. 26-1(d).) Therefore, this model has no mixing.

So, from the point of mixing, these two models are completely different. But, from the point of independence these two models have no great difference, i.e. neither shows the independence property. Both are periodic motion without chaos.

The viewpoint of this book is not to consider the initial state as a group of points. We consider only one initial configuration. The period of motion of model (A) is determined by $|p|$ to be a constant.

The above examples are obviously too simple. We do not discuss whether (26-6) and (26-7) are different in more complicated models, but at least these examples point out certain basic differences of viewpoints.

26.4. Probability and Experiments

In current literature probability usually denotes something like the Gibbs' ensemble. The discussions of equilibrium and the increase of entropy necessarily involve the evolution of probability distributions in order to see how the probability evolves from an initial distribution (or *a priori* probability) and gradually disperse to fill up the region of motion. Correlation time refers to the time of dispersion for this probability distribution.

In the author's opinion, the meaning of this probability is very unclear. Let us forget about quantum mechanics for the moment. The concepts of statistical mechanics are derived from classical mechanics. If we cannot elucidate the situation in classical mechanics, the case of quantum mechanics would be hopeless.

In one point of view, probability expresses the knowledge of the observer. If he knows more about the system, the probability distribution is more concentrated. This is obviously incorrect. The motion of a system is independent of the psychological condition of the observer. According to this view, different persons observing the same system will have the same knowledge. This is unacceptable.

Another way is to say that this probability represents the outcome of experiments. This seems to be a more reasonable view, since experiments give us a sense of objectivity. But this proposition is more or less the same as the previous one. The question is: What is the experiment? How many experiments must be performed? The answers of "macroscopic experiments" or "feasible experiments" are incorrect. In the discussion of fixed impurities and the analysis of echo phenomena in Chapter 24, the position of each impurity atom is known. Is this a feasible experiment?

The experiments discussed in the literature on statistical mechanics are usually unrelated to actual experiments. For example, the canonical ensemble represents a fixed temperature for which the experimenter should be able to measure energy to an accuracy of $O(1/\sqrt{N})$. The other large variables are also determined to $O(1/\sqrt{N})$. In practise $N \sim 10^{22}$. Therefore $N^{-\frac{1}{2}} \sim 10^{-11}$. No experiments have been performed to such accuracy. In addition if the probability tends to zero, then the event is explained to be not happening or not observed in experiments. This too is untenable, e.g. any metastable states such as diamond have a probability tending to zero. Yet the systems encountered in the laboratory are usually metastable.

One main weakness in the literature is to regard the region of motion or the equilibrium probability distribution as something known. The total energy function H is fixed, and then the equilibrium probability can be simply determined, treating every configuration with the energy E as having the same probability. This manner of analysis is a habit acquired from analysing the gaseous state. (This book has repeatedly emphasised that the determination of the region of motion is not so simple and it depends on the actual distribution of the trajectories.) This habit is a hindrance to the analysis of the logical roots of statistical mechanics.

The author thinks that Gibbs' ensemble and probability have no relation with mechanics. From the viewpoint of statistical mechanics, they are both unnecessary and unrealistic. These concepts are helpful for our abstract analysis, but they are unrelated to any discussion on the logical roots of statistical mechanics. If we want a correct analysis of these roots, we have to get rid of this concept of ensemble probabiltiy.

Modern computers can calculate the trajectory of some simple models and can calculate the various thermodynamical properties from the trajectory. (See Chapter 22 and Chapter 25.) So the problem of experiments can be separated from that of the logical roots. These computer calculations are purely objective. Of course, the understanding of these simple models does not form the basis of statistical mechanics. But if we cannot even understand these models, the future does not look bright. The numerical solution requires Newton's laws but not the concepts of ensemble or probability.

The definition of probability given in this book is quite different. Probability is looked upon as a tool for arranging the information and not a physical concept. It is not a necessity but a convenience. We must first have numbers and information before we can do statistics, classify the states, make ensembles, calculate the probability and discuss equilibrium or nonequilibrium. All the properties must come from the trajectory.

We have performed a critical analysis of the concepts of ensemble and probability as presented in the literature. Our aim is to clarify the most basic concept. But the above criticism is not meant to denigrate the achievements attained in the literature. As mentioned above, we have a strong intuition about space and the flow and dispersion of a cloud in space. The concept of ensemble gives us a new tool and quite an elegant one. We cannot avoid assumptions of mathematical

450 STATISTICAL MECHANICS

concepts, but we have to clarify our concepts and make effective use of it.

26.5. Instability of the Trajectory

To understand mixing or independence from the laws of mechanics is a very difficult task. Equations (26-6) and (26-7) seem to be rather remote from the details of the trajectory. We now discuss instability and its details. The meaning of instability is as follows. Let $s(t)$ be the trajectory. If the initial state $s(0)$ is changed by a bit to $s(0) + \delta s(0)$, then the change of the trajectory is $\delta s(t)$. If

$$\delta s(t) \propto e^{\alpha t} , \qquad (26\text{-}9)$$

with $\alpha > 0$, the trajectory is said to be unstable. If there are many different $\delta s(0)$, forming a small cloud surrounding $s(0)$, then this cloud will disperse, conforming to the requirement of mixing.

In model (A) of Sec. 26.3, if initially momentum is increased by δp, then

$$\delta x(t) = \frac{\delta p}{m} t . \qquad (26\text{-}10)$$

There is also instability, but not as serious as (26-9).

Now let us look at a more complicated example. Suppose a particle collides with a number of fixed hard spheres, as in Fig. 26-2. We consider only a model with planar motion. The configuration of this particle is determined by its position and velocity. The magnitude of the velocity is unchanged, so we only consider the direction of the velocity. Starting from the origin, if the direction makes a small change $\delta\theta(0)$ the trajectory will change, and $\delta\theta$ changes for every collision. Assume that the density of the hard spheres is very low and the mean free distance λ of the particle is much larger than the radius a of the spheres. Figure 26-2 shows the situation after two collisions. As seen from the figure

$$\delta\theta' = 2\delta\alpha = 2\lambda\delta\theta/a \cos\alpha . \qquad (26\text{-}11)$$

Therefore $|\delta\theta'| \geq (2\lambda/a)|\delta\theta|$, and after n collisions

$$|\delta\theta(t)| \geq |\delta\theta(0)| e^{t/\tau} ,$$

$$\tau \equiv \lambda \bigg/ \left(v \ln \frac{2\lambda}{a} \right) , \qquad (26\text{-}12)$$

THE ORIGIN OF THE BASIC ASSUMPTION 451

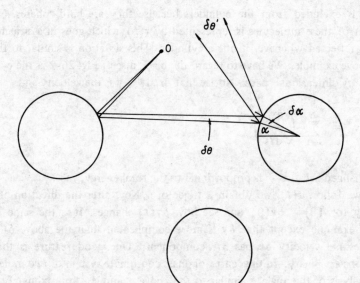

Fig. 26-2 Collision of a particle with fixed hard spheres.

where v is the speed of the particle. This result is the same as (26-9). If the time is sufficiently long, i.e.

$$t \geqslant \tau \ln(1/|\delta\theta(0)|)$$

$$= \frac{\lambda}{v} \frac{\ln|1/\delta\theta(0)|}{\ln 2\lambda/a} \quad , \tag{26-13}$$

then $|\delta\theta(t)| \sim 1$, i.e. the direction is randomised. The above analysis can be extended to the gas model with similar results. We mention it briefly here.

Suppose we have N small hard spheres of radius a and the density of the molecules is very low. Let $\mathbf{r}_i, i = 1, 2, 3, \ldots, N$ be the positions of the spheres. Let

$$\mathbf{r} \equiv (\mathbf{r}_1, \mathbf{r}_2, \ldots, \mathbf{r}_N) \quad , \tag{26-14}$$

where $\mathbf{r}$ lies in a $3N$ dimensional space. In this space there are $\frac{1}{2}N(N-1)$ cylinders:

$$|\mathbf{r}_i - \mathbf{r}_j| \leqslant a \quad , \qquad i, j = 1, 2, \ldots, N \quad , \tag{26-15}$$

and **r** is excluded from the cylinders because they are hard spheres. So the motion of these molecules is represented by $\mathbf{r}(t)$, which goes in a straight line and is reflected whenever it hits a cylinder. This situation is similar to that of the above example. We have to regard the black disc of Fig. 26-2 as the cylinder in the $3N$ dimensional space. Notice that $d\mathbf{r}/dt = \mathbf{v}$ is the velocity and

$$v^2 = \sum_{i=1}^{N} \left(\frac{d\mathbf{r}_i}{dt} \right)^2 , \qquad (26\text{-}16)$$

is unchanged because v^2 is proportional to the total energy.

If we follow $\mathbf{r}(t)$, we obtain a trajectory. Now alter the direction of $\mathbf{v}(0)$ slightly to $\mathbf{v}(0) + \delta\mathbf{v}(0)$, and see how $\delta\mathbf{v}(t)$ changes. It is the same as the above example except that $\delta\mathbf{v}$ is more complicated than the above $\delta\theta$. The infinitesimal velocity $\delta\mathbf{v}$ has $3N$ components. One good feature is that the collisions are binary. In the centre-of-mass coordinate system of two molecules, the analysis of the angle is similar to the above example. Approximately, after each collision $|\delta\mathbf{v}|$ will increase by a factor λ/a, where λ is the mean free distance. But collisions with different molecules will change the direction of $\delta\mathbf{v}$. The changes of $\delta\mathbf{v}$ in collisions with different molecules are mutually perpendicular. For example, in a collision of molecules 1 and 2, there are changes of $\mathbf{v}_1$ and $\mathbf{v}_2$, so

$$\delta\mathbf{v}_{12} = (\delta\mathbf{v}_1, \delta\mathbf{v}_2, 0, 0, \ldots, 0) . \qquad (26\text{-}17)$$

If molecules 7 and 8 collide, then only $\delta\mathbf{v}_7$ and $\delta\mathbf{v}_8$ are nonzero.

$$\delta\mathbf{v}_{78} = (0, 0, \ldots, \delta\mathbf{v}_7, \delta\mathbf{v}_8, 0, 0, \ldots, 0) . \qquad (26\text{-}18)$$

Therefore,

$$\delta\mathbf{v}_{12} \cdot \delta\mathbf{v}_{78} = 0 . \qquad (26\text{-}19)$$

This is similar to the analysis in Sec. 5.9.

So, after n collisions $\delta\mathbf{v}(t)$ becomes the sum of n vectors. These n vectors are mutually perpendicular and

$$n \sim \frac{t}{\tau_0} N ,$$

$$\tau_0 \equiv \lambda/v , \qquad (26\text{-}20)$$

THE ORIGIN OF THE BASIC ASSUMPTION

where v is the average velocity of the molecules. The magnitude of each vector is about

$$\delta v_i(t) \sim \left(\frac{\lambda}{a}\right)^{t/\tau_0} \delta v_i(0) \quad . \tag{26-21}$$

Notice that t/τ_0 is the number of collisions of each molecule. The correlation time can be estimated as $\tau_0 / \ln(\lambda/a)$ similar to (26-12). As the vectors are mutually perpendicular, they form a volume

$$\prod_i \delta v_i(t) \sim \left(\prod_i \delta v_i(0)\right)\left(\frac{\lambda}{a}\right)^{3Nt/\tau_0} . \tag{26-22}$$

The increase of this volume is exponential:

$$\left(\frac{\lambda}{a}\right)^{3Nt/\tau_0} = \exp\left[3N(t/\tau_0)\ln(\lambda/a)\right] \quad . \tag{26-23}$$

The space of **v** is momentum space $\mathbf{p} = (\mathbf{p}_1, \mathbf{p}_2, \ldots, \mathbf{p}_N)$ as $\mathbf{p} = m\mathbf{v}$. The above analysis shows how the volume expands. Notice that (26-23) is a large number of magnitude e^N. The above is only a crude analysis.[e]

The above analysis explains approximately the meaning of instability. If the trajectory is unstable, then the trajectory is not easily controlled and extremely sensitive. If the initial state differs a bit, the final outcome will be completely different. If the trajectory has the property of instability, there will be no problem in mixing. Notice that instability is a result of mechanics and does not involve the concept of probability. To use it to discuss mixing we must first assume that there is an initial probability distribution.

In this kind of analysis it is very easy to allow the assumptions of mixing and independence to creep in unnoticed. For example, in (26-13) we assume that after $|\delta\theta(t)|$ equals 1, it will be "unrelated" to $\delta\theta(0)$ and hence we can disregard $\delta\theta(0)$. In fact the above analysis does not say how $\delta\theta(t)$ changes when it becomes very large. Equation (26-12) is appropriate only when $\delta\theta(t)$ is very small. When $\delta\theta(t)$ becomes large, we are ignorant and naturally there is no reason to draw any conclusion. This analysis cannot be used as a proof of mixing or independence. Equation (26-12) or (26-13) can only be regarded as an estimate, assuming that $\delta\theta(t)$ will be unrelated to the situation at $t=0$ or independent when $\delta\theta(t)$ grow large. However, this way of analysing enables us to have a deeper understanding of the characteristics of the trajectory.

---

[e] For details, see Krylov (1979), p. 193. It is detailed but not quite rigorous.

Part of the contents of this section came from inspiration while reading Krylov's book. The discussion by Krylov is very penetrating but the views of the author do not quite agree with those of Krylov.

26.6. The General Problem of Independence

To summarise: From the point of view of physics, to understand statistical mechanics we must understand

(a) the existence of a correlation time $\tau = O(1)$. That is to say, the correlation time must not only be finite, but of $O(1)$ and not a large number of $O(e^N)$;

(b) the existence of a correlation distance $\xi = O(1)$. That is to say, a system can essentially be regarded as composed of independent parts. The correlation distance ξ must also be of $O(1)$ and cannot be a large number of order N.

These two points are the contents of the independence condition (26-7).

In recent years, the research of mechanics in the direction of random trajectory has made great advances. Many simple models (N very small, no more than 3 or 4 variables) exhibit random (i.e. independent in time) motion. These works show that random motion is a result of Newton's laws. However, up to now, the values of N investigated is still very small and is not sufficient to discuss the independence of the parts of a system, i.e. the existence of the correlation length ξ. When N is very small, the study of random motion is in the field of numerical simulation and has not reached a profound understanding of the problem. The author thinks that the research should be directed to the influence of the value of N, especially when N is around 10. Large values of $N \geqslant 10$ may produce random motion different from that of small N.

Indeed, our understanding of the basic assumption is still very primitive.

Although to derive (26-7) from the principles of mechanics seems extremely difficult, this property of independence is actually not unexpected from our daily experience. Independence is "randomness" or "chaos". This has been discussed in Chapter 3. The property of independence is so common that it should not follow only from Newton's laws. For example, if we can understand random sequences we can understand more about the basic assumption. The random sequence is generated by a computer programme and should be easier to understand.

Why a random sequence is random at all is still unknown. From the molecular motion to the throwing of a die, to the random sequence and the stock market, all these random phenomena are not understood. The cause of chaos seems to relate to the degree of complexity, but simple models can still give rise to chaos. These problems are waiting for an answer.

26.7. Difficulties Encountered in Applications

From the examples of the above sections we see that the application of statistical mechanics is an extremely difficult mathematical problem. If we cannot transform a model to one similar to the ideal gas, we cannot solve it. That is to say, we have to separate the variables of motion into mutually independent parts and then solve each part separately. In an ideal gas, each molecule is an independent part. In the crystal model, each elementary vibrating element (i.e. normal mode) is an independent part, etc.

If there are N variables, we have to look for the relation between them and put the correlated elements into a bunch, making the independence property obvious. In the various examples in this book, no matter how complicated the solution was (e.g. the exact solution of the two-dimensional Ising model), we had to separate out the independence property and the correlation property of the model, and then to solve it. At present the usual models are not tractable and have to be solved approximately or numerically.

The author thinks that the difficulty of solving problems is related to our lack of understanding of the independence property, i.e. as in the discussion of this chapter our lack of understanding of the origin of the basic assumption. The mutual interaction sets up the independence property, and is the origin of correlation. Correlation when mixed with independence makes the problem difficult to solve. From mechanics we cannot arrive at the independence property. But independence simplifies the application of the basic assumption.

At present, reliance on numerical solution is gradually increasing. As we said in Chapter 22, the numerical method is very important. But from the point of understanding the origin of statistical mechanics, this is perhaps not a good long-term solution.

Problems

1. As pointed out in Sec. 26.5, the correlation time of a gas is approximately equal to the mean free time of the molecules. This conclusion has some deficiencies. Try to find them from Prob. 23.1.

   This problem shows that in some special cases the correlation time is very long. These special cases are usually overlooked.

2. Apply the method of Sec. 26.5 to the model of crystal vibration.

   (a) Use a one-dimensional model; see Sec. 8.3. Use Eq. (8-41) as the interaction. First neglect the α' term and find the equation of motion.

   (b) Add the α' term and discuss its effect, assuming α' to be small.

   (c) Find the correlation time.

3. (a) Review Sec. 11.2. This section shows that it is not easy to determine whether a group of variables is independent.

(b) Review Secs. 12.5 to 12.8 about the central limit theorem, and the fact that the requirement of experimental observation is not strict independence but only some sort of approximate independence.

(c) Discuss the definition of entropy and the requirement of independence. What is the degree of independence necessary so that entropy has an unambiguous meaning? This question is not easy to answer.

4. Some problems of a rather philosophical nature appear in the literature of statistical mechanics, e.g. the arrow of time. From the law of mechanics we cannot determine whether time is flowing forwards or backwards. But in diffusion we know the arrow of time. In fact diffusion does not determine the arrow of time. It at $t = 0$ some solute molecules gather at a point in the solvent, after a time they will disperse. But suppose we ask: according to mechanics what is the distribution before $t = 0$? The answer is that the molecules are also dispersed. Therefore, the time direction has a symmetry, between past and future. The problem is not the direction of time flow but the dispersion of the distribution with $|t|$. Discuss this problem in the viscous liquid bottle demonstration in Chapter 24 (the dispersion of the dye molecules by turning the handle in its subsequent restoration).

5. Chaos and Prediction

Chaos and independence are properties that can be measured. We can test whether a sequence is random or not, if we calculate the correlation values from it. If we have n numbers in a random sequence, how do we calculate the next n numbers? This is a prediction problem.

According to experience the more random the number, the more difficult is the prediction. Of course if we can discover the algorithm used by the computer, then the numbers following can be calculated. This is the problem of decoding. What is the relation between the degree of randomness and the degree of difficulty of prediction? Try to discuss this.

Notice that one definition of randomness is the impossibility of prediction. This kind of definition has no relation with the discussion here.

PART VII
CONDENSATION

This part is divided into 4 chapters on topics related to the phenomena of condensation. Condensation is freezing, such as water changing to ice, or paramagnetism changing to ferromagnetism, etc. The coexistence of different phases discussed in Chapter 15 is the coexistence of different condensed states. The Ising model in Chapter 17 is one of the main models used to discuss condensation. The main point of this part is to introduce the order parameter, and discuss its properties. The coexistence of different phases can be regarded as the order parameter taking on different values. If one part of the system has a value of the order parameter different from that of the other part, then different phases coexist. For example a magnet may have half of its spins pointing up and the other half pointing down. (The order parameter here is the magnet moment.) If the temperature is high enough, the order parameter vanishes.

Chapter 27 discusses the mean field approximation. Its advantages are simplicity and clarity, expressing quite lucidly the common characteristics of various condensation phenomena. Unfortunately the approximation is rather crude, as it ignores the fluctuations of the order parameter and sometimes leads to erroneous conclusions. Chapter 28 considers the boundary between coexisting phases and analyses its fluctuation. Chapter 29 discusses cases with continuous symmetry, for which the order parameter can have continuous values. For example, given the regular arrangement of molecules in a crystal, if we displace all the molecules the thermodynamical potential is not changed. Displacement changes the order parameter and this is a continuous change. One of the results is the appearance of "soft modes", i.e. states of motion with very low energy, leading to large fluctuations of the order parameter. Chapter 30 discusses superfluidity, explaining it on the basis of condensation. The main points of this chapter are the appearance of the order parameter, the metastability of the superfluid state and the role played by the "defects" (vortices).

Chapter 27

MEAN FIELD SOLUTIONS

The mean field method is extremely important. In solving many relatively complicated problems, this is the only possible tool for analysis. Although crude, and even occasionally completely wrong, its simplicity gives it an important place in statistical mechanics. To illustrate this method, the Ising model is the best place to start. This chapter gives the method of calculation, and emphasises those points which are easily misunderstood. The mean field method assumes every element to be independent and that every element moves in a fixed environment which is determined by the average value of these independent motions.

The main model discussed here is the Ising model. We use the mean field approximation to analyse the properties of ferromagnetism and the coexistence of different phases. The various conclusions can be directly applied to the Ising gas model, thus explaining the phenomenon of the coexistence of liquid and gas. Van der Waals theory is also a mean field approximation, and we also discuss it in detail. The vector model can also be solved by the mean field method. The common features of various condensation phenomena are clearly exhibited in the mean field approximation.

27.1. Mean Field Theory

In Chapter 17 we have discussed the solution of the one-dimensional Ising ferromagnetic model. However, two-dimensional or three-dimensional Ising model are not easy to solve, so approximations become very useful. The mean field theory is the simplest approximation and quite easy to understand. Let the total energy be

$$H(s) = -\frac{1}{2} \sum_{i,j} s_i s_j J_{ij} - h \sum_i s_i \quad ,$$

$J_{ij} = J$ if i, j are nearest neighbours

$\phantom{J_{ij}} = 0$ otherwise , \hfill (27-1)

The environment of each element s_i includes the external magnetic field h and the magnetic field due to the nearest elements. The total magnetic field is

$$h'_i = h + \sum_j J_{ij} s_j \quad . \tag{27-2}$$

Because every element is changing continually, h'_i is changing too. The mean field method replaces s_i by its average value, getting a mean field

$$h' = h + Jv\langle s \rangle \quad . \tag{27-3}$$

As the average value of every s_i is the same, we have dropped the index i. The factor v in (27-3) is the total number of neighbours. Now every s_i is situated in a constant magnetic field h', so its average value is easily found to be

$$m \equiv \langle s \rangle = \frac{e^{h'/T} - e^{-h'/T}}{e^{h'/T} + e^{-h'/T}}$$

$$= \tanh\left(\frac{h + Jvm}{T}\right) \quad , \tag{27-4}$$

i.e.

$$h = T \tanh^{-1} m - Jvm \quad . \tag{27-5}$$

See Fig. 27-1 for the (h, m) curves.

If T is sufficiently large, the first term on the right-hand-side of (27-5) is larger than the second term, i.e.

$$\chi = \frac{\partial m}{\partial h} > 0 \quad , \tag{27-6}$$

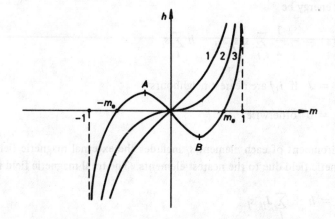

Fig. 27-1 (h, m) curves for three temperature values.

indicating paramagnetism, and m vanishes at $h = 0$ (curve 1). But if the temperature is too low, i.e.

$$T < T_c \equiv J\nu \quad , \tag{27-7}$$

one portion of the curve violates (27-6), and when h is not too large, m can assume three values (curve 3). If $T = T_c$ then χ is infinite when $h = 0$ and the curve touches the x-axis at the origin (curve 2).

When $T < T_c$ and $h = 0$ there are two nonzero solutions of m (see Fig. 27-2) i.e.

$$m = \pm m_0(T) \quad ,$$

$$m_0 = \tanh(m_0 T_c / T) \quad , \tag{27-8}$$

together with an $m = 0$ solution. When $m_0(T)$ is near 0 or T_c, the curve $m_0(T)$ is

$$m_0 \simeq 1 - 2e^{-2T_c/T} \quad , \qquad T \ll T_c \quad ,$$

$$m_0 \simeq \sqrt{3}\left(1 - \frac{T}{T_c}\right)^{1/2} \quad , \qquad T \simeq T_c \quad . \tag{27-9}$$

What is the meaning of these solutions?

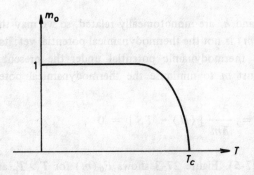

Fig. 27-2 Average value of spin as a function of temperature.

Now we return and examine the thermodynamical potential under this approximation. From the minimum of the thermodynamical potential we can explain the situation when $T < T_c$.

27.2. Total Magnetic Moment

The mean field solution can also be derived by maximising the entropy. This method works as follows. Use a simple energy function $H'(s)$ to replace $H(s)$, and use H' to calculate the entropy and $\langle H \rangle$. Then maximise the entropy under fixed $E = \langle H \rangle$, i.e. calculate the minimum of the thermodynamical potential $F = E - TS$ at constant temperature, so as to determine the best choice of H' (see Sec. 7.2(E)). Now we use

$$H'(s) = - \sum_{i=1}^{N} s_i h' \quad , \tag{27-10}$$

to calculate the entropy, regarding h' as a variational parameter. The entropy of this model is very easy to obtain:

$$S = \ln \frac{N!}{N_+! \, N_-!}$$

$$= \frac{N}{2} \left[(m+1) \ln \frac{2}{1+m} + (1-m) \ln \frac{2}{1-m} \right] \quad ,$$

$$F_0(m) = \langle H \rangle - TS \quad ,$$

$$\langle H \rangle = - \frac{1}{2} N v J m^2 - N m h \quad ,$$

$$\langle s_i \rangle = \tanh \frac{h'}{T} \equiv m \quad . \tag{27-11}$$

Notice that m and h' are monotonically related, so we may think of m as the parameter. $F_0(m)$ is not the thermodynamical potential yet; its minimum value is however the thermodynamic potential under the present approximation.

Now we adjust m to minimise the thermodynamical potential, and solve

$$\frac{\partial F_0(m)}{\partial m} = \frac{\partial}{\partial m} [\langle H \rangle - TS] = 0 \ . \tag{27-12}$$

The result is (27-5). Figure 27-3 shows $F_0(m)$ for $T > T_c$ and $T < T_c$. The solution of (27-5) is the horizontal point of the $F_0(m)$ curve, i.e. the curve of Fig. 27-2.

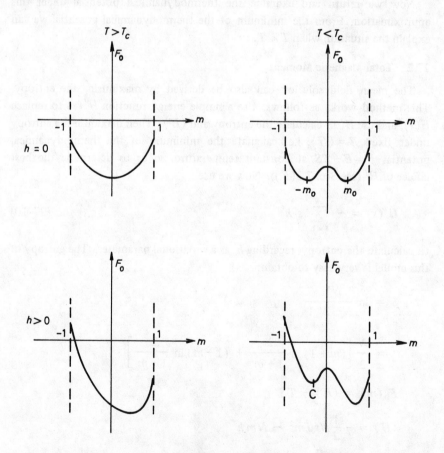

Fig. 27-3

When $T < T_c$ and h is not too large, as can be seen from Fig. 27-3, Eq. (27-5) has three solutions. But there is a solution with maximum F_0; this is the curve between AB in Fig. 27-1. The other solutions are those outside AB. If $h \neq 0$ then there is one absolute minimum, for which m and h are in the same direction. If $h = 0$, there are two minima, i.e. $\pm m_0(T)$ (see Eq. (27-8)). The point C in Fig. 27-3 is a minimum, but not an absolute minimum, and m and h are opposite in direction. This is a metastable state and we shall discuss it at the end in the problems.

Now, only if $h \neq 0$ does $F_0(m)$ have an absolute minimum. There is no question about the solution when $h \neq 0$, but when $h = 0$ and $T < T_c$ there are two minima. This can be regarded as a characteristic of ferromagnetism, i.e. when there is no magnetic field and m is nonzero the direction is not fixed. We now show the result for the total magnetic moment $M = Nm$ in Fig. 27-4, which is just Fig. 27-1 with unsuitable parts deleted.

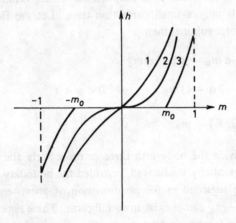

Fig. 27-4

Notice that $F_0(m)$ in Fig. 27-3, is not the thermodynamical potential for fixed total magnetic moment. Equation (27-10) is a very stringent condition, not only fixing the total magnetic moment Nm, but also fixing every $\langle s_i \rangle$ to m. In other words, the steps from (27-10) to (27-112) are: assuming that every $\langle s_i \rangle$ is the same, and that every s_i is independent, the best approximation to the true thermodynamical potential is determined by (27-12). If $\langle s_i \rangle$ is not everywhere the same, then these steps must be modified.

27.3. Thermodynamical Potential and the Coexistence of Different Phases

We have just emphasised that $F_0(m)$ in Fig. 27-3 is not yet the thermodynamical potential. What then is the thermodynamical potential? The answer is very simple: the thermodynamical potential is

$$F(h, T) = F_0(m(h, T), h, T) \quad . \tag{27-13}$$

Besides being a function of m, $F_0(m)$ is also a function of h and T (which we now display explicitly). The function $m = m(h, T)$ is the curve in Fig. 27-4, obtained by taking the minimum of $F_0(m)$.

When $h \neq 0$ there is no problem for (27-13). When $h = 0$, and $T < T_c$, $m(0, T) = \pm m_0(T)$ and there are two solutions. But $F_0(m)$ does not depend on the sign of m, and (27-13) still has no problems.

If the body is divided into two parts, one having $m = m_0$, while the other $m = -m_0$, then $F(h, T)$ is still the same. The energy of the boundary is proportional to its area while F is proportional to its volume. So the energy of the boundary is only a small correction term. Let the first part occupy a fraction α of the total volume, then

$$m(0, T) = \alpha m_0 - (1 - \alpha) m_0 \quad ,$$

$$= (2\alpha - 1) m_0 \quad , \quad 0 < \alpha < 1 \quad , \tag{27-14}$$

i.e. $-m_0 < m(0, T) < m_0$.

Of course if we divide the body into three or more parts, the thermodynamical potential is still essentially unchanged, provided the boundary surface is not too large. This can be regarded as the phenomenon of coexistence of phases, i.e. $m = m_0$ and $m = -m_0$ can coexist in equilibrium. These regions are very much like the domains in ferromagnetism, where the spins in the same domain point to the same direction, but different domains point differently. In a real ferromagnet, we have long range forces also, and it is much more complicated than the Ising model. The boundary between different domains can move, or may be fixed by impurities. This is a separate problem. Our present conclusion is that when $T < T_c$, and $h = 0$, $m(0, T)$ can take values between $\pm m_0$ and F is unchanged. Let

$$F_0'(m, T) \equiv F_0(m, h, T) + hNm \quad ,$$

$$= -\frac{1}{2} N v J m^2 - T S(m) \quad . \tag{27-15}$$

F_0' is a function of m and T only. We have not included the term $-hM$. Let the thermodynamical potential (27-13) plus hM be

$$F + hM = F'(m(h,T), T) \quad . \tag{27-16}$$

This thermodynamical potential can be regarded as a function of the total magnetic moment and temperature. Notice that the differentials of F and F' are

$$dF = -S\,dT - M\,dh \quad ,$$

$$dF' = -S\,dT + h\,dM \quad . \tag{27-17}$$

F' is the internal energy minus TS, while F is the internal energy minus hM and TS, i.e. F includes the potential energy due to the external force h. (See Chapter 16.)

According to the above analysis, if $T < T_c$ then

$$F'(m, T) = F_0'(m, T) \quad , \qquad \text{if} \quad |m| > m_0 \quad ,$$

$$= F_0'(m_0, T) \quad , \qquad \text{if} \quad |m| < m_0 \quad . \tag{27-18}$$

Hence m is the equilibrium value of M/N, i.e. $m(h, T)$. If $T > T_c$, then $F'(m, T) = F_0'(m, T)$.

Figure 27-5 is just the upper right curve of Fig. 27-3, with the bottom curve cut off by a horizontal line F'. In equilibrium,

$$\frac{\partial F'}{\partial M} = h \quad . \tag{27-19}$$

If $h = 0$, then $F' = F$ must be the minimum. Because different phases coexist and F is unchanged between m_0 and $-m_0$, so $F' = F$ must be horizontal and not convex.

The above discussion only cuts-off the convex portion of the upper right curve of Fig. 27-3. The result is simple, but the relation between F, F' and F_0 is not. The reader must note that the meanings of Fig. 27-5 and Fig. 27-3 are quite different.

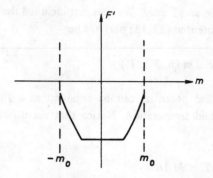

Fig. 27-5

27.4. The Ising Lattice Gas

This model is defined by the total energy H:

$$H - \mu N = -\frac{1}{2}\epsilon \sum_{i,j} n_i n_j - \mu \sum_i n_i \quad ,$$

$$N = \sum_i n_i \quad . \tag{27-20}$$

The variable n_i is the number of particles at the site i, with $n_i = 0$ or 1. The first term on the right of (27-20) is limited to neighbouring i and j. Because each n_i has two values, this model is an Ising model. Let

$$\sigma_i = 2n_i - 1 \quad , \tag{27-21}$$

then (27-20) can be transformed to the Ising ferromagnetic model:

$$H - \mu N = H' - \left(\frac{1}{8}\epsilon v + \frac{\mu}{2}\right) V \quad ,$$

$$H' = -\frac{1}{2} J \sum_{i,j} \sigma_i \sigma_j - h \sum_i \sigma_i \quad ,$$

$$J = \frac{\epsilon}{4} \quad ,$$

$$h = \frac{\mu}{2} + \frac{\epsilon v}{4} \quad . \tag{27-22}$$

Notice that now we use V as the total number of lattice points and N as the number of particles. The former number of elements N is now represented by V. With these relations between this model and the ferromagnetic model, we can translate the results of the above sections into properties of the lattice gas. The definition of the thermodynamical potential Ω is

$$e^{-\Omega/T} = \sum_s e^{-(H-\mu N)/T} \quad , \tag{27-23}$$

and the definition of the thermodynamical potential F of the ferromagnetic model is

$$e^{-F/T} = \sum_s e^{-H/T} \quad . \tag{27-24}$$

Therefore from (27-22) we know that

$$\Omega = F - \left(\frac{1}{2}\mu + \frac{1}{8}\epsilon v\right) V \quad . \tag{27-25}$$

Because $\Omega = -pV$, where p is the pressure, therefore (notice that the N in the ferromagnetic model is V here)

$$p = -\frac{F}{V} + \left(\frac{1}{2}\mu + \frac{1}{8}\epsilon v\right) = -\frac{F'}{V} + 2nh - \frac{1}{2}T_c \quad ,$$

$$T_c = Jv = \frac{1}{4}\epsilon v \quad , \tag{27-26}$$

$$n = \langle n_i \rangle = \frac{1}{2}(1 + m) \quad . \tag{27-27}$$

where F is defined from (27-16); see (27-16), (27-17), (27-18), (27-19) and Fig. 27-5. We want to know the relation between the temperature T, the pressure, and the density n. Although F' has been solved, (27-26) then is our answer; the desired relationship is, however, not yet very explicit.

First consider the case $T > T_c$. The field h can be replaced by (27-5), and F' by (27-11). Simplifying, we get

$$p = T \ln\left(\frac{1}{1-n}\right) - 2n^2 T_c \quad . \tag{27-28}$$

At low densities, i.e. $n \ll 1$, $p \approx nT$, i.e. the ideal gas law. At high densities, i.e. $n \to 1$, p tends to infinity. This is because every site can accommodate only one particle. Equation (27-28) can be rewritten as

$$p = (T - T_c) \ln\left(\frac{1}{1-n}\right) + T_c \left[\ln\left(\frac{1}{1-n}\right) - 2n^2\right] . \qquad (27\text{-}29)$$

Hence

$$\frac{dp}{dn} = \frac{(T - T_c)}{(1-n)} + T_c \frac{(1 - 2n)^2}{(1-n)} . \qquad (27\text{-}30)$$

It is easy to see that at $T = T_c$ and $n = \frac{1}{2}$, the first and second derivatives of p are all 0. This is the "critical point", corresponding to $h = 0$, $T = T_c$ and $m = 0$. The critical pressure is

$$p_c = T_c (\ln 2 - \tfrac{1}{2}) = 0.193 \, T_c . \qquad (27\text{-}31)$$

and (p_c, T_c) is the endpoint of the equilibrium vapour pressure curve (see Fig. 27-6). This curve is just the representation on the (p, T) plane of $h = 0$ and $T < T_c$. Let h in (27-26) be zero, and replace $F_0'(m_0, T)$ in (27-18) by F', then we get this curve.

$$p(T) = -\frac{F_0'(m_0(T), T)}{V} - \frac{1}{2} T_c$$

$$= -\frac{1}{2} T_c (1 + m_0^2(T)) + \frac{1}{2} T \ln \frac{4}{1 - m_0^2(T)} . \qquad (27\text{-}32)$$

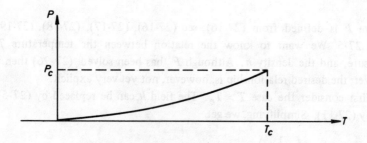

Fig. 27-6 (p, T) curve for an Ising lattice gas.

When $T < T_c$ and $h \neq 0$, we can still use (27-28). If we draw p as a function of $1/n = V/N$, then we get Fig. 27-7. The horizontal part is the pressure for the coexistence of the liquid and gas phases, i.e. the saturated vapour pressure.

The latent heat q is the heat required for one molecule to change from the liquid (l) to the gas (g) state. In this model $S(m_0)/V$ is the same as $S(-m_0)/V$, but the density $n = N/V$ is changed from $\frac{1}{2}(1+m_0)$ to $\frac{1}{2}(1-m_0)$. Therefore

$$q = T\left[\left(\frac{S}{N}\right)_g - \left(\frac{S}{N}\right)_l\right]$$

$$= TS(m_0)\left(\frac{2}{1-m_0} - \frac{2}{1+m_0}\right)\frac{1}{V}$$

$$= T\frac{2m_0}{1-m_0^2}\left[(1+m_0)\ln\left(\frac{2}{1+m_0}\right) + (1-m_0)\ln\left(\frac{2}{1-m_0}\right)\right].$$
(27-33)

This result can be obtained by differentiating (27-32), i.e.

$$\frac{dp}{dT} = \frac{q}{T\Delta v},$$

$$\Delta v = \left(\frac{1}{n}\right)_l - \left(\frac{1}{n}\right)_g.$$
(27-34)

This formula appears in Chapter 15.

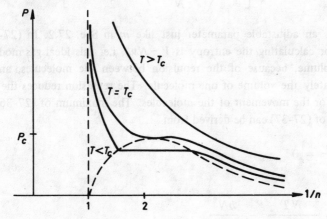

Fig. 27-7 Plot of pressure versus inverse density.

27.5. The Van der Waals Equation

In the last section we had applied the mean field approximation to the gas-liquid transition. The van der Waals equation has the same content, but with a slightly different appearance. The aim of this section is to provide further practice and introduces the law of corresponding states.

The interaction between gas molecules is repulsive at short distances and attractive at large distances. Now we replace this gas by an ideal gas model, and adjust the parameters in the model to maximise the entropy, with fixed total energy.

Let S be the entropy of the ideal gas. Under a fixed temperature and pressure, the total energy is

$$H = \frac{3}{2}NT + pV - bN^2/V \quad . \tag{27-35}$$

This formula includes the potential energy pV due to pressure and the interaction energy bN^2/V due to the attraction between the molecules. Now we maximise S at fixed H, i.e. we want to calculate the maximum of

$$S(V - Nv_0) - \frac{H}{T} \equiv -\frac{1}{T} G_0(V) \quad , \tag{27-36}$$

or the minimum of

$$G_0(V) = \frac{3}{2}NT + pV - \frac{bN^2}{V}$$

$$- T \left[N \ln \left(\frac{V - Nv_0}{N} \right) + \frac{3}{2} N \ln T \right] \quad . \tag{27-37}$$

Now V is an adjustable parameter just like m in Sec. 27.2. In (27-36) the volume for calculating the entropy is $V - Nv_0$, i.e. this ideal gas model has a smaller volume, because of the repulsion between the molecules, and v_0 is approximately the volume of one molecule. The repulsion reduces the volume available for the movement of the molecules. The maximum of (27-36) or the minimum of (27-37) can be derived from

$$\frac{\partial G_0}{\partial V} = 0 \quad , \tag{27-38}$$

i.e.

$$p = \frac{NT}{V - Nv_0} - \frac{bN^2}{V} \quad . \tag{27-39}$$

This is the van der Waals equation. (See Fig. 27-8.) We encounter no difficulty at high temperatures. At low temperatures, $\partial p/\partial V$ can be negative, i.e. volume increases with pressure. This is not reasonable.

Notice that (27-39) is obtained as the minimum of G at fixed p. But in Fig. 27-8(a) the low temperature curve has three points with $\partial G/\partial V = 0$. These three points are indicated in Fig. 27-8(b). Obviously G_0 at A is the true thermodynamical potential. We have to discard points B and C.

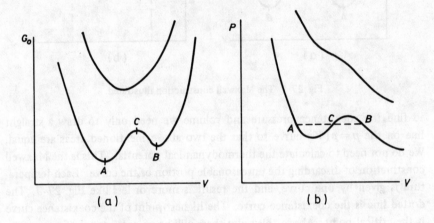

Fig. 27-8

If G_0 is the same at A and B, then there are two solutions. In this situation

$$0 = G_0(V_B) - G_0(V_A)$$

$$= \int_{V_A}^{V_B} dV \frac{\partial G_0}{\partial V} = \int_{V_A}^{V_B} dV \, [p - p(V)]$$

$$= (V_B - V_A)p - \int_{V_A}^{V_B} dV \, p(V) \quad , \tag{27-40}$$

where $p(V)$ is the expression on the right of (27-39), i.e. the curve in Fig. 27-9(b). The last term in (27-40) is the area under the curve AB, or from (27-40) the area below the straight line AB. Therefore, for the coexistence of the two phases, the shaded portions in Fig. 27-9(b) must have equal areas. So

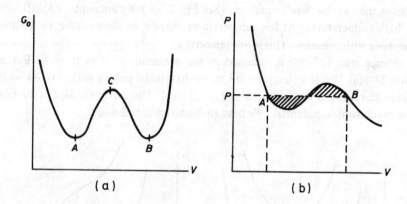

Fig. 27-9 The Maxwell construction illustrated.

to find the coexistence pressure and volume we need only to draw a straight line on the $p = p(V)$ curve so that the two above-mentioned areas are equal. We do not need to calculate the thermodynamical potential. This is the Maxwell construction of discarding the unreasonable portion of the curve. Each temperature is given by one curve, and the result is more or less like Fig. 27-7. The dotted line is the coexistence curve. The highest point of the coexistence curve is the critical point. Above this point there will be no coexistence of two states.

The characteristic of the critical point is that the first and second derivatives of the $p = p(V)$ curve are zero.

$$\frac{\partial p}{\partial V} = 0 \quad , \quad \frac{\partial^2 p}{\partial V^2} = 0 \quad . \tag{27-41}$$

From the van der Waals equation (27-39) and (27-41) we can solve for V_c and T_c, i.e. the critical volume and temperature. We get

$$V_c = 3v_0 N \;,$$

$$T_c = 8b/27v_0 \;,$$

$$p_c = b/27v_0 \;. \tag{27-42}$$

Now we use these values as units of volume, temperature and pressure and let

$$v' \equiv V/V_c \;,$$

$$T' \equiv T/T_c \;,$$

$$p' \equiv p/p_c \;. \tag{27-43}$$

MEAN FIELD SOLUTIONS 473

The van der Waals equation becomes

$$p' = \frac{8T'}{3v'-1} - \frac{3}{v'^2}. \qquad (27\text{-}44)$$

This equation does not involve any parameters. Therefore, if the van der Waals equation is correct, (27-44) can be applied to all gases. We have to measure the values at the critical point and all the other properties can be derived from (27-44). Equation (27-44) agrees surprisingly well with experiment. Many gases can be described by (27-44) to better than 10% accuracy. Equation (27-44) can be called the law of corresponding states. The reason for the success is the similarity of the interaction between gas molecules, all being the shape of Fig. 14-1.

Notice that the mean field approximation cannot accurately solve for T_c. The 10% accuracy mentioned above refers only to (27-44). The values T_c, p_c and V_c must be experimentally determined, and not taken from the solutions of the mean field approximation. We can observe this in Fig. 17-3.

27.6. Common Properties of Condensation Phenomena

In the above we have used the mean field method to discuss the models of liquid-gas phases and ferromagnetism. Although these models have formal differences, they have the same content. Hence the many different phenomena thus discussed have many common points, which we summarize as follows.

A. Coexistence of different phases

Under certain fixed conditions, there can be different equilibrium states. In the model of a ferromagnet, the different states mean magnetic moments pointing to different directions. In the liquid-gas model the different states imply different densities.

B. Critical point

Above the critical point, coexistence of different phases is not possible and there is a unique equilibrium state. In the ferromagnet, the magnetic moment disappears. In the liquid-gas model, there is no distinction between liquid and gas.

C. Order parameter

The concept of order parameter can now be naturally established. This quantity is zero above the critical temperature. Below the critical temperature, it is nonzero and can have different values for different equilibrium states. In the ferromagnet, the order parameter can be defined as the average magnetic moment. In the liquid-gas model, it can be defined as the density minus the critical density. In the ferromagnet the order parameter is a vector while in the liquid-gas model it is a scalar. Notice that the discussion above is really limited to scalar order parameters only since the magnetic moment in the ferromagnet model is limited to two directions. Vector order parameters have continually varying directions. This is the topic of Chapter 29.

The order parameter is a coordinate. Like other coordinates, it is associated with a generalised force. (See Chapter 2.) The force associated with the order parameter can be called the condensation force. The magnetic field and the pressure (or the chemical potential) are the forces associated with the magnetic moment and the density, and are the condensation forces of the above examples.

Besides the ferromagnet and the coexistence of liquid and gas, there are many condensation phenomena possessing the characteristics of A, B and C. Table 27-1 lists some common examples. Antiferromagnetism in solids shows a periodic arrangement of the direction of the spins. For example in a cubic lattice, the direction of any spin is opposite to that of its six neighbours. The order parameters for the superfluidity of ^{4}He and for superconductivity are not obvious. The state of a particle can be empty or occupied. But quantum mechanics allows a linear combination of the empty and occupied states. Hence, besides the particle population, there are other variables of motion, i.e. the amplitude of the particle. If the reader thinks in terms of spin $\frac{1}{2}$, then the "up" corresponds to the occupied state while "down" corresponds to the empty state, so spin represents the number population. But the spin can also point in other directions. These different directions correspond to the amplitude of the particle. This point will be discussed fully in Chapter 30. The situation is similar in superconductivity. The state of a pair of electrons can be occupied or empty or it can be a combination of the two.

The above common characteristics can be summarised by the simple mean field theory. Near the critical temperature, the various curves in Fig. 27-3 can be represented by the following simple expressions. When m is very small,

$$F_0(m) \approx \text{constant} + N(am^2 + bm^4 - hm)$$

$$a = a'(T - T_c) \quad , \tag{27-45}$$

where a', and b are essentially constant near T_c, and m is the order parameter.

Table 27-1 Condensation Phenomena[a]

Phenomena	Order Parameter	Example	Critical Temp. K
liquid-gas	density	H_2O	647.05
ferromagnetism	magnetic moment	Fe	1044
antiferro-magnetism	periodic magnetic moment	FeF_2	78.26
superfluid ^{4}He	amplitude of particle	^{4}He	2.17
superconductivity	amplitude of paired electrons	Pb	7.19
separation of mixture	difference of the two densities of the liquids	$CCl_4 - C_7F_{14}$	301.8
metallic alloy	periodic arrangement	CuZn	739

[a] Taken from Ma (1976), p. 6.

Equation (27-45) can be applied to any condensation phenomenon, only the values of a' and b are different. Let $h = 0$; from the condition for the minimum of $F_0(m)$, we know that $m = 0$ when $T > T_c$. When $T < T_c$, m is proportional to $(T_c - T)^{1/2}$. The derivative $\partial m/\partial h$ is proportional to $|T - T_c|^{-1}$. The heat capacity is discontinuous at $T = T_c$. These characteristics have been mentioned in the above sections. The reader can rederive it from (27-45). Equation (27-45) is the Landau theory. Its content is the same as the mean field theory. But its purpose is to discuss all the condensation phenomena within the same framework. This is a most important step.

The mean field theory does not explain all the common points of the condensation phenomena. More detailed experiments and theory show that when T is very close to T_c, the mean field approximation is not correct, e.g.

$$m \propto (T_c - T)^\beta \quad ,$$

$$\frac{\partial m}{\partial h} \propto |T - T_c|^{-\gamma} \quad . \tag{27-46}$$

The exponent β is not $\frac{1}{2}$, and γ is not 1. The values of β and γ are related to the symmetry of the order parameter. The Ising model has a symmetry of $+m$ and $-m$ while the vector model (i.e. **m** is a vector) has continuous symmetry, and the exponents in the two cases are different. If the dimensionality of the space is different, the exponents are different also. If impurities are added, the results may change. The properties near the critical point are called critical phenomena. Although this is a very interesting topic, it is too specialised and we shall not discuss it any more.

Problems

1. As seen from Fig. 27-3, when $T < T_c$ and h is not too large, $F_0(m)$ has an absolute minimum and also a local minimum. The absolute minimum is the equilibrium point. If we suddenly reverse the direction of h, what happens then?

Review the section Growth of Water Droplets in Chapter 15 and Problem 5 of Chapter 17. Try to interpret the local minimum in Fig. 27-3 as a metastable state.

2. If we allow the mean field to be different at different points, then the generalisation of (27-5) is

$$h = T \tanh^{-1} m_i - J\nu m_i - J \sum_j (m_j - m_i) \quad,$$

$$m_i = \langle s_i \rangle \quad. \tag{27-47}$$

The summation in the equation is limited to neighbouring j and i.

(a) Derive (27-47) using the methods in Secs. 27.1 and 27.2.

(b) If m_i is a slowly varying function of $\mathbf{r}_i$, show that

$$\sum_j (m_j - m_i) \simeq a^2 \nabla^2 m(\mathbf{r}) \quad, \tag{27-48}$$

where a is the distance between nearest neighbours and m is regarded as a continuous function of position. Therefore

$$h \simeq T \tanh^{-1} m - J\nu m - Ja^2 \nabla^2 m \quad. \tag{27-49}$$

The external magnetic field can also be a function of $\mathbf{r}$.

(c) Boundary Surface.

Let $h = 0$ and $T < T_c$. Assume

$$m = 0 \quad, \qquad x = 0 \quad,$$

$$m \to m_0 \quad, \qquad x \gg a \quad,$$

$$m \to -m_0 \quad, \qquad x \ll -a \quad, \tag{27-50}$$

where $\pm m_0$ are the two minimum points of $F_0(m)$, (see Secs. 27.2 and 27.3). Also assume that m is unchanged in the y, z directions. Calculate and sketch $m(x)$.

(d) Surface Tension.

Prove that the thermodynamical potential per unit area due to the boundary is

$$\sigma = \int_{-\infty}^{\infty} dx \left[\frac{1}{2a} J \left(\frac{dm}{dx} \right)^2 + f_0(m(x)) - f_0(m_0) \right] ,$$

$$f_0(m) \equiv F_0(m)/V . \qquad (27\text{-}51)$$

σ is the surface of the boundary. Use the results of (c) to get σ.

(e) The Displacement of the Boundary.

In (27-50) we imposed the condition $m(0) = 0$. If we change this condition to $m(b) = 0$, where b is any constant, is there any change to the above answer?

Discuss the meaning of b and discuss the problem of the fluctuation of the boundary. (Detailed analysis will appear in the next chapter.)

(f) Thickness of the Boundary.

Prove that if $x \gg a$,

$$m(x) - m_0 \propto e^{-x/\xi} ,$$

$$\xi \equiv a[(T_c - T)/J]^{-\frac{1}{2}} . \qquad (27\text{-}52)$$

This result is approximate for situations when $\xi \gg a$, because the assumptions of (27-48) and (27-49) require m to vary very little over a distance a.

3. Apply the mean field approximation to the vector model

$$H = - J \sum_{i,j} \mathbf{s}_i \cdot \mathbf{s}_j - \mathbf{h} \cdot \sum_i \mathbf{s}_i , \qquad (27\text{-}53)$$

where $\mathbf{s}_i$ is a unit vector and i, j are neighbouring sites on the lattice. Calculate the magnetic susceptibility and the heat capacity.

4. Apply the mean field approximation to the above vector model, except that now $\mathbf{s}_i$ is the spin $\frac{1}{2}$ operator.

5. Based on (27-44), carefully draw the various (p', v') curves and the coexistence curve.

478   STATISTICAL MECHANICS

6. Starting from the analysis of Sec. 27.5, derive the evaporation entropy for each molecule

$$\Delta s = q/T = \ln\left(\frac{3v_1' - 1}{3v_2' - 1}\right), \qquad (27\text{-}54)$$

where v_1' and v_2' are respectively the volume of each molecule of the gas the liquid, divided by the critical volume v_c. Therefore, once v_1' is measured, v_2' and Δs can be calculated. Notice that Δs is logarithmic, and is not sensitive to changes in v_1' and v_2'. As the law of corresponding states can be applied extensively, (27-54) is also very useful. Problem 15.2 can essentially be explained by this equation.

7. Suppose a class of substances have the same form of interaction $\psi(r) = \epsilon \psi(r/\sigma)$, where the form of ψ is fixed but ϵ and σ vary from substance to substance.

(a) From the method of fixed temperature, show that in suitable units for the thermodynamical variables, this class of substances have the same equation of state.

(b) Prove that if this class of substances has a critical point, then they have the same ratio of critical values

$$p_c v_c / T_c$$

where T_c, p_c and v_c are the temperature, pressure and volume per molecule at the critical point respectively. This illustrates that the law of corresponding states is universal because the interaction between gas molecules has essentially the same shape.

8. Response and Correlation.

Assume $T > T_c$, m is very small and $h \propto \delta(\mathbf{r})$. Use (27-49) to solve for $m(\mathbf{r})$. Prove that

$$G(\mathbf{r}) \propto \frac{1}{r} e^{-r/\xi},$$

$$\xi \propto (T - T_c)^{-\frac{1}{2}}. \qquad (27\text{-}55)$$

$G(\mathbf{r})$ is the response function. The correlation function $C(\mathbf{r})$ is obtained from $G = C/T$ and ξ is the correlation length. See Chapter 13 for background material.

9. The rate of scattering is particularly large near the critical point because it is proportional to the correlation length (see Sec. 11.4). Try to analyze this phenomenon.

10. Obtain the response function and correlation function for $T < T_c$, again starting from (27-49).

The last two problems show that the mean field method enables us to analyze fluctuation to a limited extent.

Near the critical point, the fluctuation is large and so is the correlation length, and this simple analysis is not too reliable.

11. Antiferromagnetic Ising Model.

Let the total energy be

$$H = J \sum_{i,j} s_i s_j - h \sum_i s_i \quad . \tag{27-56}$$

This is like the ferromagnetic model with J replaced by $-J$. The indices i and j still label neighbouring sites, and $J > 0$.

(a) Let $h = 0$ and solve the one-dimensional model.

(b) Prove: if $h = 0$, the thermodynamical potential of (27-56) is like the ferromagnetic case, that is, it is independent of the sign of J.

(c) Use the mean field approximation to solve (27-56) for $h \neq 0$. Find the heat capacity and magnetic susceptibility.

(d) What is the definition of the order parameter in this model? What is the condensation force? (It is not h.)

Chapter 28

FLUCTUATION OF THE BOUNDARY SURFACE

The mean field theory in the last chapter provided a general picture of condensation. It tells us some common properties of these phenomena, such as the similarity of the phase coexistence curve, the critical point, etc. However, we have not considered fluctuations. Fluctuations have a strong influence on condensation. If the fluctuation is too large, there will be no condensation. The characteristics of fluctuations are intimately related to condensation.

The analysis of fluctuations can be divided into three cases: (1) high temperatures, (2) low temperatures and (3) critical point. Case (1) denotes situations with temperature much higher than the critical point. The properties of the substance are not related to condensation, and the correlation length is very short. These situations can be easily analysed, and the interaction may be regarded as a relatively unimportant correction. Case (2) denotes temperatures much lower than the critical point. According to the mean field calculation, condensation has been accomplished, but fluctuations can still have very important effects. Case (3) denotes temperatures around the critical point. This situation is more complicated. Near condensation, the fluctuation amplitude is rather large, so is the correlation length, and the analysis becomes difficult. This is a topic of considerable current interest. The reader may consult the references.[a] Because it is too specialised, we do not discuss it here.

Fluctuation has a very intimate relation with the dimensionality of space and the symmetry of the order parameter. The mean field theory is unable to account for this. In Chapter 11 we had calculated the vibration amplitude of

---

[a] See Ma (1976), Stanley (1971).

atoms in a crystal and discovered that the amplitude was very large in one- or two-dimensional space, but rather small in three dimensions. These types of problems will be discussed in the next chapter.

In this chapter we shall discuss the boundary between the coexisting phases at low temperatures. This is also a problem ignored by the mean field theory. We first use a simple model of surface tension to look at the fluctuation of the boundary between liquid and gas. We then use the Ising model and the Coulomb gas model to analyse the surface of a crystal. The aim of the last section is to point out that similarity exists between the seemingly unrelated models.

28.1. The Liquid Surface Model

In Chapter 17 we solved the one-dimensional Ising model. The conclusion is that there is no ferromagnetism, and hence the mean field approximation is not appropriate for this one-dimensional model. The reason for this is very simple. The boundary points divide the body into positive and negative regions. Because the boundary points can move without change in energy, the size of each region changes continually and randomly, causing the average magnetic moment to vanish. The mean field calculation had not considered the fluctuations due to the boundary points.

This situation does not occur in the two- or three-dimensional Ising model. The boundary points become boundary lines or boundary surfaces. A negative region inside a positive region can increase by enlarging the boundary. At low temperatures the boundary is not easy to increase. This is the content of the Peierls' proof in Chapter 17.

So, in discussing the coexistence of the positive and negative states, we have to consider also the boundary. Now we look at the fluctuation of the boundary line in two dimensions. We do not use the Ising model yet, but discuss a more simplified model. Figure 28-1 shows a boundary line $z(x)$. The positive region lies below the line while the negative above. We regard the boundary line as a continuous curve, ignoring the discreteness of the lattice.

Assume that the energy of the boundary line is proportional to its length.

$$H = \alpha L ,$$

$$L = \int_0^{L_0} dx \left[1 + \left(\frac{dz}{dx} \right)^2 \right]^{1/2}$$

$$\simeq L_0 + \frac{1}{2} \int_0^{L_0} dx \left(\frac{dz}{dx} \right)^2 , \qquad (28\text{-}1)$$

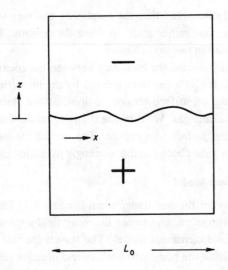

Fig. 28-1 A 2-dimensional boundary line $z(x)$.

where α is the "surface tension". If this is the boundary between liquid and gas phases, then this model is rather accurate (if there is no gravity).

The above assumes that the gradient is very small, i.e. $dz/dx \ll 1$. Let

$$z(x) = \frac{1}{\sqrt{L_0}} \sum_k \eta_k e^{ikx} ,$$

$$\eta_k \equiv \frac{1}{\sqrt{L_0}} \int_0^{L_0} dx \, z(x) e^{-ikx} . \qquad (28\text{-}2)$$

Substituting into (28-1) we get

$$L \simeq L_0 + \frac{1}{2} \sum_k k^2 |\eta_k|^2 ,$$

$$H = \alpha L = \alpha L_0 + \frac{\alpha}{2} \sum_k k^2 |\eta_k|^2 . \qquad (28\text{-}3)$$

From the law of equipartition of energy we obtain

$$\alpha k^2 |\eta_k|^2 = T . \qquad (28\text{-}4)$$

Let Δz be the difference in height between the two points x and x', and let us calculate $\langle (\Delta z)^2 \rangle$. When x and x' are very far apart, the magnitude of Δz can be regarded as the "thickness" of the boundary line.

$$\Delta z \equiv z(x) - z(x') \quad , \tag{28-5}$$

$$\begin{aligned} \langle (\Delta z)^2 \rangle &= 2 \langle z^2 \rangle - 2 \langle z(x) z(x') \rangle \\ &= \frac{2}{L_0} \sum_k \frac{T}{\alpha k^2} [1 - \cos k(x-x')] \\ &= \frac{2T}{\alpha} \int_{-\infty}^{\infty} \frac{dk}{2\pi k^2} [1 - \cos k(x-x')] \\ &= \frac{2T}{\alpha} |x-x'| \quad . \end{aligned} \tag{28-6}$$

If $|x - x'| \sim L_0$, then $\langle (\Delta z)^2 \rangle \sim (T/\alpha) L_0$. This result comes from the factor k^2 in (28-3). The displacement $k = 0$ means that all the $z(x)$ are the same, that is, a uniform displacement. Such a displacement expends no energy, that is, pushing the boundary line up or down does not change the energy. Oscillations of small k, i.e. long wavelength oscillation requires small energy. This long wavelength motion makes the height of the boundary surface fluctuate substantially. The thickness of this boundary (i.e. the square root of $\langle (\Delta z)^2 \rangle$ is thus proportional to $\sqrt{L_0}$. This boundary line is therefore "rough" and not "smooth".

Let us generalise this simple model to the boundary surface in three dimensions. Let the energy of the boundary surface be proportional to its area.

$$H = \alpha A \quad ,$$

$$A = \int d^2 r [1 + (\nabla z)^2]^{1/2}$$

$$\simeq L_0^2 + \frac{1}{2} \int d^2 r \, (\nabla z)^2 \quad . \tag{28-7}$$

This is the generalisation of (28-1) and α is the surface tension. The fluctuation of the height can be expressed by $\langle (\Delta z)^2 \rangle$. The generalisation of (28-6) is

$$\langle (\Delta z)^2 \rangle = \left(\frac{2T}{\alpha}\right) \int \frac{d^2 k}{(2\pi)^2 k^2} [1 - \cos \mathbf{k} \cdot (\mathbf{r} - \mathbf{r}')]$$

$$= \frac{T}{\pi \alpha} \int_0^{k_D} \frac{dk}{k} [1 - J_0(k|\mathbf{r} - \mathbf{r}'|)]$$

$$\simeq \frac{T}{\pi \alpha} \ln \frac{|\mathbf{r} - \mathbf{r}'|}{a} \quad,$$

$$a \simeq \frac{1}{k_D} \quad. \tag{28-8}$$

We have limited the integral within k_D and a is about the interatomic distance. Equation (28-8) is correct only when $|\mathbf{r} - \mathbf{r}'| \gg a$ and J_0 is the Bessel function.

The above result (28-8) shows that when $|\mathbf{r} - \mathbf{r}'|$ increases, $\langle (\Delta z)^2 \rangle$ increases too. Therefore, the boundary is rough. No matter how low the temperature, the thickness of the boundary surface is

$$\langle (\Delta z)^2 \rangle^{1/2} \sim \left[\frac{T}{\alpha} \ln \frac{L_0}{a} \right]^{1/2} \tag{28-9}$$

where L_0^2 is the area of the (x, y) cross-section. Notice that even if L_0 is very large (e.g. 1 km), (28-9) cannot be very large. It is not correct to regard (28-9) as infinitely large.

This model treats matter as a continuous medium and regards the boundary as an elastic line or surface. For the boundary between liquid-gas phases, this model is very accurate, except that we have not considered the effect of gravity. The gravitational force is weak but is responsible for keeping the liquid surface horizontal. Now we estimate the effect due to gravity.

When the liquid surface rises by z, it produces a pressure $\rho g z$, where ρ is the density of the liquid minus that of the gas. Hence it produces an energy $\frac{1}{2} \rho g z^2$ (per unit area) and this should be added to (28-7), which then becomes

$$H = \alpha A = \alpha L_0^2 + \frac{1}{2} \alpha \int d^2 r \left[(\nabla z)^2 + \frac{z^2}{\xi^2} \right] \quad,$$

$$\xi^2 \equiv \alpha / \rho g \quad. \tag{28-10}$$

Hence, L_0 in (28-9) should be changed to ξ.

FLUCTUATION OF THE BOUNDARY SURFACE

The above is a discussion of the boundary between liquid and gas. If we have to analyse the boundary of the surface of a crystal, then the structure of the crystal must be taken into account. The lattice arrangement of the Ising model is the simplest description of the lattice. If the discreteness of the lattice is considered, we may understand better the surface of a crystal.

The boundary line in two dimensions can be directly analysed from the Ising model (17-16). It is quite simple and can be analysed in considerable detail. The results should be similar to those above. The boundary surface in three dimensions is very complicated. If we consider the lattice effect, then the boundary surface at low temperatures will not be rough. We first study the boundary line in the next section.

28.2. The Boundary Surface of the Ising Model

From the inequality (17-27), when $T < J$ the average value of s_i is very close to 1. If we take the s_i on the boundary as -1, then nearly all the elements will be -1. Now we take the elements at the top of the square lattice as -1, and those at the bottom as $+1$, then we certainly get a boundary line in between (Fig. 28-2). If T is not very close to T_c, the elements above the boundary line are nearly all -1, and those below the line nearly all $+1$. Let us calculate the thermodynamical potential of the boundary line. The energy of the boundary line is its total length times $2J$. The total length is

$$L = L_0 + \sum_{k=1}^{L_0} |y_k| \ . \tag{28-11}$$

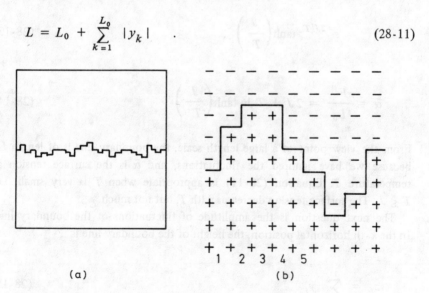

Fig. 28-2 (a) Boundary line for a square lattice. (b) An enlargement of a portion of (a).

Here, y_k is the length of the k-th vertical boundary line counting from the left. In Fig. 28-2(b) $y_1 = 2, y_2 = 1, y_3 = 0, y_4 = -2, y_5 = -1, \ldots$ etc. Therefore the thermodynamical potential F'' of the boundary line is

$$F'' = -T \ln Z'' \ ,$$

$$Z'' = \sum_{y_1 = -\infty}^{\infty} \sum_{y_2 = -\infty}^{\infty} \ldots \sum_{y_{L_0} = -\infty}^{\infty} e^{-2JL/T} \, \delta(y_1 + y_2 + \ldots + y_{L_0}) \ .$$

(28-12)

The δ function in the equation fixes the right end of the boundary line to be in the middle of the right edge. Since L_0 is a large number, the δ function can be ignored (with an error of $O(\ln L_0)$ in F''). Neglecting the δ function, Z'' can easily be calculated:

$$Z'' = \zeta^{L_0} \ ,$$

$$\zeta = e^{-2J/T} \left(1 + 2 \sum_{y=1}^{\infty} e^{-2Jy/T} \right)$$

$$= e^{-2J/T} \tanh\left(\frac{J}{T}\right) \ ,$$

(28-13)

$$\bar{\alpha} \equiv \frac{F''}{L_0} = 2J + T \ln \tanh\left(\frac{J}{T}\right) \ .$$

(28-14)

From the view point of a large length scale, the boundary line is of length L_0 because we have ignored the fluctuations, and $\bar{\alpha}$ is the surface tension at temperature T. Equation (28-11) is appropriate when T is very small, i.e. $T \lesssim J$. The surface tension decreases with T, but not much.

The next question is the amplitude of fluctuations of the boundary line. In the k-th horizontal position, the height of the boundary line is

$$z_k = \sum_{j=1}^{k} y_j \ .$$

(28-15)

FLUCTUATION OF THE BOUNDARY SURFACE

Hence, the difference in height between two points on the boundary line is

$$\Delta z \equiv z_{k+n} - z_k = \sum_{j=k+1}^{k+n} y_j . \qquad (28\text{-}16)$$

If $n \ll L_0$, each y_j will be independent variables. Using the central limit theorem we get the distribution of Δz:

$$\rho(\Delta z) \simeq \frac{1}{\sqrt{2\pi}\,\sigma} e^{-(\Delta z)^2/2\sigma^2}$$

$$\sigma^2 = n \langle y^2 \rangle$$

$$= \frac{2n\, e^{-2J/T}}{(1 - e^{-2J/T})^2}$$

$$= n/[2 \sinh^2 (J/T)] . \qquad (28\text{-}17)$$

Hence this boundary line fluctuates violently. The amplitude of fluctuation in the middle is about

$$\frac{\sqrt{L_0}}{\sinh (J/T)} . \qquad (28\text{-}18)$$

Although $\sqrt{L_0}$ is much smaller than L_0, it is still not a microscopic length scale. Therefore, this boundary line is a wiggly and rough curve, not a smooth and straight line.

The above analysis did not consider the "overhanging" or isolated regions in Fig. 28-3. Nevertheless, the conclusion that the boundary line is rough is not altered even if we include these effects.

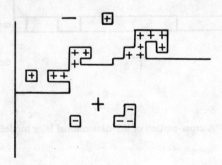

Fig. 28-3

28.3. Crystal Surface — the Coulomb Gas Model

It was pointed out some thirty years ago that at low temperatues the surface of a crystal should be smooth. When the temperature exceeds a certain value, the surface will be roughened. If we use the Ising model for analysis, this temperature should be approximately equal to the critical temperature of the two-dimensional Ising ferromagnet.[b] The argument goes as follows.

Figure 28-4 is a cross-section of a three-dimensional Ising model. The elements above the middle layer are all $-$, and those below all $+$. The elements in the middle layer can be positive or negative, because the upper neighbours are $-$ and the lower ones are $+$, and their effects tend to cancel each other, so that the change in energy of the elements in the middle layer is determined by that layer itself only. Therefore, the elements in the middle layer form a two-dimensional Ising model. If the temperature is lower than T_{c2}, the critical temperature of the two-dimensional Ising model, the elements in the middle layer will condense into $+$ or $-$ values. Of course, there will still be some fluctuation. But this boundary surface is smooth apart from this small fluctuation. The thick line in Fig. 28-4 represents the boundary surface. Of course, the layers next to the boundary line also have the above property of not being influenced by their upper or lower neighbours. Therefore, fluctuation is not limited to the middle layer. However, the above analysis is essentially very reasonable. If $T \gg T_{c2}$ the boundary surface should be smooth. This analysis also points out that the model of Sec. 28.1 (see Eq. (28-7)) has no smooth boundary surface because it has no separated layers. The existence of separated layers is a necessary condition for a smooth boundary surface.

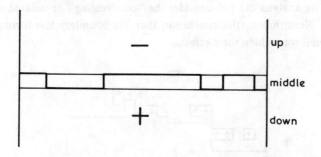

Fig. 28-4 A cross-section of a 3-dimensional Ising model.

[b] Burton, Cabrera and Frank (1951).

The Ising model is too complicated for this analysis. We now add some effects of the separated layers to the model described by Eq. 28-7, which will then allow a crude analysis of the roughening transition.

The following analysis replaces the model of Sec. 28.1 with a Coulomb gas model (see Sec. 19.5). How do charges come in? The reason is that the $(\nabla z)^2$ in Eq. (28-7) has an intimate relationship with the electrostatic interaction. If we regard z as the electric potential, then $-\nabla z$ is the electric field $\mathbf{E}$ and (28-7) is just the energy of the electric field.

$$H \propto \int d^2 r \, E^2 \quad , \tag{28-19}$$

and the minimum of the electrostatic energy is determined by $\nabla \cdot \mathbf{E} = 0$, i.e. $\nabla^2 z = 0$. Any interaction involving z can be regarded as the interaction with charges, e.g. if we add to H a term

$$\int d^2 r \, q(r) \, z(r) \quad , \tag{28-20}$$

then $q(r)$ can be regarded as the charge distribution because $z(r)$ is the electric potential. Although the following steps are a bit more complicated, the introduction of the concept of "charges" is not unnatural. Now the question is how to expose the difference between continuous z and discrete z.[c]

Now rewrite (28-7) as

$$H = \alpha L_0^2 + \frac{\alpha}{2} \sum_{\mathbf{R}} \sum_{\mathbf{a}} [z(\mathbf{R}) - z(\mathbf{R}+\mathbf{a})]^2 \quad , \tag{28-21}$$

where $\mathbf{R}$ denotes horizontal lattice sites and $\mathbf{a}$ the horizontal vectors. The thermodynamical potential is

$$F = -T \ln Z \quad ,$$

$$Z = \prod_k \sum_{z(\mathbf{R})=-\infty}^{\infty} e^{-H/T} \quad , \tag{28-22}$$

and $z(\mathbf{R})$ are now integers. The distance between each layer is 1. Equations (28-21) and (28-22) only put (28-7) on the discrete lattice. This is not the

---

[c] For the following method, see Chui and Weeks (1976), Weeks (1980).

Ising model but a new one. If we replace the summation in (28-22) by integration over z, then this model would be no different from (28-7). The discreteness of the lattice in the horizontal direction has no great influence, as will become obvious later.

The difference between summation and integration can be computed using the Poisson sum rule in Chapter 16. (See Eq. (16-49).) The value of z extends from $-\infty$ to ∞. This formula can be written as

$$\sum_{n=-\infty}^{\infty} f(n) = \sum_{q=-\infty}^{\infty} f_q = \int_{-\infty}^{\infty} dz\, f(z) + \sum_{q=1}^{\infty} (f_q + f_{=q}) ,$$

$$f_q \equiv \int_{-\infty}^{\infty} dz\, e^{-i2\pi qz} f(z) . \qquad (28\text{-}23)$$

For detailed derivation, see Chapter 16.

Every summation in (28-22) can be evaluated by the above formula. We get

$$Z = Z_0 \prod_{\mathbf{R}} \sum_{q(\mathbf{R})=-\infty}^{\infty} \left\langle \exp\left\{-i2\pi \sum_{\mathbf{R}} q(\mathbf{R})\, z(\mathbf{R})\right\} \right\rangle_0 , \qquad (28\text{-}24)$$

where Z_0 is the value of Z obtained by considering $z(\mathbf{R})$ as continuous, and $\langle \ldots \rangle_0$ is the average evaluated with respect to this continuous-valued model.

$$Z_0 \equiv \prod_{\mathbf{R}} \int dz(\mathbf{R})\, e^{-H/T} ,$$

$$\langle A \rangle_0 \equiv \prod_{\mathbf{R}} \int dz(\mathbf{R})\, A\, e^{-H/T}\, Z_0^{-1} . \qquad (28\text{-}25)$$

If $z(\mathbf{R})$ is a continuous variable, then its distribution is normal because (28-21) is quadratic in z. Then

$$\left\langle \exp\left\{-i2\pi \sum_{\mathbf{R}} q(\mathbf{R})\, z(\mathbf{R})\right\} \right\rangle_0$$

$$= \exp\left\{-\frac{(2\pi)^2}{2} \sum_{\mathbf{R}} \sum_{\mathbf{R}'} q(\mathbf{R})\, q(\mathbf{R}')\, \langle z(\mathbf{R})\, z(\mathbf{R}') \rangle_{0c}\right\} . \qquad (28\text{-}26)$$

This is the property of the normal distribution. (See Eqs. (12-32), (12-38) and the discussion in Chapter 12.) The calculation of $\langle z(\mathbf{R}) z(\mathbf{R}') \rangle_{0c}$ is similar to (28-8). If $|\mathbf{R}-\mathbf{R}'|$ is very large, then

$$\langle z(\mathbf{R}) z(\mathbf{R}') \rangle_{0c} \simeq -\frac{T}{2\pi\alpha} \ln \frac{|\mathbf{R}-\mathbf{R}'|}{a} + \text{constant} \quad . \quad (28\text{-}27)$$

Substituting into (28-26) and (28-24), we get

$$Z = Z_0 \prod_{\mathbf{R}} \sum_{q(\mathbf{R})=-\infty}^{\infty} \exp\left\{-\frac{1}{2T'} \sum_{\mathbf{R},\mathbf{R}'} q(\mathbf{R}) q(\mathbf{R}') u(\mathbf{R}-\mathbf{R}')\right\} ,$$

$$u(\mathbf{R}-\mathbf{R}') = -\ln \frac{|\mathbf{R}-\mathbf{R}'|}{a} ,$$

$$T' \equiv \frac{\alpha}{2\pi T} \quad . \quad (28\text{-}28)$$

This model is just the two-dimensional Coulomb gas model in Sec. 19.5, except that the charge can be any integer, and the temperature T' is inversely proportional to T and proportional to the surface tension α. Here $q(\mathbf{R})$ is the charge at $\mathbf{R}$. If all the $q(\mathbf{R})$ are zero, then we are back at the model of Eq. (28-7). Hence the influence of the separated layers can be ascribed to the influence of these charges. Notice that because of the symmetry of $+q$ and $-q$, the total charge is zero. To calculate $\langle (\Delta z)^2 \rangle$, in addition to (28-8) we also have to consider the influence of these charges. Now $\langle z(\mathbf{R})z(\mathbf{R}') \rangle_c$ can be obtained by the following steps: Replace $q(\mathbf{R})$ and $q(\mathbf{R}')$ by $q(\mathbf{R}) + \lambda$ and $q(\mathbf{R}') + \lambda'$, where λ and λ' are infinitesimal. Hence the increase in the thermodynamical potential can be computed from (28-24).

$$\frac{\partial^2 F}{\partial \lambda \partial \lambda'} = -\frac{4\pi^2}{T} \langle z(\mathbf{R}) z(\mathbf{R}') \rangle_c ,$$

$$\langle (\Delta z)^2 \rangle \equiv 2\langle z^2 \rangle - 2\langle z(\mathbf{R}) z(\mathbf{R}') \rangle . \quad (28\text{-}29)$$

Now λ and λ' can be regarded as charges added at $\mathbf{R}$ and $\mathbf{R}'$. Therefore (28-29) is the interaction energy between these two charges (divided by $\lambda\lambda'$). Hence the correlation value of $z(\mathbf{R})$ and $z(\mathbf{R}')$ becomes the interaction energy between the two charges.

From the analysis of Chapter 19, at low temperatures, i.e. T' very small, positive and negative charges will pair up and the dielectric constant is finite. The interaction between the charges is still logarithmic, i.e.

$$\langle z(\mathbf{R})\, z(\mathbf{R'}) \rangle_c \propto -\ln \frac{|\mathbf{R}-\mathbf{R'}|}{a} ,$$

$$T' < \tfrac{1}{4}, \quad \text{i.e.} \quad T > 2\alpha/\pi . \tag{28-30}$$

The value $\tfrac{1}{4}$ is obtained from Eq. (19-28) and the smallest q is 1. If $T' > \tfrac{1}{4}$, then single ions exist, producing "screening" phenomena, therefore

$$\langle z(\mathbf{R})\, z(\mathbf{R'}) \rangle_c \propto \frac{e^{-|\mathbf{R}-\mathbf{R'}|/b}}{\sqrt{|\mathbf{R}-\mathbf{R'}|}} ,$$

$$T' > \tfrac{1}{4}, \quad \text{i.e.} \quad T < 2\alpha/\pi , \tag{28-31}$$

$$\frac{1}{b^2} = \frac{4\pi n}{T'} . \tag{28-32}$$

Here n is the density of the $+1$ or -1 charges (see (19-34)). Charges with $|q| > 1$ contribute little to the process.

Hence at high temperatures, $T > 2\alpha/\pi$, the conclusion of roughening is unchanged. At low temperatures, when $T < 2\alpha/\pi$, the boundary surface is smooth and the correlation distance of the fluctuation is b. The charge density n has yet to be calculated, but we omit it here.

The above analysis utilises the Poisson summation formula to write the influence of the separated layers as the influence of a group of charges. We then use the model of electrostatic interaction to obtain the solution. This way of transferring one model to another and the looking for a solution is a common technique. This electrostatic model shall reappear in the next two chapters.

The roughness of the surface of a crystal affects the rate of growth of the crystal very much. The literature is plentiful on this aspect.[d] The reader should notice that the above model is not adequate to deal with the structure of the surface of a solid, which is a major problem that can only be understood from the molecular interaction.

[d] See Chui and Weeks (1980), Krumbhaar and Binder (1979).

Problems

1. Review Problem 27.2. The z in (28-1) and (28-7) are equivalent to b in that problem. Discuss the meaning of the scale ξ there for the model in Sec. 28.1.

2. (a) Obtain the thermodynamical potential F of model (28-1) starting from (28-3). The potential of each k element is also very easy to calculate.

   (b) Let $\bar{\alpha} = F/L_0$. This can be called the surface tension. This is macroscopic surface tension and is different from the microscopic one (see (28-1) and (28-3)). Discuss the meanings of $\bar{\alpha}$ and α, and estimate their difference using water as an example.

   (c) Compare this result with that in Sec. 28.2.

3. From the Ising boundary model derive the continuous boundary surface model.

   (a) Starting from model (28-11), take points at

   $$k = 0, b, 2b, 3b, \ldots, mb, \ldots, \qquad (28\text{-}33)$$

   and let $b \gg 1$. If we join these points by straight lines, they become a boundary. Of course the detailed change between mb and $(m+1)b$ is lost. Prove that Z'' in (28-12) can be written as

   $$Z'' = \prod_{m=1}^{L'} \sum_{\eta_m = -\infty}^{\infty} \zeta(\eta_m)\, \delta(\eta_1 + \eta_2 + \ldots + \eta_{L'}) ,$$

   $$L' = L_0/b , \qquad (28\text{-}34)$$

   $$e^{2jT} \zeta(\eta) = \prod_{k=1}^{m} \sum_{y_k = -\infty}^{\infty} e^{-2J|y_k|/T} \delta(\eta - y_1 - \ldots - y_N) . \qquad (28\text{-}35)$$

   (b) Using the central limit theorem, prove that

   $$\ln \zeta(\eta) \simeq b \ln \zeta - \frac{1}{2} \ln(2\pi b \langle y^2 \rangle) - \eta^2/(2b \langle y^2 \rangle) . \qquad (28\text{-}36)$$

   For the meaning of ζ see (28-13) and (28-14), and for $\langle y^2 \rangle$ see (28-17).

   (c) Define the effective total energy as

   $$H'(\eta_1, \eta_2, \ldots, \eta_{L'}) = -T \sum_{m=1}^{L'} \ln \eta(\eta_m) . \qquad (28\text{-}37)$$

   Prove that the thermodynamical potential obtained from H' is the same as that from $H = 2JL$ in (28-11). This is used to check the correctness of the above procedure.

494 STATISTICAL MECHANICS

(d) In (28-36), all are constants except the last term which contains η. Prove that

$$H' = \text{constant} + \frac{\alpha}{2}\int dx \left(\frac{dz}{dx}\right)^2$$

$$\alpha \equiv \frac{T}{\langle y^2 \rangle} \, . \tag{28-38}$$

Here, z is regarded as a smooth function of $x = mb$, $dx = b$ and $\eta/b = dz/dx$. The integral in the above equation is only an abbreviation for the summation of m.

Therefore, the model in Sec. 28.2 is essentially like that in Sec. 28.1, except that the scales are different. The scale is (28-38) or (28-1) is b, while that of (28-11) is 1. Notice that here $b \gg 1$, $dz \gg 1$ and the constant term in (28-38) is not $\propto L_0$.

(e) Notice that α in (28-38) is a complicated function of T (see (28-14)). The constant term in (28-38) is very important. If we want to calculate the exact $\bar{\alpha}$, this constant must be determined exactly. The reader must understand the meaning and origin of H'. Try to discuss the importance. From (28-38), b does not seem to appear, but without b the meaning of (28-38) would be problematic.

4. Use (28-10) to calculate the value of ξ for water, and also use (28-9) to calculate the thickness of the water surface at room temperature.

5. Discuss the fluctuation of the shape of a water droplet.
   (a) Generalise (28-7) to discuss a surface which is nearly spherical.
   (b) Use the spherical harmonics to analyse the shape of a water droplet.

Chapter 29

MODELS WITH CONTINUOUS SYMMETRY

This chapter discusses three examples — the planar vector model (the XY model), the crystal model and the quantum vector model, using rather crude analysis to discuss the fluctuation of the order parameter. The amplitude of fluctuation is related to the dimensionality of space as well as the symmetry of the system. These examples demonstrate such a relationship. Lastly we discuss the common features of these models — i.e. condensation and continuous symmetry. Under this condition of condensation, low energy, long wavelength soft modes can appear. In one- or two-dimensional models, these soft modes can soften and break the condensation. Besides the smooth soft modes, "defects" are also an important fluctuations. We shall use the two-dimensional XY model as an example to discuss the origin and effects of defects.

29.1. The Planar Vector Model

We first look at the simplest vector model whose total energy is given by

$$H = -\frac{1}{2} \sum_{i,j} J_{ij}\, \mathbf{s}_i \cdot \mathbf{s}_j \quad,$$

$$J_{ij} = J\,, \quad \text{if } i,j \text{ are neighbours} \quad,$$
$$\phantom{J_{ij}} = 0\,, \quad \text{otherwise} \quad, \hspace{4em} (29\text{-}1)$$

where $\mathbf{s}_i$ is a unit planar vector, i.e. $\mathbf{s}_i = (\mathbf{s}_{ix}, \mathbf{s}_{iy})$ and $\mathbf{s}_{ix}^2 + \mathbf{s}_{iy}^2 = 1$. This is the so-called XY model. As with the superfluid model in the next chapter, we now look at the fluctuations of the direction of the vector. First simplify (29-1) a bit and let

$$s_{ix} = \cos \theta_i ,$$

$$s_{iy} = \sin \theta_i ,$$

then $s_i \cdot s_j = \cos(\theta_i - \theta_j)$. (29-2)

Now assume that the neighbouring angles are nearly the same, then

$$\cos(\theta_i - \theta_j) \simeq 1 - \frac{1}{2}(\theta_i - \theta_j)^2 .$$

Let **R** denote the position of the vectors, then (29-1) can be written as

$$H = \frac{1}{4} J \sum_{\mathbf{R}} \sum_{\mathbf{a}} [\theta(\mathbf{R}) - \theta(\mathbf{R} + \mathbf{a})]^2 + \text{constant} , \quad (29\text{-}3)$$

where $\mathbf{R} + \mathbf{a}$ denotes the nearest neighbours of $\mathbf{R}$. Finally let

$$\theta(\mathbf{R}) = \frac{1}{\sqrt{N}} \sum_{\mathbf{k}} \theta_{\mathbf{k}} e^{i\mathbf{k} \cdot \mathbf{R}} ,$$

$$\theta_{\mathbf{k}} \equiv \frac{1}{\sqrt{N}} \sum_{\mathbf{R}} \theta(\mathbf{R}) e^{-i\mathbf{k} \cdot \mathbf{R}} . \quad (29\text{-}4)$$

Substituting in (29-3) we get

$$H = \frac{1}{2} \sum_{\mathbf{k}} J_{\mathbf{k}} |\theta_{\mathbf{k}}|^2 ,$$

$$J_{\mathbf{k}} \equiv J \sum_{\mathbf{a}} |1 - e^{-i\mathbf{k} \cdot \mathbf{a}}|^2 = 2J \sum_{\mathbf{a}} (1 - \cos \mathbf{k} \cdot \mathbf{a}) . \quad (29\text{-}5)$$

When k is very small, $J_{\mathbf{k}} \propto k^2$. From (29-5), we see that each $\theta_{\mathbf{k}}$ is independent, and therefore normally distributed. All the average values are readily computed. The equipartition of energy gives

$$\langle |\theta_{\mathbf{k}}|^2 \rangle = T/J_{\mathbf{k}} . \quad (29\text{-}6)$$

From this we can calculate

$$\langle (\Delta\theta)^2 \rangle \equiv \langle (\theta(\mathbf{R}) - \theta(0))^2 \rangle$$

$$= \frac{2}{N} \sum_\mathbf{k} \frac{T}{J_\mathbf{k}} (1 - \cos \mathbf{k} \cdot \mathbf{R}) \quad . \tag{29-7}$$

If R is very large, then the small k terms are important. Write $J_\mathbf{k}$ as the small k approximation $J_\mathbf{k} \approx \alpha k^2 a^2$, then

$$\langle (\Delta\theta)^2 \rangle = \frac{2T}{N\alpha} \sum_\mathbf{k} \frac{1}{a^2 k^2} (1 - \cos \mathbf{k} \cdot \mathbf{R}) \quad . \tag{29-8}$$

This calculation has been performed for both one and two dimensions in the last chapter (see (28-5) and (28-8)), the result being

$$\langle (\Delta\theta)^2 \rangle = \begin{cases} \dfrac{2TR}{\alpha a} \, , & d=1 \, , \\[6pt] \dfrac{T}{\pi\alpha} \ln \dfrac{R}{a} \, , & d=2 \, , \\[6pt] \dfrac{T}{\pi^2 \alpha} \left(k_D a - \dfrac{\pi a}{4R} \right) , & d=3 \, , \end{cases} \tag{29-9}$$

where d is the dimension of space. The integral for $d = 3$ is not difficult. We only have to note that $1/k^2$ is the solution of $-\nabla^2$, i.e. the solution of the electrostatic equation:

$$-\nabla^2 \phi = \delta(\mathbf{R}) \quad . \tag{29-10}$$

Except for some minor details, (29-9) is the solution of (29-10). The factor $1/k_D$ is approximately the size of the lattice. It can be easily seen from (29-9) that the fluctuation of the angle when $d \leq 2$ increases with R like the fluctuation of the liquid surface in Sec. 28.1. The correlation function of various s_i can be calculated from (29-9):

$$C(\mathbf{R}) \equiv \langle \mathbf{s}(\mathbf{R}) \cdot \mathbf{s}(0) \rangle = \langle \cos(\theta(\mathbf{R}) - \theta(0)) \rangle$$

$$= \mathrm{Re} \, \langle e^{i(\theta(\mathbf{R}) - \theta(0))} \rangle$$

$$= e^{-\langle (\Delta\theta)^2 \rangle / 2} \quad . \tag{29-11}$$

The last step uses the normal distribution of θ (see the cumulant theorems (12-32) and (12-38)). From (29-9) we get

$$C(\mathbf{R}) = \begin{cases} e^{-(T/\alpha)R/a} , & d = 1 , \\ (R/a)^{-T/2\pi\alpha} , & d = 2 , \\ e^{-Tk_D a/\pi^2 \alpha} \left[1 + \dfrac{\pi}{4k_D R} \right] , & d = 3 , \end{cases} \quad (29\text{-}12)$$

The above assumes that $R \gg a$.

From these results we can see that when $d \leqslant 2$ no matter how small T is, provided $T \neq 0$, $C(R)$ decreases with R and tends to zero. Only when $d = 3$ does $C(R)$ tend to a constant. That is to say, when $d \leqslant 2$ there is no collective arrangement. Notice that when $d = 2$, $C(R)$ vanishes very slowly and on any reasonable scale, $C(R)$ does not tend to zero and this has been mentioned in the last chapter. The factor $\ln(R/a)$ in (29-9) cannot be very large. Therefore when $d = 2$ we cannot say that condensation does not occur.

According to (29-12), when $d = 2$ or 3, collective arrangement occurs at any temperature. But why do we find no trace of the transition temperature in (29-12)? The reason is that we have assume $\Delta\theta$ to be a small quantity. If $|\Delta\theta| \gtrsim \pi$, then the above calculation is meaningless. The angles π and $-\pi$ are the same direction, and θ and $\theta + 2n\pi, n = \pm 1, \pm 2, \ldots$ are the same. Therefore the conclusions of (29-9) and (29-12) are valid for low temperature only. The above conclusion of no collective arrangement does not follow from a rigorous proof, but a rigorous proof exists and we shall mention it in Sec. 29.4. (The one-dimensional model can be easily solved and the conclusion of (29-12) is correct. The reader may attempt to solve it.)

We now look at another example, the crystal. At first sight this is quite different from the XY model, but in fact they are very similar.

29.2. Density Fluctuations of a Crystal

In a crystal the atoms are periodically arranged, and hence the density is inhomogeneous. X-ray scattering measures the scattering intensity, which is proportional to

$$\langle |\rho_\mathbf{k}|^2 \rangle = |\langle \rho_\mathbf{k} \rangle|^2 + \langle |\rho_\mathbf{k}|^2 \rangle_c \quad . \quad (29\text{-}13)$$

The experimental analysis is like that in Sec. 11.4. Because of the periodic arrangement of the crystal, when k equals certain values, i.e. the reciprocal vector,[a] $\langle \rho_k \rangle$ is nonzero. We first examine the relation between $\langle \rho_k \rangle$ and the vibration displacement $\mathbf{u}$:

$$\langle \rho_k \rangle \equiv \int d^3 r \, e^{-i\mathbf{k} \cdot \mathbf{r}} \langle \rho(\mathbf{r}) \rangle$$

$$= \sum_\mathbf{R} \langle e^{-i\mathbf{k} \cdot (\mathbf{R} + \mathbf{u}(\mathbf{R}))} \rangle . \tag{29-14}$$

In the equation $\mathbf{R}$ denotes the average position of the atom and $\mathbf{R} + \mathbf{u}(\mathbf{R})$ its actual coordinate. The displacements $\mathbf{u}(\mathbf{R})$ are the variable of motion. Now we ignore quantum mechanics, so $\mathbf{u}(\mathbf{R})$ are coordinates executing simple harmonic motion:

$$\mathbf{u}(\mathbf{R}) = \frac{1}{\sqrt{N}} \sum_{\lambda \mathbf{q}} e^{i\mathbf{q} \cdot \mathbf{R}} \eta_{\mathbf{q}\lambda} \, \mathbf{e}_{\lambda \mathbf{q}} . \tag{29-15}$$

The probability distribution of each $\eta_{\mathbf{q}\lambda}$ is normal, and

$$\langle \eta_{\mathbf{q}\lambda} \rangle = 0 \quad , \quad \langle |\eta_{\mathbf{q}\lambda}|^2 \rangle = \frac{T}{m \omega_{\mathbf{q}\lambda}^2} . \tag{29-16}$$

The cumulants of the normal distribution vanish after the second one. Hence if we use the cumulant theorems of Chapter 12 (see (12-32) and (12-38)) we get

$$\langle e^{-i\mathbf{k} \cdot \mathbf{u}(\mathbf{R})} \rangle = e^{-\frac{1}{2} \langle (\mathbf{k} \cdot \mathbf{u}(\mathbf{R}))^2 \rangle}$$

$$\equiv e^{-W(k)} , \tag{29-17}$$

and W has been computed in Chapter 11. (See (11-21).) Using the Debye model we get

$$W(k) = \frac{k^2 T \omega_D}{4\pi^2 n m c^3} \quad , \quad d = 3 \quad , \tag{29-18}$$

---

[a] The reciprocal vector $\mathbf{G}$ represents the periodic structure and $e^{i\mathbf{G} \cdot \mathbf{r}}$ has the same period as the crystal.

where n is the density of the atoms N/V, m is the mass of the atom, c is the speed of sound and ω_D is the Debye frequency. When $\mathbf{k} = \mathbf{G}$, a reciprocal vector,

$$e^{i\mathbf{G}\cdot\mathbf{R}} = 1 ,$$

$$|\langle \rho_G \rangle|^2 = N^2 e^{-2W(G)} . \tag{29-19}$$

As temperature rises, this intensity decreases; e^{-2W} is the so-called Debye-Waller factor and represents the decay of the scattering due to vibration.

We now calculate (29-13)

$$\langle |\rho_\mathbf{k}|^2 \rangle = \sum_\mathbf{R} \sum_{\mathbf{R}'} e^{-i\mathbf{k}\cdot(\mathbf{R}-\mathbf{R}')} \langle e^{-i\mathbf{k}\cdot(\mathbf{u}(\mathbf{R})-\mathbf{u}(\mathbf{R}'))} \rangle$$

$$= N \sum_\mathbf{R} e^{-i\mathbf{k}\cdot\mathbf{R}} e^{-\frac{1}{2}\langle(\mathbf{k}\cdot\Delta\mathbf{u})^2\rangle} ,$$

$$\Delta\mathbf{u} \equiv \mathbf{u}(\mathbf{R}) - \mathbf{u}(0) . \tag{29-20}$$

We use the cumulant theorem in the last step, in a manner similar to (29-17). The calculation of $\langle(\mathbf{k}\cdot\Delta\mathbf{u})^2\rangle$ is like that of $\langle(\Delta z)^2\rangle$ in Sec. 28.1 or that of $\langle(\Delta\theta)^2\rangle$ in this chapter, (see (29-8) and (29-9)).

$$\langle(\mathbf{k}\cdot\Delta\mathbf{u})^2\rangle = \frac{2k^2 T}{nmc^2} \int \frac{d^d q}{(2\pi)^d} \frac{1}{q^2}(1 - \cos\mathbf{q}\cdot\mathbf{R})$$

$$= \begin{cases} \dfrac{2T}{\alpha}R , & d = 1 , \\[2mm] \dfrac{T}{\pi\alpha} \ln\left(\dfrac{R}{a}\right) , & d = 2 , \\[2mm] \dfrac{T}{\pi^2\alpha}\left(\dfrac{\omega_D}{c} - \dfrac{\pi}{4R}\right) , & d = 3 . \end{cases}$$

$$\alpha \equiv \frac{nmc^2}{k^2} . \tag{29-21}$$

MODELS WITH CONTINUOUS SYMMETRY 501

This is similar to (29-9). The factor $1/q^2$ in the integral comes from $1/\omega_{q\lambda}^2$ in (29-16) as $\omega_{q\lambda}^2 = c^2 q^2$. This q^2 is $-\nabla^2$ in another guise. The energy of an elastic body is proportional to the square of the gradient of the displacement. If the whole body undergoes a uniform displacement, i.e. $q = 0$, then the energy is not changed. When q is very small the energy is small too. Because of the vibration of small q, i.e. long waves, the amplitude **u** is large. From (29-18) and (29-12) we know that when $d \leqslant 2$ the amplitude is so large that the periodic structure is lost. Of course when $d = 2$, $\ln(R/a)$ cannot be too large and the periodicity is not totally lost.

Now substitute (29-21) into (29-20) and let **k** = **G**. We obtain

$$\langle |\rho_G|^2 \rangle = \begin{cases} 2Nn\alpha/T, & d = 1, \\ e^{-2W(G)}[N^2 + O(N^{5/3})], & d = 3, \end{cases}$$

$$\alpha \equiv nmc^2/G^2. \tag{29-22}$$

The case is special when $d = 2$:

$$\langle |\rho_G|^2 \rangle = \sum_R \left(\frac{R}{a}\right)^{-T/2\pi\alpha}$$

$$\simeq \left(1 - \frac{T}{4\pi\alpha}\right) N^{2-(T/2\pi\alpha)}, \quad T < 4\pi\alpha,$$

$$\simeq \left(\frac{T}{4\pi\alpha} - 1\right) N, \quad T > 4\pi\alpha. \tag{29-23}$$

Therefore, the intensity of X-ray scattering by a three-dimensional solid (scattering of **k** = **G**) is $\sim N^2$, plus a small correction of $N^{5/3}$, and e^{-2W} decreases with T. (See (29-18) and (29-19).) The proportionality to N^2 represents the periodic structure of the crystal. Crystals in one-dimensional space have no periodic structure, because $\langle |\rho_G|^2 \rangle$ is only proportional to N. For crystals in two dimensions, at high temperature $\langle |\rho_G|^2 \rangle$ does not exhibit periodic structure, but when $T < 4\pi\alpha$ it is intermediate between N^2 and N.[b]

[b] For a more detailed analysis, see Imry and Guntler (1971).

29.3. Quantum Vector Model

The above examples do not involve quantum mechanics. Now we look at the simplest quantum ferromagnet model, whose Hamiltonian is

$$H = -\frac{1}{2} J \sum_{\mathbf{R}} \sum_{\mathbf{a}} \mathbf{s}(\mathbf{R}) \cdot \mathbf{s}(\mathbf{R} + \mathbf{a}) \quad, \tag{29-24}$$

where $\mathbf{R}$ is the position of a lattice site, $\mathbf{R} + \mathbf{a}$ denotes a neighbouring site, and $\mathbf{s}(\mathbf{R})$ is the spin $\frac{1}{2}$ operator, whose three components are

$$s_x = \frac{1}{2}\begin{pmatrix} 0 & 1 \\ 1 & 0 \end{pmatrix}, \quad s_y = \frac{1}{2}\begin{pmatrix} 0 & -i \\ i & 0 \end{pmatrix}, \quad s_z = \frac{1}{2}\begin{pmatrix} 1 & 0 \\ 0 & -1 \end{pmatrix}.$$

$$\tag{29-25}$$

In addition there are two convenient combinations s_+ and s_-:

$$s_\pm \equiv s_x \pm i s_y \quad,$$

$$s_+ = \begin{pmatrix} 0 & 1 \\ 0 & 0 \end{pmatrix}, \quad s_- = \begin{pmatrix} 0 & 0 \\ 1 & 0 \end{pmatrix}. \tag{29-26}$$

Each $\mathbf{R}$ is associated with a set of these operators.

The ground state of H is easily seen. If all the spins point in the same direction, H is minimum. Now pick a particular direction, the direction $-z$, i.e. the "downward" direction. Let all the spins point downwards. We call this state Φ_0:

$$H\Phi_0 = E_0 \Phi_0 \quad,$$

$$E_0 = -\frac{1}{4} J d N \quad, \tag{29-27}$$

where d is the dimensionality of space, $2d$ is the number of nearest neighbours of each lattice site (assumed to be a cubic lattice) and N is the total number of lattice sites, as in the above examples. Notice that s has three components, i.e. s is a vector in three dimensions, while the dimension of the space d of the lattice can be chosen at will.

We now look at the excited state of H. The simplest state is obtained by flipping one of the spins, and we get

$$\varphi(\mathbf{R}) \equiv s_+(\mathbf{R}) \Phi_0 \quad . \tag{29-28}$$

There are N such excited states, but they are not eigenvectors of H, i.e. not the stationary states. However $H\varphi(\mathbf{R})$ does not involve any other types of excited states (it only involves states with one up spin). Therefore, a linear combination of these states can give the stationary state. The simplest such combination is

$$\sum_{\mathbf{R}} \varphi(\mathbf{R}) = S_+ \Phi_0 \quad ,$$

$$S_+ \equiv \sum_{\mathbf{R}} s_+(\mathbf{R}) \quad . \tag{29-29}$$

This combination is in fact another ground state, not an excited state. Since H is a scalar, it is invariant under a rotation of the coordinates, i.e. H has rotational symmetry:

$$[H, \mathbf{S}] = 0 \quad ,$$

$$\mathbf{S} \equiv \sum_{\mathbf{R}} \mathbf{s}(\mathbf{R}) \quad . \tag{29-30}$$

The operator $\mathbf{S}$ is the total angular momentum. Hence, applying S_+ on Φ_0 gives a state of the same energy, and applying it again we get another and there is a total of $N+1$ ground states.

Now let

$$\varphi_{\mathbf{k}} \equiv \frac{1}{\sqrt{N}} \sum_{\mathbf{R}} e^{i\mathbf{k} \cdot \mathbf{R}} \varphi(\mathbf{R}) \quad . \tag{29-31}$$

This state is an eigenstate of H and its energy $\omega_{\mathbf{k}}$ (over E_0) is

$$(H - E_0) \varphi_{\mathbf{k}} = \omega_{\mathbf{k}} \varphi_{\mathbf{k}} \quad ,$$

$$\omega_{\mathbf{k}} = \frac{1}{2} J \sum_{\mathbf{a}} (1 - \cos \mathbf{k} \cdot \mathbf{a}) \quad . \tag{29-32}$$

This result is obtained by directly substituting (29-31) into (29-24). Notice that H is unchanged by a uniform displacement of the lattice. (Assume cyclic boundary, i.e. no boundary.) Hence the eigenstates of H must be eigenstates of the displacement operator, i.e. plane waves. Equation (29-31) is the state of one magnon, because there is only one spin turned up. This magnon can be treated as a particle with energy ω_k and momentum $\mathbf{k}$, and (29-28) can be looked upon as a magnon at $\mathbf{R}$.

At low temperatures, the number of magnons is very small, and this model becomes an ideal gas. (We neglect the situations when two or more magnons are at the same point.)

The magnons are bosons because $[s(\mathbf{R}), s(\mathbf{R}')] = 0$ if $\mathbf{R} \neq \mathbf{R}'$ (i.e. if there is no overlapping). The number of magnons is not conserved and hence its chemical potential is zero. The number distribution is like that of the phonons in a crystal. The total energy is

$$E - E_0 = \sum_\mathbf{k} \frac{\omega_\mathbf{k}}{e^{\omega_\mathbf{k}/T} - 1}$$

$$= V(2\pi)^{-d} \int d^d k \, \frac{\omega_\mathbf{k}}{e^{\omega_\mathbf{k}/T} - 1} \, . \tag{29-33}$$

When one spin is flipped, the total angular momentum in the z direction is increased by 1, and the total number of magnons is

$$N' = \sum_\mathbf{k} (e^{\omega_\mathbf{k}/T} - 1)^{-1} \, ,$$

$$|\langle \mathbf{S} \rangle| = \frac{N}{2}\left(1 - \frac{2N'}{N}\right) \, . \tag{29-34}$$

As N' increases the total magnetic moment decreases. From (29-33) and (29-34) we can calculate the thermodynamical properties.

If k is very small, then (29-32) becomes

$$\omega_\mathbf{k} \simeq \lambda k^2 \, ,$$

$$\lambda \equiv \frac{1}{4} J a^2 \, . \tag{29-35}$$

If $\omega_\mathbf{k} \ll T$, then

$$(e^{\omega_\mathbf{k}/T} - 1)^{-1} \simeq T/\omega_\mathbf{k} \, . \tag{29-36}$$

We can estimate N':

$$\frac{N'}{N} \simeq \frac{1}{n} (2\pi)^{-d} \int d^d k \left(\frac{T}{\lambda k^2}\right)$$

$$\simeq \left(\frac{T}{n\lambda}\right) L , \quad d = 1 ,$$

$$\simeq \frac{T}{n\lambda} \ln\left(\frac{L}{a}\right) , \quad d = 2 ,$$

$$\simeq \frac{T k_D}{2\pi \lambda n} , \quad d = 3 , \quad (29\text{-}37)$$

where $k_D \sim 1/a$ is the upper limit of the k integration, L is the size of the system and $n = N/V$. We can repeat the procedure of the above two sections. When $d \leq 2$, no matter how T is, if L is sufficiently large, N'/N can exceed $\frac{1}{2}$. If N'/N is too large then the total magnetic moment disappears. (See (29-34).) If $d = 2$, $\ln(L/a)$ cannot be too large. Therefore if T is sufficiently small, ferromagnetism can be maintained. When $d = 3$ there is no problem of ferromagnetism at low temperatures. When $d = 1$ ferromagnetism does not exist.

In this model the fluctuation of the spins is described by the magnons, avoiding direct reference to the directions of the vectors. Indeed we can ignore quantum mechanics and arrive at the conclusion (29-37) quickly. In (29-36) we have already ignored quantum mechanics. The aim of this section is to show that conclusions such as (29-37) have nothing to do with quantum mechanics.

29.4. Continuous Symmetry and Soft Modes

From the fluctuation of the liquid surface to the fluctuation of the spins we have repeatedly encountered the same results. Obviously these examples reflect a common phenomenon.

There is a common point in the models of these examples, that is, the total energy function H has a certain continuous symmetry and in the condensed state this continuous symmetry is destroyed because of freezing.

Symmetry means that H is invariant under a certain change of the variables. For example if we rotate the coordinates, the energy of the vector model is unchanged; if we translate the coordinates, the interaction between the atoms is unchanged, if we move the liquid surface uniformly upwards or downwards, the energy is unchanged. Symmetry, through the basic assumption of statistical mechanics, becomes the homogeneity of the equilibrium state. Translational invariance results in homogeneity of the spatial distribution; there cannot be a nonuniform density. This is because the determination of the region of motion in configuration space is related to the total energy only.

Hence, condensation violates this homogeneity. The trajectory of motion in phase space has been limited to the part of the region of motion determined by the basic assumption, and it cannot go beyond it. Thus, there is a special direction (e.g. the total magnetic moment) or a nonuniform density (e.g. the solid-liquid interface).

We repeat once more that condensation is not limited to regular condensed states. In amorphous magnets the frozen spins are random. According to our definition, this frozen state still has a preferred direction. If we rotate the coordinates the state will be different. Although the vector sum of all the spins is zero, each spin has its own average value pointing to a certain direction. We can take any three mutually perpendicular directions of the spin as the coordinate axes attached inside the body. These directions are frozen in and unchanged. All solids, not only crystals, violate the homogeneity of space. When the positions of the atoms are frozen in, translational symmetry is destroyed, whether it is a crystal or a random arrangement. We can pick the positions of four atoms not on the same plane to determine the position and orientation of this solid.

What is continuous symmetry? The opposite of continuous is discrete. The symmetry of the Ising ferromagnet model is discrete. If we change all s_i to $-s_i$, the energy is unchanged. The change of this variable is not continuous. The symmetry of the vector model is continuous: the direction of the coordinate axes can be continuously rotated. Coordinate translation is also a continuous change.

The freezing of the continuous symmetry results in the existence of many ground states. For example the ferromagnet in the above example has $N+1$ ground states, i.e. the magnetic moment can condense in any one direction. Likewise the solid can stay in any position. Due to the existence of so many ground states, it is not hard to perceive that there will be many excited states with very low energy. The explanation is as follows.

A system is made up of many parts, e.g. the above ferromagnet model can be thought of as composed of many individual ferromagnets. In the ground state, all the spins point in the same direction. Now rotate the individual spin by a small amount. If all the rotation angles are the same, the energy is unchanged. But if the various rotation angles are slightly different, changing smoothly from one part to another, then the energy will be slightly increased. This is a low energy excited state. This can be explained by writing down the equations from quantum mechanics. Let each spin be rotated about the x-axis, the operator for a rotation of θ is, for small θ,

$$e^{-i\theta S_x} \simeq 1 - i\theta S_x = 1 - i\theta \sum_{\mathbf{R}} s_x(\mathbf{R}) \quad . \tag{29-38}$$

If each spin is rotated by a different angle, then the operator is

$$U \equiv 1 - i \sum_{\mathbf{R}} \theta(\mathbf{R}) s_x(\mathbf{R}) \quad . \tag{29-39}$$

If each θ is the same, then $UHU^{-1} = H$, i.e. $U\Phi_0$ is also the ground state. H is the sum of many terms which individually have their own symmetry. Each term has only a few spin operators, like

$$H_{\mathbf{R}} \equiv -J\mathbf{s}(\mathbf{R}) \cdot \mathbf{s}(\mathbf{R} + \mathbf{a}) \quad , \tag{29-40}$$

in (29-24). If these two spins are rotated by the same angle, this term is unchanged, but if they are rotated differently, i.e. $\theta(\mathbf{R}) \neq \theta(\mathbf{R} + \mathbf{a})$, then this term will change. How much is the change? This depends on the original relative orientation of the two spins. If they are originally perpendicular, then the change of this term is proportional to the difference of $\theta(\mathbf{R})$ and $\theta(\mathbf{R} + \mathbf{a})$. If they are originally parallel, then it is proportional to the square of the difference between the two angles. Let $\theta(\mathbf{R}) \propto e^{-i\mathbf{k} \cdot \mathbf{R}}$, then

$$UH_{\mathbf{R}} U^{-1} - H_{\mathbf{R}} = O(ka)$$

$$\text{or } O((ka)^2) \quad . \tag{29-41}$$

Equation (29-41) will hold so long as $H_{\mathbf{R}}$ does not involve spins too far apart, and not just for the special form (29-40). Therefore our conclusion is

$$\langle \Psi | (H - E_0) | \Psi \rangle = O(ka) \quad \text{or} \quad O(k^2 a^2) \quad ,$$

$$\Psi = (U - 1)\Phi_0 \quad ,$$

$$\theta(\mathbf{R}) \propto e^{-i\mathbf{k} \cdot \mathbf{R}} \quad , \tag{29-42}$$

where **k** can have many different small values. Hence we have proved the existence of the low energy excited states. The above analysis is not limited to spin models, or quantum mechanics. However, it is restricted to model with short range interaction. We have deliberately displayed the factor of a in ka in (29-41) and (29-42). If a is too large, the result is not certain to be correct. If we are dealing with models with electrostatic interaction, this result is then unreliable.

The above result of the appearance of low energy excited states (energy $\to 0$ as $k \to 0$) when the continuous symmetry is frozen has the same meaning as the Goldstone theorem in quantum field theory.

The phonons for the vibration of a crystal and the above-mentioned magnons are this kind of low energy excitations. So is the surface tension wave in a liquid surface. We called these the "soft modes", as their wavelength is long and their frequency low.

If the distribution of the order parameter is random, like the spins in an amorphous magnet, then the soft modes may not have a definite wavelength. Its form is determined by the distribution of the spins and the interaction energy. In amorphous solids the soft modes may not be plane waves. There is as yet little work in this direction.

Therefore, for both classical and quantum mechanical models, if there is a freezing in the continuous symmetry, soft modes appear. If the amplitude of these soft modes is large, they can break the condensation. In one and two dimensions, the role of long wavelength is quite important, and the soft modes can soften the order parameter. The above examples all reflect this phenomenon.

This phenomenon of condensation softening can be treated rigorously. The conclusion is essentially the same as the above examples, but the rigorous analysis yield many other important details.[c]

29.5. Defects in the Condensation

The above discusses the smooth fluctuation of the order parameter. This is the characteristic of continuous symmetry. The change is smooth and the energy is low. The amplitude of the fluctuation can be regarded as a smooth and continuous function of space. Both $\theta(\mathbf{R})$ in the XY model and $u(\mathbf{R})$ in the crystal model can be regarded as continuous functions of $\mathbf{R}$ because their short distance change (of the order of a lattice spacing) is very small.

---

[c] See Hohenberg (1957), Mermin and Wagner (1966). These are easy to read and are very important papers.

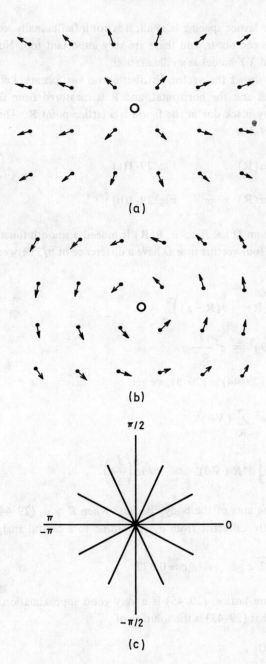

Fig. 29-1 (a), (b) Vector distributions $\theta(\mathbf{R})$ on a plane. $\alpha(\mathbf{R})$ is shown in (c). Notice that π and $-\pi$ have the same meaning.

Although the lattice spacing is small, it is not infinitesimally small. Discontinuous situations can occur, and these are very important too. Now we use the two-dimensional XY model as an illustration.

Figure 29-1 shows the vector distribution on two planes. Let $\alpha(\mathbf{R})$ be the angle between $\mathbf{R}$ and the horizontal, and $\mathbf{R}$ is measured from the origin O in the figure. Every black dot in the figure is a lattice point $\mathbf{R}$. The vector distribution in Fig. 29-1 is

$$\theta(\mathbf{R}) = \alpha(\mathbf{R}) \quad , \qquad \text{Fig. 29-1(a)}$$

$$\theta(\mathbf{R}) = \alpha(\mathbf{R}) + \frac{\pi}{2} \quad , \qquad \text{Fig. 29-1(b)} \quad . \tag{29-43}$$

In regions far from O i.e. $R \gg a$, $\theta(\mathbf{R})$ is indeed a smooth function. But near O it is not. The four vectors near O have a difference of $\pi/2$ between neighbours. If $R \gg a$, then

$$\frac{1}{2} \sum_{\mathbf{a}} \left(\theta(\mathbf{R}) - \theta(\mathbf{R}+\mathbf{a}) \right)^2$$

$$\simeq a^2 (\nabla \theta)^2 = a^2 \frac{1}{R^2} \quad . \tag{29-44}$$

If we substitute (29-44) in (29-3), we get

$$H' = \frac{J}{2} a^2 \sum_{\mathbf{R}} (\nabla \theta)^2$$

$$= \frac{J}{2} \int d^2 R \, (\nabla \theta)^2 \simeq \pi J \ln\left(\frac{L}{a'}\right) \quad , \tag{29-45}$$

where πL^2 is the area of the body. Because when $R \simeq a$, (29-44) is incorrect, a' will be slightly different from a. According to a careful analysis, if we use

$$\frac{a}{a'} = 2\sqrt{2} \, e^c \quad , \qquad c = 0.577 \quad , \tag{29-46}$$

then for a square lattice, (29-45) is a very good approximation.[d] The reader may recognise that (29-43) is the solution of

$$\nabla^2 \theta = 0 \quad , \tag{29-47}$$

---

[d] See Kosterlitz (1974), p. 1049.

not too near the origin. This is equivalent to saying that, were it not for the defect near the origin, (29-43) would minimise the energy

$$H \simeq \frac{1}{2} J \int d^2 r \, (\nabla \theta)^2 \quad . \tag{29-48}$$

Therefore, apart from the defect near the origin, (29-43) is quite "perfect". Notice that going round this defect once, $\theta(\mathbf{R})$ in (29-43) changes by 2π:

$$\oint \nabla \theta \cdot d\mathbf{R} = 2\pi \quad . \tag{29-49}$$

These are expressions commonly seen in fluid mechanics and electrostatics. If $\theta(\mathbf{R})$ is regarded as the velocity potential in fluid mechanics, then $\nabla \theta$ is the velocity of flow and the configuration of θ in Fig. 29-1 is a vortex. In two-dimensional electrostatics a charge produces an electric field proportional to $1/R$. Therefore, (29-44) can also be seen as the result of having a charge. Hence the effect of this defect is easy to analyse.

Of course, (29-45) points out that the energy of a vortex or charge increases like $\ln L$. But if there is a pair of opposite charges or vortices, then energy is only related to their distance apart (see Fig. 29-2).

$\theta(\mathbf{R})$ in Fig. 29-2 is

$$\theta(\mathbf{R}) = \alpha(\mathbf{R}) - \alpha(\mathbf{R} - \mathbf{R}') \quad . \tag{29-50}$$

The energy of a pair of vortices is

$$H' = 2\pi J \ln\left(\frac{R'}{a'}\right) \quad . \tag{29-51}$$

If there are many vortices, then

$$H' = -\frac{1}{2} \sum_{i,j} q_i q_j \ln\left(\frac{|\mathbf{R}_i - \mathbf{R}_j|}{a'}\right) \quad , \tag{29-52}$$

where $\mathbf{R}_i$ are the position of the vortex centre or the position of the charge and

$$\frac{1}{\sqrt{2\pi J}} q_i = \frac{1}{2\pi} \oint \nabla \theta \cdot d\mathbf{R} \quad , \tag{29-53}$$

is the integral around $\mathbf{R}_i$, representing the vortex strength or the charge. Now we summarise the meaning of (29-52) as follows.

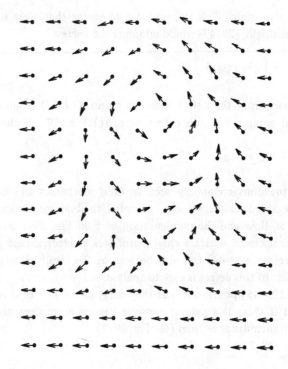

Fig. 29-2 The energy of two opposite vortices depends on their separation.

At every $\mathbf{R}_i$, we cut off a small hole of radius a' and on the plane punctured by these holes we solve $\nabla^2 \theta = 0$. The solution of this equation is a smooth function, and this is substituted into (29-48) to calculate the energy, thus obtaining (29-52). There is a condition for each small hole, i.e. the integral (29-53) must be an integer. There is an additional condition that the total charge is zero.

$$\sum_i q_i = 0 \ . \tag{29-54}$$

Equation (29-52) is the energy of a group of defects. This energy is not infinite because a' is not zero, i.e. the lattice spacing is not zero.

As $\theta(\mathbf{R})$ is the electric potential, the influence of the charges on the electric potential has been discussed in Chapter 19. At low temperatures, positive and negative charges pair together, leading to a finite dielectric constant. The fluctuation and correlation of $\theta(\mathbf{R})$ follow mainly the results of Sec. 29.1.

If the temperature is high, the charges will separate, and the correlation length for the fluctuation $\theta(\mathbf{R})$ is finite, hence condensation collapses. The temperature where this takes place is $T_0 \simeq q^2/4 = \pi J/2$,[e] according to the rough estimate of Chapter 19.

So, it can be seen that defects can cause the condensation to collapse. Notice that at high temperatures, neighbouring vectors will not have directions which are nearly the same and the concept of defects will not be very useful. When nowhere is "perfect", what is the point of talking about "defects"? The above charge model assumes that defects are few, and $\theta(\mathbf{R})$ is an otherwise perfect and smooth function, only with holes at a few places. The paired charges or separated charges are built upon this assumption of near perfection. Nevertheless, the separation temperature $T_0 \simeq \pi J/2$ for the paired charges is a rather high temperature compared to J. At such a high temperature, is the above assumption still valid? If we use the mean field method of the last chapter to calculate the critical temperature, the answer is $T_c = 2J$. (The calculation is very simple, and is left as an exercise for the reader.) This is close to $T_0 = \pi J/2$, and is somewhat worrying. But, perhaps by luck, the density of the charges at $T = T_0$ is still very low. Generally speaking the energy of each charge is about $\epsilon = q^2 \ln l/a$ and the average density is $(a/l)^2$, where l is the average distance between charges. Hence

$$\left(\frac{a}{l}\right)^2 \sim e^{-\epsilon/T} \sim \left(\frac{a}{l}\right)^{q^2/T}. \tag{29-55}$$

Therefore, if we want $a/l \gtrsim 1$, then we must have $T \gtrsim q^2 = 4T_0$. Near $T = T_0$, the density of the charges should not be very large. Perhaps this rough estimate is not sufficient to dispel all doubts. Later numerical simulation provided better evidence.

The above is for the two-dimensional XY model. In three dimensions the planar holes would become tubes, i.e. the defects are vortex lines. The term "charges" will not be appropriate and must be changed to line current, and $\nabla\theta$ denotes the direction of the magnetic field. A vortex line can wind into a circle forming a vortex ring, or the ends of the line may be positioned on the boundary of the body. In this way, defects involve not just a few lattice points but a large number of them, and the energy is certainly very high. Therefore, they should not appear in the equilibrium state. But, this vortex line is metastable and once formed will not disappear easily. To eliminate a vortex we have

---

[e] For a more careful analysis, see Kosterlitz (1974).

514 STATISTICAL MECHANICS

to rearrange many elements. (This is also true in two dimensions.) These will be rediscussed in the next chapter on superfluidity.

The defects in crystals are much more complicated, and the simplest is the line dislocation (Fig. 29-3). There is an extra line in the crystal shown in the figure. There are many other types of defects and we shall not discuss these here. The reader can consult books on material science. The theory of defects also appears in other models. The reader can look up the recent literature.[f]

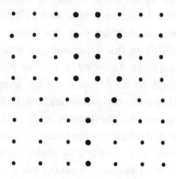

Fig. 29-3 A line dislocation in a crystal.

Problems

1. Solve the one-dimensional XY model directly from (29-1). Do not use (29-3).

   (a) Calculate the thermodynamic potential, entropy and heat capacity, and sketch them versus T.

   (b) Calculate the correlation function $C(R)$. [For definition see (29-11).]

   (c) Compare the above result with the solution of (29-3).

   (d) If the angle between two neighbouring vectors is greater than a right angle, we say that there is a defect between these two elements. With this definition of defect, calculate the defect density in this model.

2. In the above problem, would we encounter difficulty when we define entropy? How can we resolve this?

   Notice that s_i is a continuous variable. All continuous variables require quantum mechanics for the definition of entropy.

3. Review Chapter 27. Use the mean field approximation for the XY model (29-1). Calculate the critical temperature T_c.

---

[f] E.g. Young (1979), Nelson and Halperin (1980).

4. Add a magnetic field to (29-1):

$$-\sum_i \mathbf{h}_i \cdot \mathbf{s}_i \quad . \tag{29-56}$$

Generalise the analysis of Problems 27.2 and 27.8 to the XY model, assuming that $\mathbf{h}_i$ is very small.

(a) Generalise (27-47). Consider the two special cases of $T \ll T_c$ and T near T_c. Notice that when $T < T_c$, the order parameter points to a fixed direction.

(b) Calculate the correlation function $C(R)$ for $T \ll T_c$, and for T near T_c. Compare it with the analysis of Sec. 29.1.

5. Apply the analysis of the above three problems to the three-dimensional vector model, i.e. $\mathbf{s}_i$ is a vector in three dimensions, and space has dimension d. This is the so-called classical Heisenberg model.

6. Bose gas and gauge transformation.

(a) Review Problem 3.8. The discussion in Chapter 3 neglected the interaction. Now we add an interaction term

$$H' = \frac{1}{2} \sum_{i,j} u(\mathbf{r}_i - \mathbf{r}_j) \quad , \tag{29-57}$$

where $\mathbf{r}_i, i = 1, 2, \ldots, N$ are the positions of the particles.

(b) Let $\Psi(\mathbf{r}_1, \ldots, \mathbf{r}_N)$ be a wavefunction of the system. Define the gauge transformation as

$$\Psi' = e^{-i\alpha N} \Psi \quad , \tag{29-58}$$

where α is the gauge rotation angle. Prove that the total energy is unchanged under the gauge transformation, i.e.

$$e^{i\alpha N} H e^{-i\alpha N} = H \quad . \tag{29-59}$$

Here N is regarded as an operator and this equation says that particle number is conserved.

(c) For the moment neglect H' and consider the kinetic energy:

$$H_0 = \sum_{\mathbf{k}} \epsilon_{\mathbf{k}} N_{\mathbf{k}} \quad , \qquad \epsilon_{\mathbf{k}} = k^2/2m \quad , \tag{29-60}$$

$N_{\mathbf{k}}$ is the number of particles in the state $\mathbf{k}$.

This can be regarded as a group of simple harmonic oscillators, because N_k is a positive integer. State **k** can be looked upon as a simple pendulum, with energy

$$\epsilon_k N_k = \tfrac{1}{2} \epsilon_k (p^2 + q^2 - 1) \quad , \tag{29-61}$$

where p and q are the momentum and position of the simple pendulum. The configuration is therefore represented by a point on the (q,p) plane.

Prove that the gauge transformation is a coordinate transformation on the (q,p) plane:

$$e^{i\alpha N_k}(q,p)\,e^{-i\alpha N_k}$$
$$= (q \cos\alpha + p \sin\alpha,\; p \cos\alpha - q \sin\alpha) \quad . \tag{29-62}$$

We can write it more simply as

$$e^{i\alpha N_k} a_k e^{-i\alpha N_k} = a_k e^{-i\alpha} \quad , \tag{29-63}$$

$$a_k \equiv \frac{1}{\sqrt{2}}(q + ip) \,, \qquad N_k = a_k^* a_k \quad . \tag{29-64}$$

The operator a_k is the annihilation operator, while a_k^* is the creation operator.

(d) Problem (3.8) shows that at low temperatures, N_0 is very large. Hence the state $\mathbf{k}=0$ can be treated classically. The expectation value $\langle a_0 \rangle$ is a large number of order $\sqrt{N_0}$ and this can be regarded as a condensation phenomenon, i.e. the Bose-Einstein condensation.

Apply the analysis of soft modes in Sec. 29.4 to this model.

(e) Suppose H' cannot be neglected, but condensation still occurs, i.e. $N_0 = O(N)$. Use the analysis in Sec. 29.4 to discuss the low energy excited states.

We shall rediscuss the bosons in the next chapter.

7. Discuss the quantum antiferromagnetic model, i.e. changing J in (29-24) to $-J$. This model is not so simple as the ferromagnetic one. The reader can consult books on solid state physics.

8. (a) Compute $W(k)$ for $d = 1, 2$, (see (29-17) and (29-18)) and compare with (29-21).

(b) Find the position distribution of any one atom:

$$\langle \delta(\mathbf{r} - \mathbf{u}(\mathbf{R})) \rangle \quad .$$

(c) Suppose a copper film (with one layer of copper atoms) has a size $L = 1$ km. Estimate the value of $\langle u^2 \rangle$.

9. Suppose N atoms form a one-dimensional chain with N' impurity atoms distributed randomly in between, with $N \gg N'$. The total energy function is like problem (8.12), but the mass m' of the impurity atoms is much larger than the mass m of the other atoms.

(a) Calculate the influence of the impurity atoms on the thermodynamic properties.

(b) Discuss the influence of the impurity atoms on the amplitude of atomic vibrations.

10. In a d-dimensional crystal, N' atoms are replaced by impurity atoms which are distributed randomly. The last problem deals with $d = 1$.

Discuss the influence of the impurity atoms on the various conclusions of Sec. 29.2.

11. Review the fundamentals of fluid mechanics and electrostatics, paying special attention to the application of the complex variable theory.

(a) Let $z = x + iy$, and let $f(z)$ be a function of z. Suppose that in a certain neighbourhood, $f(z)$ is an analytic function of z, i.e. df/dz exists. Prove that in that neighbourhood

$$\left(\frac{\partial^2}{\partial x^2} + \frac{\partial^2}{\partial y^2} \right) f = 0 \qquad (29\text{-}65)$$

Let $f = \phi + i\psi$. Then ϕ and ψ satisfy $\nabla^2 \phi = 0$ and $\nabla^2 \psi = 0$.

(b) Prove that the equipotential lines of ϕ and ψ are orthogonal i.e. $\nabla \psi \cdot \nabla \phi = 0$.

(c) Let $f(z) = \ln z$. Discuss how to use this function to analyse vortices and charges.

Chapter 30
THEORY OF SUPERFLUIDITY

Superfluidity denotes frictionless, continuous flow in the absence of external force. This phenomenon occurs at low temperatures, e.g. ^{4}He liquid ($T < 2.17$ K), and ^{3}He liquid (T lower than about 10^{-3} K). The electrons of many metals exhibit this phenomenon at low temperatures, with the transition temperature depending on the kind of metal; and this superfluidity of electrons is called superconductivity. The superfluidity of ^{3}He is rather complicated. Because ^{3}He and electrons are both fermions, the theory of ^{3}He is like the theory of superconductivity. On the other hand ^{4}He are bosons, and the situation is simpler. All the phenomena of superfluidity share certain basic features. This chapter does not attempt to give a systematic introduction to the properties of liquid He and superconductors. Our aim is to analyse the basic principles of superfluidity which is a rather mysterious property. Here we shall use the simplest model to explain it.

Superfluidity involves two phenomena: condensation and metastability, the former being a necessary condition for the latter. The state of flow is a metastable state, and over a short period of time (which may be years) it is effectively an equilibrium state and the thermodynamical potential is a minimum so that the flow is not stopped by perturbations (provided the perturbations are not too large). Our aim for discussing the metastable state is to understand the process causing the flow velocity to decrease. The thermodynamical potential of the flowing state is higher than that of the static state and is not an absolute minimum. But if the flowing state is to make a transition to the static state, it must first overcome a potential barrier. (See Chapter 21 for the discussion of the metastable state.) We must understand what this potential barrier is, only then will we have an initial understanding of superfluidity.

If there is no condensation, the stability of the flow will not occur. The mystery of superfluidity lies in this order parameter. Indeed, what is being condensed? We shall use a ferromagnetic model to explain this point. We first establish a lattice gas model. The content of this model is like the spin $\frac{1}{2}$ ferromagnetic model in the last chapter. Because we are quite familiar with the ferromagnetic model, we can use it to analyse the property of the lattice gas, just as we analysed the coexistence of liquid and gas using the Ising ferromagnet in Chapter 27. Here the fluid model is the quantum lattice gas, an offspring of the quantum spin $\frac{1}{2}$ model. The order parameter is the magnetic moment, which is nothing mysterious.

After establishing the order parameter, we can discuss superfluidity. We first simplify the model to a planar vector model, i.e. the XY model in the last chapter. Then we discuss the order parameter in the volume of a torus. The winding number is then introduced. It represents the change of the angle of the order parameter divided by 2π when circling the torus once. This winding number n is a conserved quantity if we do not consider the condensed defects. The superfluid flow velocity is proportional to n. The defects are just the vortices. The content of the last part of this chapter is to investigate how vortices change n. From this we can understand the essence of the theory of superfluidity. The detailed analysis rests on the two-dimensional model (i.e. the model of ^{4}He adsorbed on a smooth surface). The experiences we gain in Chapter 19 and the last chapter will be very helpful here. The theory is simpler in two dimensions but the experiment is more difficult. Both experiment and theory appeared only recently and the bibliography is rich. But the aim of this chapter is not to introduce the literature to the reader. Many other interesting problems, like the various sound waves, the thermodynamic properties at low temperatures, have not been mentioned here. Neither is superconductivity, even though it is an important and interesting science. There are many similarities with the basic principles of ^{4}He, even though electrons are fermions and are charged. Another important topic is the behaviour of superfluidity and other properties near the critical point. We do not discuss this either. Hence the content of this chapter is quite minimal, discussing the metastability of the condensation and the flow. We shall make a careful analysis of these two important points.

30.1. Quantum Lattice Gas

Let us set up a quantum lattice gas model now. Let $n(\mathbf{R})$ be the particle number on the site $\mathbf{R}$, with value 0 or 1.[a] Let

$$n(\mathbf{R}) - \tfrac{1}{2} = s_z(\mathbf{R}) \quad , \tag{30-1}$$

where s_z is the z component of a spin $\tfrac{1}{2}$ operator. Hence the lattice gas model has a content similar to the spin $\tfrac{1}{2}$ model.

The state with zero particle number is the vacuum, i.e. the state with all spins pointing to $-z$ and we call it Φ_0,

$$\sum_{\mathbf{R}} n(\mathbf{R}) \Phi_0 = 0 \quad . \tag{30-2}$$

The state $s_+(\mathbf{R}) \Phi_0$ describes a single particle at $\mathbf{R}$. If the particle is a plane wave, then its wavefunction is

$$\psi_\mathbf{k} = \frac{1}{\sqrt{N}} \sum_{\mathbf{R}} e^{i\mathbf{k}\cdot\mathbf{R}} s_+(\mathbf{R}) \Phi_0 \quad . \tag{30-3}$$

The operator $s_+(\mathbf{R})$ increases the value of $s_z(\mathbf{R})$ by 1, in other words, it adds a particle at $\mathbf{R}$. Now we have to write down a reasonable Hamiltonian operator H. We first ask: what is the kinetic energy? The kinetic energy of a particle should be $k^2/2m$. But now the particle is restricted to a discrete lattice, it is not easy to find a reasonable kinetic energy operator. Now we construct an approximate form. Let

$$K \equiv \frac{1}{4ma^2} \sum_{\mathbf{R}} \sum_{\mathbf{a}} [s_+(\mathbf{R}) - s_+(\mathbf{R}+\mathbf{a})] [s_-(\mathbf{R}) - s_-(\mathbf{R}+\mathbf{a})] \quad . \tag{30-4}$$

If $ka \ll 1$, operating with K on (30-3) will give

$$K \psi_\mathbf{k} \simeq \frac{k^2}{2m} \psi_\mathbf{k} \quad . \tag{30-5}$$

In (30-4) $\mathbf{a}$ is the neighbouring vectors of the site, and we take the lattice to be cubical. Equation (30-4) can also be written as

$$K = -\frac{1}{2} J \sum_{\mathbf{R}} \sum_{\mathbf{a}} \mathbf{s}(\mathbf{R}) \cdot \mathbf{s}(\mathbf{R}+\mathbf{a})$$

$$+ \frac{1}{2} J \sum_{\mathbf{R}} \sum_{\mathbf{a}} s_z(\mathbf{R}) s_z(\mathbf{R}+\mathbf{a}) + JN' d \quad , \tag{30-6}$$

---

[a] This n is not the winding number which we later discuss.

where N' is the total number of particles, $J \equiv (ma^2)^{-1}$. Hence K is very much like the exchange interaction in the ferromagnetic model. We consider the case when the particle number is unchanged, i.e. N' is a constant. Now we simply discuss the following model.

$$H = K + U$$

$$= -\frac{1}{2} J \sum_{\mathbf{R}} \sum_{\mathbf{a}} \mathbf{s}(\mathbf{R}) \cdot \mathbf{s}(\mathbf{R}+\mathbf{a}) + \text{constant} ,$$

$$U \equiv -\frac{1}{2} J \sum_{\mathbf{R}} \sum_{\mathbf{a}} n(\mathbf{R}) \, n(\mathbf{R}+\mathbf{a}) \quad . \tag{30-7}$$

That is to say, this gas has kinetic energy K and potential energy U, which comes from the attraction between neighbouring particles. Notice that $n = s_z + \frac{1}{2}$. Therefore U is the second term on the right of (30-6) plus a constant. (N' is a constant.) Hence this model is exactly the same as the ferromagnetic model. Lastly, we need a momentum operator. Let

$$\mathbf{P} = -\frac{i}{2} \sum_{\mathbf{R}} \sum_{\mathbf{a}} \frac{1}{2} [s_+(\mathbf{R}) + s_+(\mathbf{R}+\mathbf{a})]$$

$$\cdot \frac{1}{a} [s_-(\mathbf{R}+\mathbf{a}) - s_-(\mathbf{R})] \frac{\mathbf{a}}{a} \quad . \tag{30-8}$$

This operator acting on the $\psi_{\mathbf{k}}$ of (30-3) gives, for $ka \ll 1$,

$$\mathbf{P} \psi_{\mathbf{k}} \simeq \mathbf{k} \psi_{\mathbf{k}} \quad . \tag{30-9}$$

Therefore $\mathbf{P}$ can essentially be regarded as the momentum operator. In the problems at the end of the chapter, we shall discuss some details of K and $\mathbf{P}$.

The aim of introducing this model is to point out several important features of quantum mechanics.

The Ising ferromagnet is not a spin $\frac{1}{2}$ vector model. The variable of motion of the Ising model has two values, like the spin $\frac{1}{2}$ model, but is otherwise quite different. Spin $\frac{1}{2}$ is a vector. Although each spin $\frac{1}{2}$ has two states, it is still a vector. Creating a three-dimensional vector from two states is the result of quantum mechanics. Without quantum mechanics, this is impossible. However, if we have many spin $\frac{1}{2}$ together pointing in the same direction, then the vector property of this ensemble can be analysed by the concepts of classical mechanics.

The Ising lattice gas model is a disguised form of the Ising ferromagnetic model, and the quantum lattice gas model (30-7) is a disguise of the spin $\frac{1}{2}$ model. The Ising gas is a classical model and its state is determined by the distribution of the particles just like the value of each spin of the Ising ferromagnet. The state of the quantum lattice gas model is specified by the distribution of the particles, just like the value of each $s_z(\mathbf{R})$ in the spin $\frac{1}{2}$ model. But, besides s_z, there are also s_x and s_y. That is to say, in quantum mechanics, all operators are variables of motion. If there are two states $n = 1$ and 0, then besides n, there must be two other variables s_x and s_y.

The spin $\frac{1}{2}$ model can have a condensed state when the total spin **S** condenses in any direction. The components S_x, S_y and S_z can all be nonzero. For the same reason, the quantum lattice gas model can have condensed states, and have nonzero S_x and S_y. ($S_z = N' - \frac{1}{2}N$ where N' is the total number of particles and N is the total number of lattice sites.) These nonzero S_x and S_y have no equivalent in a classical gas. But from the ferromagnetic model, this is a straightforward conclusion. The condensation of S_x and S_y is a necessary condition for superfluidity.

Model (30-7) can be regarded as a prototype ^{4}He system. The size of the lattice a is the size of an atom. There is a restriction of at most one atom per lattice site, representing the short distance repulsion between atoms. U is a special attractive force, rigged to make up a simple form of H even though it may not be entirely realistic.

Notice that this is a model for bosons. Although each lattice site can accommodate at most one atom, this is not due to the fermionic character but due to the hard-sphere repulsion. The wavefunction is symmetric.

30.2. The Ground State and Low Temperature Model

Now we analyse model (30-7), which is just a spin $\frac{1}{2}$ ferromagnetic model and is like (29-24) of the last chapter. The only condition is that

$$S_z = N' - \frac{N}{2}, \qquad (30\text{-}10)$$

is a constant, because the total number of particles N' is a constant.

$$N = V/a^3, \qquad (30\text{-}11)$$

is the number of lattice sites, V is the total volume and a^3 is the volume of a lattice cell. We first look at the ground state, which is the state with all the spins parallel. The direction of the spins, because of (30-10), is limited on a ring; see Fig. 30-1.

THEORY OF SUPERFLUIDITY 523

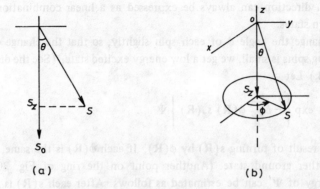

Fig. 30-1 (a) The ground state with all spins pointing downward;
(b) shows the ground state obtained by rotating (a).

The total length of **S** is $\frac{1}{2}N$, i.e. N spin $\frac{1}{2}$ linked together. The vector **S** makes an angle θ with the $-z$ direction. This angle is determined by the values of S_z and $|\mathbf{S}|$. The azimuthal angle ϕ can take any value.

The wavefunction of this ground state is very easy to calculate. The vector $\mathbf{S}_0$ in Fig. 30-1(a) denotes Φ_0, i.e. the zero particle state with all the spins pointing downwards. To obtain the ground state of N' particles, we only need to turn $\mathbf{S}_0$ by an angle θ:

$$\Psi = e^{-i\phi S_z} e^{-i\theta S_y} \Phi_0 . \tag{30-12}$$

We first rotate by θ about the y-direction, then by ϕ about the z-direction, then we get the direction of **S** in Fig. 30-1(b). Ψ is the ground state wavefunction. These rotations can be performed immediately because each spin can be rotated independently. The reader can prove that

$$\Psi = \prod_{\mathbf{R}} (u + v s_+(\mathbf{R})) \Phi_0 ,$$

$$u \equiv e^{i\phi/2} \cos \theta/2 ,$$

$$v \equiv -e^{i\phi/2} \sin \theta/2 . \tag{30-13}$$

In (30-13), $|u|^2$ is the probability of having no particle at **R**, while $|v|^2$ is that of finding one particle at **R**. Because of a delicate combination of u and v, S_x and S_y are not zero and the angle ϕ is therefore defined. From the viewpoint of the spin model this combination is not surprising. The state of the spin pointing to a certain direction can always be expressed as a linear combination of the up and down states.

If we change the angle ϕ of each spin slightly, so that the change of ϕ for neighbouring spins is small, we get a low energy excited state. (See the discussion of Sec. 29.4.) Let

$$\Psi' \equiv \exp\left\{-i \sum_{\mathbf{R}} \phi(\mathbf{R}) \, s_z(\mathbf{R})\right\} \Psi \quad . \tag{30-14}$$

This is the result of turning $s(\mathbf{R})$ by $\phi(\mathbf{R})$. If each $\phi(\mathbf{R})$ is the same, then Ψ is just another ground state. (Another point on the ring of Fig. 30-1(b).)

The energy of Ψ' can be estimated as follows. After each $s(\mathbf{R})$ is rotated, what is the change of $s(\mathbf{R}) \cdot s(\mathbf{R} + \mathbf{a})$? If all the spins are parallel, this product is $\frac{1}{4}$. Now $s(\mathbf{R})$ has been turned by $\phi(\mathbf{R})$ and $s(\mathbf{R} + \mathbf{a})$ by $\phi(\mathbf{R} + \mathbf{a})$, and the change of this product is

$$\frac{1}{4} \sin^2 \theta \, \cos[\phi(\mathbf{R}) - \phi(\mathbf{R} + \mathbf{a})] \quad . \tag{30-15}$$

Because the rotation is around the z-axis, only the components on the xy plane is influenced, which is the reason for the factor $\sin^2 \theta$ in (30-15). Therefore the change of the total energy is

$$\langle \Psi' | H | \Psi' \rangle - \langle \Psi | H | \Psi \rangle \equiv H' \quad ,$$

$$H' = -\frac{J'}{2} \sum_{\mathbf{R}} \sum_{\mathbf{a}} \cos[\phi(\mathbf{R}) - \phi(\mathbf{R} + \mathbf{a})] \quad ,$$

$$J' = \frac{1}{4} J \sin^2 \theta \quad . \tag{30-16}$$

The angle θ is determined by $S_z = N' - \frac{1}{2}N$ and $|\mathbf{S}| = N/2$:

$$\sin^2 \theta = 1 - (2N'/N - 1)^2$$

$$= 4n(1-n) \quad ,$$

$$n \equiv N'/N \quad . \tag{30-17}$$

Therefore,

$$J' = J n(1-n) \quad . \tag{30-18}$$

Equation (30-16) is just the planar vector model or the XY model discussed in the last chapter. The oscillation of the angle ϕ is a kind of the soft modes discussed. The arrangement of the spins is a condensation, and this model has rotational symmetry with respect to the z axis. The low temperature properties of this model can be analysed by the Hamiltonian H' in (30-16). Notice that (30-16) is no longer a quantum model and the variables $\phi(\mathbf{R})$ are regarded as classical variables. Of course this model H' is derived from the quantum model (30-7). It is the low temperature version of (30-7) and only considers the motion of angle ϕ of the various spins.

H' can be written in a better form. Because $J \equiv 1/ma^2$ and

$$\sum_{\mathbf{R}} \simeq \frac{1}{a^3} \int d^3 R \quad,$$

$$\cos[\phi(\mathbf{R}) - \phi(\mathbf{R} + \mathbf{a})] \simeq 1 - \tfrac{1}{2}(\mathbf{a} \cdot \nabla \phi)^2 \quad, \tag{30-19}$$

we have

$$H' \simeq \frac{\rho}{2m} \int d^3 R \, (\nabla \phi)^2 + \text{constant} \quad, \tag{30-20}$$

$$\rho \equiv \frac{1}{a^3} n(1-n)$$

$$= \frac{N'}{V}\left(1 - \frac{N'}{V} a^3\right) \quad. \tag{30-21}$$

If the lattice spacing is very small, then ρ is the density of the particles. The reader should be reminded that (30-20) is the ferromagnetic model in a different guise.

30.3. State of Flow and Winding Number

Suppose we have a torus of length L and cross-section A, with the z axis along the torus. We put our lattice gas in this torus. The volume is $V = AL$ and the energy is H'. The use of (30-20) is simpler for our purpose.

If ϕ is a constant, then H' assumes the lowest value, so ϕ is the ground state. Now we calculate the other minima of the energy H'. Setting the derivative of H' with respect to $\phi(\mathbf{R})$ to zero, we get

$$\nabla^2 \phi = 0 \quad. \tag{30-22}$$

Of course ϕ = constant is a solution, but

$$\phi = \phi_n \equiv \frac{2\pi n}{L} z , \qquad n = \pm 1, \pm 2, \ldots , \qquad (30\text{-}23)$$

are also solutions of (30-20), because $\phi + 2n\pi$ is the same as ϕ, since ϕ is an angle. The z-axis is along the torus and $z = 0$ and $z = L$ are the same point. Hence $\phi(x, y, L) - \phi(x, y, 0)$ must be a multiple of 2π. Substituting (30-23) into (30-20), we obtain

$$H' = \frac{\rho V}{2m}\left(\frac{2\pi n}{L}\right)^2 \equiv E_n . \qquad (30\text{-}24)$$

E_n is a minimum value of H' and not a maximum because if we put $\phi = \phi' + \phi_n$ into (30-8) (ϕ' very small) we get E_n + a positive number.

We call n the winding number, whose meaning is obvious from Fig. 30-2. Figure 30-2 shows a torus, the cross-section being a circle of radius $\frac{1}{2} \sin \theta \equiv s_\perp$, the z coordinate is along the torus and the length is L. The coordinates ($s_\perp \cos\phi$, $s_\perp \sin\phi$) describe a curve winding the torus. But ($s_\perp \cos\phi$, $s_\perp \sin\phi$) is just (s_x, s_y). Therefore Fig. 30-2 is the distribution of the order parameter (represented by (30-23)) along the torus, and n is the winding number of this curve:

$$\text{Winding number} = \frac{1}{2\pi} \int_0^L dz \, \frac{\partial}{\partial z} \phi(x, y, z) . \qquad (30\text{-}25)$$

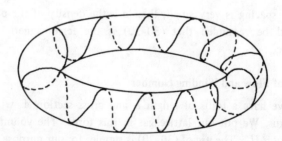

Fig. 30-2

According to the present model the state represented by ϕ_n is stable. The winding number n is unchangeable. As seen from Fig. 30-2, a winding line on on the torus cannot change its value of n except by breaking the line and reconnecting. That is to say n is a conserved quantity. No matter how the soft modes oscillate, if ϕ is a continuous function, its winding number is unchanged. Although $E_n > 0$, ϕ_n represents a stable state. When the temperature is nonzero, the thermodynamic potential F_n can be calculated from

$$H' = E_n + \frac{\rho}{2m} \int d^3R \, (\nabla \phi')^2 ,$$

$$\phi = \phi_n + \phi' . \tag{30-26}$$

The answer is obviously $F_n = E_n + F_0$. This F_0 is the thermodynamic potential for the ground state $n = 0$. The state ϕ_n is flowing. We need (30-8) to calculate its total energy. Now

$$s_\pm(\mathbf{R}) = \frac{1}{2} \sin\theta \, e^{\pm i\phi(\mathbf{R})} , \tag{30-27}$$

(see Fig. 30-1(b)). Notice that Ψ' of (30-14) can be written as

$$\Psi' = \prod_{\mathbf{R}} [u(\mathbf{R}) + v(\mathbf{R}) s_+(\mathbf{R})] \, \Phi_0 , \tag{30-28}$$

where u and v are obtained by replacing ϕ by $\phi(\mathbf{R})$ in the u and v of (30-13). Equation (30-27) represents the average value of $s_\pm(\mathbf{R})$ with respect to Ψ', i.e.

$$s_+ = v^* u ,$$

$$s_- = u^* v . \tag{30-29}$$

Equation (30-8) can be simplified as

$$\mathbf{P} \simeq \rho \int d^3R \, \nabla \phi , \tag{30-30}$$

where we have made use of

$$\frac{1}{4a^3} \sin^2\theta = \rho , \tag{30-31}$$

see (30-17) and (30-19). Therefore, putting ϕ_n into (30-30), we get the momentum along the torus

$$P_z = \rho V(2\pi n/L) \equiv m\rho V v_z ,$$

$$v_z \equiv (2\pi n/L)/m . \tag{30-32}$$

This is a stable total momentum and represents superfluidity, and v_s is the so-called superfluid velocity. The winding number n cannot change because $\sin\theta$ is nonzero (see Fig. 30-1), i.e. we have a condensation of $s(\mathbf{R})$. This is a macroscopic thermodynamical result. A slight increase of temperature or the addition of some impurities does not affect this stable equilibrium state. A superfluid is not the same as a fluid motion with the viscosity tending to zero. The viscosity, though tending to zero, is still nonzero and the addition of a few impurities will influence the flow velocity, slowing it down. This kind of normal flow with vanishing viscosity is therefore not a metastable state but an unstable state, and is completely different from the superfluid state.

The constancy of the winding number n comes about because we only consider smooth $\phi(\mathbf{R})$, regarding it as a continuous function of $\mathbf{R}$. If we also consider discontinuous situations, i.e. the defects discussed in Chapter 5, then n would not remain unchanged indefinitely. As this model is the XY model, the defects are the vortices. In the next section, we shall see how vortices change the above stable flow into a metastable or even an unstable state.

30.4. Stability of Superfluidity

The last section concludes that superfluidity is maintained by the conservation of the winding number n. To destroy the conservation of n is to destroy superfluidity.

Now we look at the influence of defects, starting from a one-dimensional model. (The above model is three-dimensional but to change it to one or two dimensions is very easy.) The conservation of n depends on the continuity of $\phi(\mathbf{R})$, i.e. the fact that the curve in Fig. 30-2 is continuous. But $\mathbf{R}$ is not continuous, instead it represents a set of discrete points. In one dimension, these points arrange in a chain like a necklace and the curve in Fig. 30-2 is really a chain of discrete points. If the difference of ϕ for neighbouring elements is close to π, then this curve can be regarded as broken. The probability of breaking is about

$$e^{-\epsilon/T}, \qquad \epsilon = \frac{\rho a}{2m}\left(\frac{\pi}{a}\right)^2, \qquad (30\text{-}33)$$

i.e. $(\partial\phi/\partial z)^2\, dz \sim (\pi/a)^2 a$. There are about N_π breaks where

$$N_\pi = N e^{-\epsilon/T} = \frac{L}{a} e^{-\epsilon/T} . \qquad (30\text{-}34)$$

Assume that $N_\pi \gg 1$, i.e. the breaks happen repeatedly. Of course after this curve is broken, it will be reconnected. But the angle difference before and after the break may be zero or $\pm 2\pi$; that is to say, n can be unchanged or it can increase or decrease. This n is no longer a conserved quantity and superfluidity disappears correspondingly.

If N_π is very small, i.e. T is very small, then L/a will not be large. Superfluidity can be regarded as a metastable state, i.e. n changes very slowly. The time scale of breaking is τ.

$$\tau \simeq \tau_0/N_\pi \ , \tag{30-35}$$

$$\begin{aligned}\frac{dn}{dt} &\simeq -\frac{1}{T}\left(\frac{\partial E_n}{\partial n}\right)\frac{1}{\tau} \\ &\simeq \frac{n}{\tau_0}\frac{4\pi\rho}{maT}e^{-\epsilon/T} \ .\end{aligned} \tag{30-36}$$

The factor $1/\tau_0$ can be regarded as the reaction rate at high temperatures. This is only a rough estimate.

Now consider the two-dimensional situation. The defects in two-dimensional space are vortices, and can be regarded as charges. $\nabla\phi(\mathbf{R})$ is regarded as the electric field. From Chapter 19 and the last chapter, we know that these charges will pair up as dipoles at low temperatures, and their interactions can be represented by a dielectric constant ϵ. When dipoles are formed, the electric field is smaller by a factor ϵ:

$$\text{Electric field} \ \propto \ \frac{1}{\epsilon}\left(\frac{\Delta\phi}{L}\right) \ , \qquad \Delta\phi = 2\pi n \ .$$

Therefore, the energy of the electric field is proportional to

$$\frac{1}{\epsilon}\left(\frac{\Delta\phi}{L}\right)^2 \ . \tag{30-37}$$

The difference between having and not having defects is in the value of ϵ. If there are no defects, $\epsilon = 1$. If there are defects, then $\epsilon > 1$. Equation (30-24) can be rewritten as

$$E_n = \frac{\rho_s V}{2m}\left(\frac{2n\pi}{L}\right)^2 \ , \qquad \rho_s \equiv \rho/\epsilon \ . \tag{30-38}$$

530 STATISTICAL MECHANICS

Therefore, at low temperatures, the flow velocity can be maintained, but with a slightly lower value. The ρ in (30-32) will now be replaced by ρ_s.

The parameter ρ_s is the elasticity of the order parameter, i.e.

$$\text{energy} \propto \rho_s \left(\frac{\Delta \phi}{L}\right)^2 , \qquad (30\text{-}39)$$

where $\Delta \phi / L$ is the degree of twisting of **s**. We can call ρ_s the elasticity. When the elasticity is not zero, the order parameter can withstand a twist.

If the temperature is too high, $T > T_0$ where

$$T_0 \simeq \frac{q^2}{4} = \frac{\pi \rho}{2m} , \qquad (30\text{-}40)$$

then the dipoles will separate and become free charges, $\epsilon \to \infty$ and $\rho_s \to 0$. The superfluid state will no longer exist. Notice that the relation between the unit of charge q and ρ/m is $q = 2\pi\rho/m$ (see (30-20), (29-48) and (29-53)).

The above discussions can be illustrated by some diagrams. Figure 30-3 shows something like a bicycle wheel. The tyre is our two-dimensional model with circumference L. The radial lines are the equipotential lines of ϕ. If there are no vortices in the tyre, this figure represents ϕ_n, $n = 16$. The total momentum can be obtained from (30-32). Of course we can regard the picture as a big vortex with a big hole in the middle and superfluid flows in the tyre. The strength of the big vortex is $16q$.

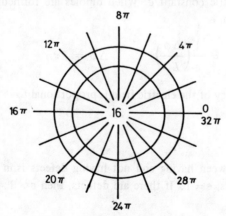

Fig. 30-3

The radial lines of Fig. 30-3 are the equipotential lines and can also be regarded as the electric lines emanating from the central charge nq, with the electric field pointing radially outwards. This is a rather convenient viewpoint. The strength of the electric field is

$$\frac{nq}{(L/2\pi)} = \frac{2\pi nq}{L} = qmv_s . \qquad (30\text{-}41)$$

Here $L/2\pi$ is the radius of the wheel assuming that the width of the tyre is much smaller than L.

If free charges appear, i.e. $T > T_0$, then the free charges screen the electric lines as in Fig. 30-4. There will no longer be electric lines inside the tyre and there is no flow.

If $T < T_0$, the charges pair up to form dipoles, and $16/\epsilon$ electric lines can still penetrate.

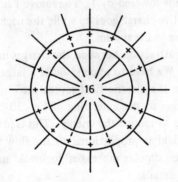

Fig. 30-4

Figure 30-5 is an enlarged portion of the tyre. Some electric lines penetrate through, while some meet the dipoles and the electric field will be less between the dipoles. On the average the electric field is reduced by a factor ϵ. (See the discussion at the end of Chapter 19.) The velocity is also smaller by a factor ϵ.

We now examine whether the state of flow is stable when $T < T_0$. This situation is much more complicated than the one-dimensional case. We only perform a crude analysis here. We shall not consider the detailed motion of the vortices, but look at the problem from the point of view of the process of energy decrease.

To change the flow velocity we must change the winding number n. As seen from Fig. 30-5 we have to change the number of electric lines inside the system.

532 STATISTICAL MECHANICS

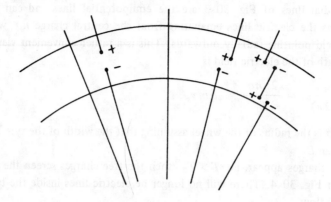

Fig. 30-5

As we decrease 1 line, n is lowered by 1. To remove 1 line we have to break up a dipole, so that the positive charge goes up while the negative charge goes down. In this way n will decrease in the system.

Of course there is an attraction between the charges in the dipole; otherwise they will not be paired. We have to expend quite a large energy to break them up. But once broken up, the electric field represented by the electric lines will attract the negative charges to the inner ring of Fig. 30-5 and push the positive charges to the outer ring to reduce the energy. This is just a barrier penetration problem. This electric field is proportional to the flow velocity. Hence if the flow velocity is large, the dipoles are easier to break and n decreases rapidly. Now let us look into the details.

The energy of the dipole is the interaction energy of $+q$ and $-q$ plus the potential in the external electric field

$$E = \frac{2\pi n q}{L \epsilon} = \frac{m v_s q}{\epsilon} , \qquad (30\text{-}42)$$

i.e.

$$u(r) = \frac{1}{\epsilon} q^2 \ln \frac{r}{a} - qEr , \qquad (30\text{-}43)$$

where r is the distance between $+q$ and $-q$. Figure 30-6 shows $u(r)$, whose maximum Λ occurs at $r = r_c$,

$$r_c = \frac{L}{2\pi n} = \frac{1}{m v_s} . \qquad (30\text{-}44)$$

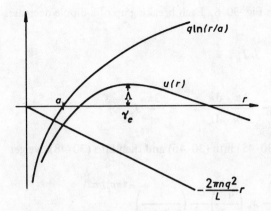

Fig. 30-6 Plot of $u(r)$.

Therefore if the flow velocity is large, r_c is small. The value of Λ is

$$\Lambda = \frac{q^2}{\epsilon}\left(\ln\frac{r_c}{a} - mv_s r_c\right)$$

$$= \frac{2\pi\rho_s}{m}\left(\ln\frac{1}{mv_s a} - 1\right). \tag{30-45}$$

If r is much smaller than r_c, then the dipoles are bound very strongly. To separate one pair is extremely difficult, and they must surmount the barrier Λ. We can estimate how many free charges flow out per unit time and unit length. We use the results in Chapter 21. (See (21-12).) The flow rate is

$$J_\pi \simeq \frac{DN_\pi}{V}\frac{e^{-\Lambda/T}}{(\Delta r)}, \tag{30-46}$$

$$(\Delta r)^2 \simeq \frac{T}{(\partial^2 u/\partial r^2)_c}$$

$$= \frac{\epsilon T}{q^2}r_c^2, \tag{30-47}$$

where D is the diffusion coefficient of the charges, N_π/V is the density of the dipoles, V is the total area of the system and Δr represents the width of the barrier peak in Fig. 30-6. Each breaking up of a dipole decreases n by 1. Hence

$$\frac{dn}{dt} = -LJ_\pi ,$$

$$\frac{dv_s}{dt} = -\frac{2\pi}{Lm}\frac{dn}{dt} = -\frac{2\pi}{m}J_\pi . \qquad (30\text{-}48)$$

Substituting (30-45) into (30-46) and then into (30-48), we get

$$\frac{dv_s}{dt} \simeq -v_s \left(\frac{q^2}{\epsilon T}\right)\frac{D}{a^2}\left(\frac{1}{mv_s a}\right)^{-2\pi\rho_s/mT} . \qquad (30\text{-}49)$$

The above assumes $N_\pi/V \sim (1/a^2)\exp(-q^2/T)(q^2/\epsilon) = 2\pi\rho_s/m$. The diffusion coefficient must be calculated separately. This is related to the surface on which the ^{4}He is adsorbed and the fluid mechanics of the vortices.

The most important conclusion of the above is that superfluidity in two dimensions is a metastable state. The larger the flow velocity the faster it decreases. But at low temperatures and small velocities, it is rather stable. This is due to the last factor in (30-49) caused by $e^{-\Lambda/T}$.

The solution of (30-49) for long times is (assume $mv_s a \ll 1$):

$$v_s \sim \frac{1}{ma}\left(\frac{t}{\tau}\right)^{-\alpha T} ,$$

$$\alpha = \frac{m}{2\pi\rho_s} ,$$

$$\frac{1}{\tau} \sim \frac{D}{\alpha T a^2} . \qquad (30\text{-}50)$$

At low temperatures, v_s disappears very slowly.

If mv_s is very close to $1/a$, then the barrier in Fig. 30-6 will disappear and there is then no superfluidity.

The situation is even more complicated in three dimensions. The defects are vortex lines. It involves a chain of atoms (because it is a line) and the energy is quite large. Hence it will be stable or metastable by itself and will not appear or disappear suddenly in the equilibrium state. Superfluidity is quite stable in three dimensions. If originally there are some vortex lines on the boundary, a large flow velocity will elongate them forming large circles, or push them to the other side of the tube to decrease the winding number. These are very interesting fluid mechanics problems. There was much work on this aspect some ten years ago, much earlier than the work in two dimensions.[b] The disappearance of superfluidity in three dimensions is usually due to the extremely high flow velocity. The critical velocity is related to the details of the shape of the container. The barrier for the metastable state is very high because the energy of the vortex lines is large. The energy of the molecules in the equilibrium state is not sufficient to overcome this barrier.

Problems

The requirements for this chapter are high and greater effort on the part of the reader is necessary.

1. The properties of helium are discussed in considerable details in the reference books (e.g., Woo (1972), Wilks (1967), Putterman (1974), etc.). The reader can read them for himself.

2. (a) Prove (30-5). Notice that

$$[s_+(\mathbf{R}), s_-(\mathbf{R}')] = 2s_z(\mathbf{R})\delta_{\mathbf{R}\mathbf{R}'} ,$$

$$[s_z(\mathbf{R}), s_\pm(\mathbf{R}')] = \pm s_\pm(\mathbf{R})\delta_{\mathbf{R}\mathbf{R}'} . \qquad (30\text{-}51)$$

(b) Prove (30-9).

3. When N'/N is very small (N' is the number of particles, see (30-10)), prove that this model is very much like the ideal gas model. Let

$$\psi(\mathbf{R}) \equiv a^{-3/2} s_-(\mathbf{R}) , \qquad \psi^*(\mathbf{R}) \equiv a^{-3/2} s_+(\mathbf{R}) , \qquad (30\text{-}52)$$

then

$$[\psi(\mathbf{R}), \psi^*(\mathbf{R}')] = \delta(\mathbf{R}-\mathbf{R}') ,$$

$$a^{-3} \delta_{\mathbf{R}\mathbf{R}'} \simeq \delta(\mathbf{R}-\mathbf{R}') . \qquad (30\text{-}53)$$

Prove that $\psi^*(\mathbf{R})\psi(\mathbf{R})$ is approximately the number operator of the particles. Notice that this model is still not the same as the ideal gas model because each lattice site can only accommodate one particle, so there is still some interaction between the particles.

[b] See the book by Wilks (1967), and also Putterman (1974), and Anderson (1966).

4. If the reader finds difficulty following the derivations in Sec. 30.2, he should review the quantum mechanics of spin $\frac{1}{2}$ and the transformation of axes:

$$e^{-i\theta s_\alpha} = \cos\frac{\theta}{2} - 2is_\alpha \sin\frac{\theta}{2} , \qquad \alpha = x, y, z . \qquad (30\text{-}54)$$

In (30-12) Φ_0 denotes the state with all the spins pointing downwards. Therefore the state of each spin can be represented by $\begin{pmatrix} 0 \\ 1 \end{pmatrix}$.

5. Derive (30-15) and (30-16) from (30-14).

6. The derivation of (30-20) from (30-18) involves the computation of functional derivatives. To differentiate a function with respect to $\phi(x)$ we need to remember the formula

$$\frac{\delta}{\delta\phi(x)} \phi(x') = \delta(x - x') . \qquad (30\text{-}55)$$

If we run into difficulty, the way out is to divide x into small sections and treat $\phi(x)$ of each section as a variable.

Calculate $\delta F/\delta\phi(x)$ for the following:

(a) $\quad F = \int dx\, dx'\, \phi^2(x)\, \phi^2(x')\, A(x, x') , \qquad (30\text{-}56)$

(b) $\quad F = \int dx\, \phi^2(x) \left(\frac{d\phi}{dx}\right)^2 , \qquad (30\text{-}57)$

(c) $\quad F = \int dx\, [\phi(x)\, h(x) + (\nabla^2 \phi(x))^2]$

$\qquad\qquad + \left(\int dx\, \phi^2(x)\right)^2 . \qquad (30\text{-}58)$

7. The discussion of this chapter is limited to the low temperature case. If T is close to the critical temperature, then the model in Sec. 30.2 is not suitable.

(a) Use the mean field approximation to solve model (30-7). Condition (30-10) must still be satisfied. Calculate the critical temperature T_c, the heat capacity and the order parameter $\langle s \rangle$, assuming that the order parameter is everywhere the same.

(b) Assuming that the order parameter is inhomogeneous:

$$\langle s_-(\mathbf{R}) \rangle = \psi(\mathbf{R}) . \qquad (30\text{-}59)$$

Prove that $\psi(\mathbf{R})$ is approximately the solution of

$$-\nabla^2 \psi - A\psi + B\psi^3 = 0 , \qquad (30\text{-}60)$$

(assume that $T_c - T$ is very small). Starting from (30-7), calculate A and B. Refer to Prob. 27.2 and Prob. 29.3. Use the method in Sec. 27.2 to obtain expressions like (27-11) and (27-12), including

$$F_0(\psi) = \langle H \rangle - TS$$

$$\propto \int d^3R \left[\frac{1}{2} |\nabla \psi|^2 - \frac{1}{2} A |\psi|^2 + \frac{B}{4} |\psi|^4 \right] ,$$

(30-61)

(27-12) is equivalent to (30-60). From $\delta F_0 / \delta \psi = 0$ we immediately get (30-60).

(c) Use (30-60) to replace (30-21) and discuss the various conclusions of Sec. 30.3. The concept of winding number is still applicable.

8. The above model has been used to analyse the superconductivity of fine metal wires. Although we have not discussed the theory of superconductivity, the above model can nevertheless be used. The basic concept of this chapter can also be applied, but the details are different.

The constants A and B should be computed from the model of the metal. Discuss the change of the winding number n. Notice that $|\psi|$ in this model can change, i.e. the curve of Fig. 30-2 is not limited to the surface of the torus. But if n changes, the curve must cut through the torus at a certain place. That is to say, during the process where n changes to $n-1$, $|\psi|$ must decrease to zero at some point and then increase again. This process has been carefully analysed. (See Langer and Ambegaokar (1966).) The basic line of thinking is as follows:

(a) The phase space is the set of ψ. Each ψ is a function of z, which is the coordinate along the torus. We assume that ψ does not change with x or y.

(b) Equation (30-60) determines the minimum of F_0 in this space.

$$\psi_n \propto e^{i2\pi n z/L} ,$$

(30-62)

is a minimum of F_0 and is the absolute minimum at ψ_0.

(c) From ψ_n to ψ_{n-1} we must overcome points with large $F_0(\psi)$, i.e. overcome a barrier. The lowest point of the barrier is a saddle point. Fortunately the solutions of (30-60) includes the minima, the maxima, and the saddle points. Let the saddle point be ψ'. Hence the height of the barrier is

$$\Lambda = F_0(\psi') - F_0(\psi_n) .$$

(30-63)

(d) From the discussion of Chapter 21 we can estimate the reaction rate of the process n to $n-1$.

$$\frac{1}{\tau} = \frac{1}{\tau_0} e^{-\Lambda/T} .$$

(30-64)

The factor τ_0 is not easily computed, but the other factor $e^{-\Lambda/T}$ comes from the barrier.

The reader can attempt the problem first and then consult the reference material. If the reader wants to do more extensive reading, he can refer to Putterman (1974).

9. Phonon gas.

The main emphasis of this chapter is on superfluidity. The thermodynamic properties of bosons are very interesting too, but have not been discussed. The reader can derive some results as follows.

(a) Review Prob. 29.6. Prove that the Hamiltonian can be written as

$$H = \sum_{\mathbf{k}} \epsilon_{\mathbf{k}} a_{\mathbf{k}}^* a_{\mathbf{k}} + \frac{1}{2V} \sum_{\mathbf{k},\mathbf{p},\mathbf{q}} u_{\mathbf{q}} a_{\mathbf{k}+\mathbf{p}}^* a_{\mathbf{p}-\mathbf{q}}^* a_{\mathbf{p}} a_{\mathbf{q}} ,$$

$$u_{\mathbf{q}} \equiv \int d^3 r \, e^{i\mathbf{q}\cdot\mathbf{r}} u(r) . \tag{30-65}$$

(See (29-57) to (29-64).)

(b) Suppose the interaction u is very weak. At low temperatures $N_0 = a_0^* a_0$ is nearly the same as N, and the other $a_{\mathbf{k}}$ can be treated as small quantities. Hence H is approximately equal to

$$H \simeq \sum_{\mathbf{k}} (\epsilon_{\mathbf{k}} + 2n u_{\mathbf{k}}) a_{\mathbf{k}}^* a_{\mathbf{k}}$$

$$+ n \sum_{\mathbf{k}} u_{\mathbf{k}} (a_{-\mathbf{k}}^* a_{\mathbf{k}}^* + a_{-\mathbf{k}} a_{\mathbf{k}}) + \text{constant} ,$$

$$n \equiv N/V . \tag{30-66}$$

(c) Notice that state $\mathbf{k}$ is only correlated with $-\mathbf{k}$. Hence model (30-66) can be easily solved. Calculate the zero-point energy and the wavefunction. The zero-point wavefunction of each pair of $(\mathbf{k}, -\mathbf{k})$ states is

$$\Psi_0 = \sum_m A_m (a_{-\mathbf{k}}^* a_{-\mathbf{k}}^*)^m \Phi_0 . \tag{30-67}$$

Φ_0 is the zero particle state, i.e. the vacuum. Calculate A_m.

(d) Prove that the excited states can be regarded as an ideal phonon gas. The phonons are defined by the excited states of $(\mathbf{k}, -\mathbf{k})$ states. Let

$$\alpha_{\mathbf{k}} \equiv u_{\mathbf{k}} a_{\mathbf{k}} - v_{\mathbf{k}} a_{-\mathbf{k}}^* . \tag{30-68}$$

This is the annihilation operator of the phonon and α_k^* is the creation operator. Here u_k and v_k are constants. From

$$[\alpha_k, \alpha_k^*] = 1 , \qquad \alpha_k \Psi_0 = 0 , \qquad (30\text{-}69)$$

calculate u_k and v_k and prove that

$$H = \sum_k \omega_k \alpha_k^* \alpha_k + \text{constant} ,$$

$$\omega_k^2 \equiv 2n u_k \epsilon_k + \epsilon_k^2 . \qquad (30\text{-}70)$$

If k is very small, then

$$\omega_k \simeq ck ,$$

$$c^2 \equiv n u_0/m . \qquad (30\text{-}71)$$

This is the motivation for the term phonon. The above were obtained some forty years ago by Bogoliubov.

(e) Use (30-70) to calculate the thermodynamical properties.

Notice that in older textbooks or literature it is usual to connect superfluidity with the energy of the phonon, and taking $c > 0$ as the origin of superfluidity. When the flow velocity is lower than c, no phonons are emitted and energy cannot be dissipated. This view is incorrect. (Solids have phonons too, but no superfluidity.) This chapter emphasises that the conservation of the winding number and the elasticity of the order parameter are the roots of superfluidity. The reader should take special note of this.

REFERENCES

Abraham, F. F. (1979) *Phys. Rep.* **53**, No. 2.

Alexandrowicz, Z. (1975) *J. Stat. Phys.* **13** 231.

Abrikosov, A. A., L. P. Gorkov and I. E. Dzyaloshinski (1963) *Methods of Quantum Field Theory in Statistical Physics* (Prentice-Hall, New Jersey).

Anderson, P. W. (1966) *Rev. Mod. Phys.* **38** 298.

Anderson, P. W., B. I. Halperin and C. Varma (1972) *Phil. Mag.* **25** 1.

Arnold, V. I. and A. Avez (1968) *Ergodic Problems of Classical Mechanics* (W. A. Benjamin, New York).

Berne, B. J., (1977), ed., *Statistical Mechanics* (Plenum Press, New York).

Beth, E., and G. E. Uhlenbeck (1937) *Physica* **4** 915.

Binder, K. (1979) *Monte Carlo Methods in Statistical Physics* (Springer-Verlag, Berlin).

Brout, R., and P. Carruthers (1963) *Lectures in the Many Electron Problem* (Gordon and Breach, New York).

Burton, W. K., N. Cabrera and F. C. Frank (1951) *Phil. Trans. Roy. Soc. London* **243A** 299.

Chui, S. T. and J. D. Weeks (1976) *Phys. Rev.* **B14** 4978.

Dasgupta, C., S. Ma and C. K. Hu (1979), *Phys. Rev.* **B20** 3837.

Dashen, R., S. Ma and H. J. Bernstein (1969) *Phys. Rev.* **187** 345.

Dashen, R. and S. Ma (1971) *Phys. Rev.* **A4** 700.

Dundon, J. M. and J. M. Goodkind (1974) *Phys. Rev. Lett.* **32** 1343-6.

Dyson, F. J. (1967) *J. Math. Phys.* **8** 1538.

Feigenbaum, M. J. (1981) *J. Stat. Phys.* **19** 25.

Fisher, M. E. (1978) *Physica* **106A** 28.

Forster, D. (1975) *Hydrodynamic Fluctuations, Broken Symmetry and Correlation Functions* (W. A. Benjamin, New York).

Gibbs, J. W. (1960) *Elementary Principles of Statistical Mechanics* (Dover Publications, New York).

Griffiths, R. B. (1972) in *Phase Transition and Critical Phenomena* Vol. 1, C. Domb and M. S. Green, eds. (Academic Press, New York).

Grimes, C. C. and G. A. Adams (1979) *Phys. Rev. Lett.* **42** 795.

Imry, Y. and L. Gunther (1971) *Phys. Rev.* **B3** 3939.

Hahn, E. L. (1950) *Phys. Rev.* **80** 580.

Hohenberg, P. C. (1967) *Phys. Rev.* **158** 383.

Huang, K. (1963) *Statistical Mechanics* (J. Wiley, New York).

Kikuchi, F. (1951) *Phys. Rev.* **81** 988.

Kittel, C. (1956), (1966) *Introduction to Solid State Physics*, 3rd ed. (J. Wiley, New York).

Kosterlitz, J. M. and D. J. Thouless, (1973) *J. Phys.* **C6** 1181.

Kosterlitz, J. M. (1974) *J. Phys.* **C1** 1046.

Krylov, N. S. (1979) *Works on the Foundations of Statistical Physics* (Princeton University Press, Princeton).

Langer, J. S., and V. Ambegaokar (1967) *Phys. Rev.* **164** 498.

Landau, L. D. and E. M. Lifshitz (1980) *Statistical Physics* (Pergamon Press, London).

Lebowitz, J. L. and E. H. Lieb (1969) *Phys. Rev. Lett.* **22** 631.

Lifshitz, E. M. and L. P. Pitaeviskii (1980) *Statistical Physics* (Pergamon Press, London).

Ma, S. K. (1976) *Modern Theory of Critical Phenomena* (Benjamin, New York).

Ma, S. K. (1981a) *J. Stat. Phys.* **26** 221.

Ma, S. K. and M. Payne (1981) *Phys. Rev.* **B24** 3984.

Malmberg, J. H., C. B. Wharton, R. W. Gould, and T. M. O'Neils, (1968) *Phys. Fluids* **11** 1147.

Mattis, D. C. (1965) *The Theory of Magnetism* (Harper and Row, New York).

Matsubara, T. (1955) *Prog. Theor. Phys.* (Kyoto) **14** 351.

Mairovitch, H., (1977) *Chem. Phys. Lett.* **45** 389.

Mermin, N. D. and H. Wagner (1966) *Phys. Rev. Lett.* **17** 1133.

Nelson, D. R. and B. I. Halperin (1980) *Phys. Rev.* **B21** 5312.

O'Neil, T. M. and R. W. Gould (1968) *Phys. Fluids* **11** 134.

Onsager, L. (1944) *Phys. Rev.* **65** 117.

Osgood, E. B. and J. M. Goodkind (1967) *Phys. Rev. Lett.* **18** 894.

Peierls, R. (1936) *Camb. Phil. Soc.* **32** 477.

Peierls, R. (1955) *Quantum Theory of Solid*. (Oxford Clarendon Press, London).

Peierls, R. (1979) *Surprises in Theoretical Physics* (Princeton University Press, Princeton).

Prohofsky, E. W. and J. A. Krumhansl (1964) *Phys. Rev.* **A133** 1403.

Putterman, S. J. (1974) *Superfluid Hydrodynamics* (North-Holland, Amsterdam).

Rudnick, J. (1970) Ph. D. Thesis, University of California, San Diego.

Sinai, Ya. G. (1977) *Introduction to Ergodic Theory* (Princeton University Press, Princeton).

Stanley, H. E. (1971) *Introduction to Phase Transition and Critical Phenomena* (Oxford University Press, London).

Tabor, D. (1979) *Gases, Liquids and Solids* 2nd ed. (Cambridge University Press, Cambridge).

Tolman, R. C. (1962) *Principles of Statistical Mechanics* (Oxford University Press, London).

Wannier, G. H. (1966) *Statistical Physics* (J. Wiley, New York).

Weeks, J. D. (1980) in *Ordering in Strongly Fluctuating Condensed Matter Systems,* T. Riste, ed. (Plenum, New York).

Wilks, J. (1961) *Third Law of Thermodynamics* (Oxford University Press, London).

Wilks, J. (1967) *Liquid and Solid Helium* (Oxford University Press, London).

Yang, C. N. (1952) *Phys. Rev.* **85** 809.

Yang, C. N. and T. D. Lee (1952) *Phys. Rev.* **87** 404.

Young, A. P. (1979) *Phys. Rev.* **B19** 1855.

Zemansky, M. W., (1957) *Heat and Thermodynamics,* 4th ed. (McGraw-Hill, New York).

References from Chinese Books

Lee, Y. Y. (1967) "University Physics".
大學物理學，東華書局

Ma, S. K. (1981b) "Physics", Vol. 2, 2.
熱力學，氣體運動論，統計力學，聯經出版事業

Tang, C. T. (1971), ed. "Materials".
氦和多體物理，中華書局

Woo, C. W. (1972) "Helium and Many Body Physics".
材料，科學圖書社

Wu, T. Y. (1979) "Thermodynamics, Kinetic Theory and Statistical Mechanics".
物理，中國物理學會

Yang, W. C. (1979) "Probability".
機率論，正中書局

INDEX

Adiabatic compressibility coefficient 366
Adiabatic demagnetisation 280
Adiabatic process 13, 24, 103, 400
Amorphous magnet 6, 324
Amorphous states 312, 323, 411
Amorphous substance 6
Antiferromagnetism 6, 474, 479
Anharmonic vibration 150
Average values 167

Barrier penetration 370, 532
Binary alloy 307
Binomial distribution 112, 201
Black body radiation 41
Boiling point 265
Boltzmann distribution 37, 41
Boltzmann equation 346
Bose-Einstein condensation 516
Boson 35, 41
Bubble chamber 273

Coincidence, method of 426
Canonical ensemble 130
Caratheodory theorem 28
Carnot heat engine 24
Central limit theorem 200
Chaos 185, 456
Characteristic function 199
Charge density wave 64
Chemical reactions 12, 30, 43
Chemical potential 11, 94

Chemical potential for photon gas 42
Classical distribution 42
Clausius-Clapeyron equation 264
Cloud chamber 273
Coefficient of friction 368
Collisions with impurities 354
Concentrations, law of 43
Condensation 266
Condensation forces 474
Conditional distribution 181
Conductor 6
Correlation function 182, 189
   Fourier transform of 229
Correlation length 231 237, 332, 430
Correlation time 190, 390, 431
Correlation value 171
Corresponding states 470
Coulomb gas 342, 343
Cumulant 203
Cumulant expansion theorem 207
Curie's law 282
Critical point 262, 472
Cyrstal 6
Crystal, atom in 183
Crystal, density fluctuation of 498
Crystal, imperfect 150

Debye-Huckel theory 330
Debye model 98
Debye-Waller factor 500
Decoding 456

Defects 508
Dense fermion gas 363
Dense gas 46
Density correlation function 230, 332
Density expansion 250
Density of states 49
Density wave 64
Detailed balance 30, 150
Diamagnetism 283
Diatomic molecule 150
Dielectric constant 333
Dielectric constant of plasma gas 365
Different time correlation 193, 194
Diffusion distance 318
Diffusion equation 216
Diode model 369
Dipole interaction 160
Dispersion 167
Distribution 166
Distribution function 20, 444
Distribution region 66

Effective energy 223
Effective interaction energy 315
Efficiency 24
Einstein's law 116
Elasticity of order parameter 530
Electron cyrstals 351
Electronic dipole interaction 247
Energy distribution 32
Energy, total 13
Ensembles 130
Entropy 9, 13, 14, 19, 67, 128, 403, 443
Entropy of mixing 2 gases 22, 23
Equal time correlation 192, 193
Equation of state 28
Equilibrium, local 5
Equilibrium state 2, 3, 8, 17
Equipartition of energy, law of 80, 98
Ergodicity 442
Evaporation 263
Exchange interaction 291
Exclusion principle 31, 236
Exponential distribution 196
Extensive quantity 69

Fermi distribution 38
Fermi gas model, see free electron model
Fermi momentum 49
Fermi surface 49, 236

Fermi-Thomas model 341
Fermions 35
Fermion gas 240
Ferromagnetism 6, 8, 463, 464
Ferromagnet model 502
First law of thermodynamics 13
First-order transition 273
Fluctuation 164, 166
Fluctuation-dissipation theorem 215
Fluid 6
Fokker-Planck equation 377
Free electron model 48
Free flight time 5
Free particle 31
Frequency spectrum 239
Frozen entropy 436

Gap, energy 57
Gas dilute 231
Gas ideal 19, 35, 230
Gauge transformation 516
Generalised coordinates 12
Glassy slate 6, 312
Goldstone theorem 508
Grand canonical ensemble 130, 248

H-theorem 363
Hard sphere interaction 232, 233
Heat 12, 13, 403
Heat capacity 38, 53, 99
Heat of evaporation 269
Heat of melting 269
Heat of sublimation 268
Heisenberg model 294, 515
Holes 51
Homogeneity of space 506

Ice point 8
Ideal gas, correction to 231
Impurity atoms 260
Impurities, mobile 312, 318
Impurities, stationary 312, 318
Independence 180, 181, 445, 454
Independence, mutual 185
Infinite integrals 232
Information theory 429
Instability 159, 243, 450
Internal energy 118
Insulator 6, 60
Invariant quantities 68, 111

INDEX 547

Irreversible process 10, 412, 436
Ising antiferromagnetic model 307
Ising lattice gas 466
Ising model 117, 131, 281, 432
Ising variable 297
Isothermal process 17

Joint distribution 181

Kondo problem 324
Kosterlitz-Thouless model 337

Lagrange multipliers, method of 89
Landau damping 422
Landau diamagnetism 286, 296
Landau theory 475
Large numbers, law of 69, 78
Latent heat 263
Law of crystallisation 397
Leonard-Jones potential 247
Levinson theorem 258
Lindemann formula 270
Line dislocation 514
Long range interaction 117
Longitudinal wave 98

Macroscopic 4
Macroscopic limit, see thermodynamics limit
Magnetic fluid 12, 227
Magnetic moment 11, 227
Magnetic susceptibility 227, 281
Magnon 35, 504
Mean field approximation 243
Mean field theory 458
Mean free path 5, 73, 103
Mean free time 5, 359
Metal surface, model of 341, 342
Metastable state 161, 372, 434
Melting 397
Melting point 270
Mixing 445
Mixture 323
Molecular motion 4
Monte Carlo method 30
Monte Carlo simulation 391

Nonequilibrium phenomena 4
Nonequilibrium problem 313
Nonequilibrium states 41
Normal distribution 115, 198

Normalization 168
Nuclear cooling 280

Observation time 2, 8, 114, 134, 186, 190, 214, 281, 405, 428
One-dimensional gas 145
Open system 229
Open system, rule for 127
Order parameter 473
Osmosis 312
Osmotic pressure 313

Pairwise independence 188
Paramagnetism 6, 277, 460
Partition function 125
Partition line 302
Partition point 298
Partition surface 303
Pauli susceptibility 283
Peak integration 89
Peierl's instability 48, 62
Periodic structure of atoms 55
Phase equilibrium 261
Phase space 33, 66
Phase shift 254, 255
Phase transitions 8, 149, 243
Phonon 35, 102
Photon gas 42
Planar vector model, see XY model
Planck distribution 42
Plasma echo 422
Plasma oscillations 349, 364
Plasma gas 341
Poisson distribution 183
Poisson sum rule 290
Pomeranchuk cooling 295
Pressure 11
Pressurisation, cooling by 295
Probability definition 167, 175
Projection 81
Propagation function 229

Quantum distribution 42
Quantum lattice gas 520
Quantum numbers 31
Quantum vector model, see ferromagnet model
Quenching 322

Random motion 214

Random sequence 391
Randomness 180, 189, 437
Region of motion 114, 219
Remanence 325
Response functions 228, 237
Reversible process 10, 12, 435, 436
RKKY interaction 324
Rotation of molecules 139
Rules of calculation 122

Sample set 174
Saturation, degree of 267
Saturated vapor pressure 263
Scattering operator 36
Scattering probability 112
Screening 331
Screening length 331
Shortwavelength approximation,
  see WKB approximation 334
Second law of thermodynamics 24, 405
Second order transition 273
Second sound 102, 353
Second virial coefficient 250
Semiconductor 60
Soft modes 508
Solid 6
Sound waves 103
Smooth process 11
Spin echo 419
Spin glass 6, 324
Spin glass phase 325
Spin wave 35
Star, reaction occurring
  inside 47
Statistics by classification 91
Statistics by summation 91
Statistical model of the atom,
  see Fermi-Thomas model
Statistical weight 168
Sticking coefficient 265
Stimulated emission 43
Stirling's formula 78
Superconductor 6
Superfluid 6
Superfluidity 528
Superheated liquid 273
Supersaturated vapour 267

Surface tension 266, 477
Susceptibility, differential 226
Symmetry 506
System, subsystem 88

Temperature 9, 13
Temperature, absolute 19
Temperature critical 47
Thermal electron radiation 63
Thermodynamics 10, 12
Thermodynamic limit 154
Thermodynamic potentials 120
Third law of thermodynamics 408, 435
Time correlation function 190
Time delay in collisions 250
Tin, states of 410
Transfer matrix, method 310
Transition temperatures 8
Transverse wave 98
Triple point 263
Two-state cluster model 380

van der Waals equation 250, 470
Vapour pressure curve 264
Vector model 294
Vibration 94
Vibration of nuclei 139
Virial coefficient 246
Virial expansion 251
Virial theorem 259
Viscous liquid bottle 416
Viscosity 362, 368
Vortex 511, 529

White dwarf 63
Winding number 526
WKB approximation 33
Work 11, 403

XY model 495, 513

Yang-Lee theorem 156

Zero-sound 351
Zero-point energy 50, 97
Zero-point pressure 51
Zero-point vibration 269